The Successful Internship

PERSONAL, PROFESSIONAL, AND CIVIC DEVELOPMENT IN EXPERIENTIAL LEARNING

D0226125

Praise for The Successful Internship

A must read, full of "best practices," the book will not only help students apply classroom knowledge in the field, it will help students acquire the non-academic skills necessary for successful performance in the world of work.

John S. Duley, Emeritus, Michigan State University

Where most guides to the internship process stay on the nuts-and-bolts level and treat the placement merely as a path to a good job, Sweitzer and King insist that we pay attention to the multiple forms of learning. They flesh out their very sound practical advice with helpful references about theories on everything from learning styles to citizenship, from adult development to social change. I already own two editions of the book, but will add this new one to my collection: It's a gem.

David Thornton Moore, New York University

Whether you are a total beginner or even an experienced caring person, this book will enable you to become more effective and make a larger difference in the lives of others.

Allen E. Ivey. Emeritus, University of Massachusetts, Amherst

The authors tackle this critical educational experience from an interdisciplinary perspective. Woven throughout the book are their unique developmental stages, through which they suggest all interns will pass. In addition, the book pushes students to understand how their work relates to living in a democratic and social world, one in which we have responsibility toward ourselves, the clients with whom we work, and the communities in which we live.

Edward Neukrug, Old Dominion University

The authors have combined their sophisticated experience with some of the best theory and practice in the field of experiential education.

Garry Hesser, Augsburg College

Internships today are much more than apprenticeships; they are multidimensional problem solving experiences for students. Sweitzer and King have clearly and thoroughly provided practical resources on multiple dimensions—academic, personal, organizational, social, and civic—as well as on the traditional professional dimension of engaged internship learning. This edition guides the intern to be a 21st century problem solver by focusing on high impact educational practices and being an engaged learner, and provides effective tools for managing the internship's ups and downs.

Dwight E. Giles, Jr., University of Massachusetts, Boston

The Successful Internship

**PERSONAL, PROFESSIONAL, AND
CIVIC DEVELOPMENT IN
EXPERIENTIAL LEARNING**

FOURTH EDITION

H. Frederick Sweitzer
University of Hartford

Mary A. King
Fitchburg State University

BROOKS/COLE
CENGAGE Learning·

Australia • Brazil • Japan • Korea • Mexico • Singapore • Spain • United Kingdom • United States

BROOKS/COLE
CENGAGE Learning

The Successful Internship:
Personal, Professional, and
Civic Development in
Experiential Learning,
Fourth Edition
H. Frederick Sweitzer and
Mary A. King

Publisher: Jon-David Hague

Assistant Editor: Sean M. Cronin

Editorial Assistant:
Amelia L. Blevins

Senior Brand Manager:
Elisabeth Rhoden

Senior Market Development
Manager: Kara Kindstrom

Manufacturing Planner:
Judy Inouye

Rights Acquisitions Specialist:
Don Schlotman

Art and Cover Direction,
Production Services, and
Composition: PreMediaGlobal

Cover Image: Jennifer Gottschalk

For product information and technology assistance, contact us at
Cengage Learning Customer & Sales Support, 1-800-354-9706.
For permission to use material from this text or product,
submit all requests online at **www.cengage.com/permissions.**
Further permissions questions can be e-mailed to
permissionrequest@cengage.com.

Library of Congress Control Number: 2012954572

ISBN-13: 978-1-285-07719-2

ISBN-10: 1-285-07719-9

Brooks/Cole
20 Davis Drive
Belmont, CA 94002-3098
USA

Cengage Learning is a leading provider of customized learning solutions with office locations around the globe, including Singapore, the United Kingdom, Australia, Mexico, Brazil, and Japan. Locate your local office at **www.cengage.com/global**.

Cengage Learning products are represented in Canada by Nelson Education, Ltd.

To learn more about Brooks/Cole, visit **www.cengage.com/brookscole**

Purchase any of our products at your local college store or at our preferred online store **www.cengagebrain.com**.

Printed in the United States of America
2 3 4 5 6 7 17 16 15 14

Remembering Lisa
...with affection and great respect for the memory of Lisa Gebo, long-time editor at Brooks/Cole and long-time colleague and friend,

And

Honoring Our Students
...whose voices over the years inform and echo throughout the pages of this book.

We dedicate this book with love to the young adults in our lives.

From Fred
To my goddaughters, Meagan Prescott and Kelly Thomas, who have grown into the kind of women I would be proud to have as daughters and as friends.

From Mary
To my son, Patrick Zimmermann, who inspires my life with his heart, soul, and indomitable spirit.

The Koru

The Koru is a symbol used by the Maori culture
in New Zealand to represent new beginnings,
growth, and harmony. This symbol also represents
a fern slowly unfolding toward the light.

 Contents

Section Four: Crescendos 329

Chapter 12 Riding High: The Competence Stage 331

 Foreword

In this fourth edition, the authors continue to increase the capacity of faculty, staff, and students to see the transformative potential of an internship as a vehicle for personal, academic, and civic learning and development. This book also embodies best practices in internships that are consistent with the roots of experiential education as well as new developments in cognitive science and student engagement.

In many ways, the science of learning has caught up with intuition and practice knowledge. Early in the 20th century, John Dewey (and others) was in the forefront of education reform that would prepare students holistically for productive lives as individuals, as workers, and as civic participants. "The great waste in school," Dewey wrote in *The School and Society* in 1899, "comes from [the students'] inability to utilize the experience [he/she] gets outside...while on the other hand [the student] is unable to apply in daily life what...is learn[ed] in school." Dewey wanted education that would lead to "embodied intelligence," or what we might call transformative learning. What Dewey meant was a kind of education in which what the student learns becomes part of who they are—that they are fundamentally changed in the process of education. It is that transformative potential that comes about through experiential education provided through internships.

Unlike in Dewey's time, we now know a lot more about how students learn. Developments in the cognitive sciences and developmental psychology over the last quarter century have produced an empirically based science of learning (Bransford et al, 2000). In sum, we know that "The learner is not a receptacle of knowledge, but rather creates his or her learning actively and uniquely; learning is about making meaning for each individual learner by establishing and reworking patterns, relationships, and connections; direct experience decisively shapes individual understanding (i.e., 'situated learning'); learning occurs best in the context of a compelling 'presenting problem'; and beyond stimulation, learning requires reflection" (Ewell, 1997).

So, based on the cognitive sciences, what should our teaching and learning practices look like? They should incorporate approaches that emphasize application and experience; that emphasize linking established concepts to new situations; that emphasize interpersonal collaboration; and that consistently develop cross-disciplinary skills (Ewell, 1997). They should look like high-quality internships.

The problem, as Derek Bok (Bok, 2008) has pointed out, is that (quite ironically) educators are not applying the empirical findings on learning in their practice. This book helps to fill that gap, providing a critical resource for faculty in any discipline, students in any major, and staff in any area of the campus to design, implement, and participate in effective internships.

The authors have done an excellent job of having this edition reflect the most recent research deepening our understanding of learning and the importance of experience. The emphasis on engagement in learning, drawing from the research based on the results from the National Survey of Student Engagement (NSSE), has led to an emphasis on engaged learning and high impact practices. High impact practices such as internships, service-learning, and community-based learning are ones "that educational research suggests increase rates of student retention and student engagement" (Kuh, 2008) in learning. We know from NSSE data that "…such experiences make learning more meaningful, and ultimately more useful because what students know becomes a part of who they are" (NSSE, 2002). Through experience, they are transformed as learners.

Research has also shown us what it takes for an experience to be high impact. What the authors have done is to offer students and faculty both a theoretical justification and practical methods for operationalizing those principles in the context of an academic internship. High impact learning comes about because of the quality of "real-world application" (Kuh, 2006). It is this quality that contributes to "deep learning" as an outcome: students are more able to "attend to underlying meaning as well as surface content, integrate and synthesize different ideas, discern patterns of evidence, apply knowledge in different situations, and view issues from multiple perspectives" (Laird, 2008). Or as a student in a community-based course I taught reflectively wrote in his journal, "I have become more aware of my surroundings, have learned to look more deeply into the words of writers, and have learned to formulate my own opinions. Perhaps what I like best about this class is that it synthesized all of my years of book learning and applied it to why I was here in the first place. Lately I've been having difficulty justifying the cost of my education versus what I really learned about what is necessary in living…. I don't think I ever really *knew* what it meant until I was trying to incorporate my experiences [in the community] with the many readings we worked with" (Saltmarsh, 2000, p. 52). This is the kind of learning that comes from experience linked to rigorous academic study, and the authors offer an internship experience with academic integrity, community relevance, and civic development.

Finally, this new edition is timely—we face 21st-century transdisciplinary "wicked problems" that require innovative solutions, collaborative public problem solving that requires working with people from different cultural backgrounds and different life experiences and perspectives, and ways to democratically draw on the assets of the knowledge and experience that everyone contributes to learning to building a wider public culture of democracy. Internships are a potentially powerful tool in preparing students to meet these challenges.

When asked about the kind of learning that was needed for the 21st century, John Abbot, the Director of Britain's Education 2000 Trust, replied, "people worldwide need a whole series of new competencies... But I doubt that such abilities can be taught solely in the classroom, or be developed solely by teachers. Higher order thinking and problem solving skills grow out of direct experience, not simply teaching; they require more than a classroom activity. They develop through active involvement and real life experiences in workplaces and the community" (Marchese, 1996). Twenty-first century learning requires the transformative power of experiential education, and the authors have helped bring that power to the world of internships.

John Saltmarsh, Ph.D.
Co-Director, The New England Resource Center
    for Higher Education
Professor, Higher Education Administration
University of Massachusetts, Boston
November 2012

References

Bok, D. (2008). *Our underachieving colleges: A candid look at how much students learn and why they should be learning more.* Princeton University Press.

Bransford, J. (2000). *How people learn: Brain, mind, experience, and school.* National Academies Press.

Ewell, P. T. (1997). Organizing for learning: A new imperative. *AAHE bulletin, 50,* 3–6.

Kuh, G. D. (2008). High-impact educational practices. *What they are, who has access to them, and why they matter.* Association of American Colleges and Universities.

Laird, N., et al. (2008). The effects of discipline on deep approaches to student learning and college outcomes. *Research in higher education,* 49:6, September.

Marchese, T. (1996). The search for next-century learning: An interview with John Abbott. *AAHE bulletin,* March. 3–6.

National Survey of Student Engagement (NSSE). (2002). *From promise to progress: How colleges and universities are using student engagement results to improve collegiate quality.* Indiana University.

Saltmarsh, J. (2000). Emerson's prophecy. *Connecting past and present: Concepts and models for service-learning in history,* 43–60.

 Preface

Internships have been a part of the landscape of higher education for a long time. In the helping, teaching, and health professions, they are the coin of the realm; virtually no program at the undergraduate or graduate level is without at least one major field experience. Internships are on the rise in a number of other disciplines as well, including many traditional liberal arts fields. As internships spread to a wider cross section of disciplines, they are drawing more attention campus-wide. This book is, first and foremost, intended as a guide for students and faculty who want to get the highest return for their effort in the form of learning (*academic, career*) and development (*professional, personal, civic*). And although it is less transparent to the reader, the book also attempts to locate academic internships within these emerging and continuing trends in higher education:

- *Engaged Learning*, which tells us that students learn best when they make active connections to and with the material. This long-held belief in education is being confirmed by emerging neurobiological research that helps us understand what actually happens in the brain when people learn;

- *Experiential Learning*, which tells us that experience is a powerful teacher, but it is far more successful when it is guided and structured to maximize learning;

- *Education for Civic and Democratic Engagement*, which tells us that every person and every profession has a responsibility to the communities in which they live and to take an active role in participatory democracy; and, that colleges ought to focus deliberately on this aspect of education; and

- *High-Impact Educational Practices*, which are educational approaches that have been shown over the last ten years to be effective in promoting student engagement. There is also a growing body of literature on the pedagogy of these practices—that is, how to maximize their impact.

In short, the internship is a nearly boundless opportunity for learning. Like any such opportunity its success depends on the ability of faculty and program staff to structure and guide the learning and of students' ability and willingness to engage the material and the experience. We also believe that success depends on the ability of the faculty, program staff, and students to attend to and engage in the *lived* experience of the internship—the current of thoughts, concerns, and emotions that run beneath the surface and more than occasionally rises to the top. Perhaps because of the sheer number of hours spent at the internship, the intensity of the work, the prospect of gaining (or not) entry into a desired field of work, or all of these factors, the *affective* dimension of the internship is more salient for students than in traditional academic experiences; faculty and students ignore it that their peril.

Our experience in listening to the affective, lived experience and guiding our students through it, as well as listening to faculty, program staff and students at many other institutions and in many other fields, led us to conceive of a set of developmental stages, a progression of concerns through which interns tend to pass in predictable order, although not at a predictable pace. Knowledge of that progression and of the tasks necessary to move through it helps students, faculty, program staff, as well as site supervisors in a number of ways.

Over the last several years, that same process of listening has brought us to a revision of the stages. In previous editions of the book, we have posited five stages of an internship: Anticipation, Disillusionment, Confrontation, Competence, and Culmination. *Disillusionment* was a term used to connote a major *crisis in confidence* for the intern, a sense that things are going terribly wrong. *Confrontation* was the term we used for engaging and addressing that crisis and resolving major problems in the internship. While this sequence seemed quite common in many internships, especially those in high stakes, emotionally challenging, highly personal and interpersonal settings, in some of those placements and in other instances there were two important differences that we decided to address in this edtion. The first is that significant problems can occur *before* or *after* the Disillusionment stage. And, some interns were moving through the internship without experiencing disillusionment. That does not mean they had no problems; on the contrary, interns can and do encounter problems at every stage. But they do not all experience the pervasive sense of disappointment, frustration, and anger that we associated with the disillusionment stage.

Although the original stage model (DSM-1) recognized those two differences as realities of the experience, it did not account for them in the framework. This led us to reconsider the five-stage conceptual model to respond to the question: *What does the experience of disillusionment look like in each of the stages?* We have, then, shifted our perspective from a focus on the experiences *some* were having (to varying degrees) to one on which *all* were having. We now posit four stages: Anticipation, Exploration, Competence, and Culmination. In each stage, we emphasize the need to become and remain *engaged* in the internship process. *Engagement*, as we use the term, has two dimensions: engagement as a stance toward the interns' experience and engagement with the particular tasks and challenges of each stage. Each stage of an internship brings with it certain challenges, or even crises. When interns engage those challenges, they meet them head on and take the actions they need to move through them—not around them. *Disengagement*, on the other hand, means acting passively, namely, letting things take their course, hunkering down, waiting it out, or withdrawing emotionally and even physically. Extreme levels of disengagement can result from a combination of external and internal circumstances. Some circumstances challenge the limits of an intern's knowledge, skills, and internal resources. Some circumstances are unfair, or even tragic. A disengaged response is perhaps reasonable and expected under such circumstances; such a response, whether to major or minor challenges, can also lead interns to feel, over time, like they are in a hole and can't get out; this is the *experience of disillusionment*.

This book has been adopted by a wide variety of helping professions (which were its original audience) as well as by other academic disciplines and professions at both graduate and undergraduate levels. Because the internship also serves different roles across academic and student affairs programs, student use of this book varies. For example, in some cases, the students are in a culminating internship, and most of the specific skill development will have been accomplished earlier in their academic work, while in other cases, internships and field experiences are woven into the entire program, and skill development proceeds in tandem. Some students will use this book as part of an on-campus or online seminar that accompanies their experience (that is how we use it). But others use it as a self-guide or a resource, selecting chapters based on what is important at the moment and communicating with instructors individually though not necessarily meeting in groups.

Because of this diversity, we have tried to give interns and instructors as much flexibility as possible in using the book. We have included concepts and examples from a range of professions. We have included resources for further exploration so that students and instructors can build on those areas that are of interest and relevance. We have also provided a wide range of reflective questions at the end of each chapter, anticipating that students and instructors will choose from among them. Some of those questions work best for individual reflection and some are designed for group work. We have also tried to create a book that can be augmented or supplemented with more discipline- and skill-specific assignments, publications, and instruction.

Using This Book for Optimal Effectiveness

Our experience has been that students go through the stages in a predictable order but often at different rates of *speed* and *emotional intensity*.

Seminar classes tend to cover chapters in sequential order and at particular times in the internship with a couple of caveats. First, we believe that students know best what they need to read, so if they want to jump into the Essentials chapters when they open the book, we honor that. We also listen carefully to what individual students are saying in class, online, and in their logs or journals and suggest that a particular student reread a chapter or skip ahead to future ones as needed.

The book is organized into four sections and the chapters in each section coalesce around the themes of the sections. The chapters, though, do not need to be read in keeping with the calendar of the internship. For example, the first section is *Foundations* and the chapters that comprise this section (1–4) are sometimes read before the internship or field experience begins, as part of a prerequisite seminar or as preparatory reading. Chapter 3 in this section is the first of the Internship Essentials chapters (*Tools for Staying Engaged*). The second section is titled *Beginnings*, and the chapters that comprise this section (5–7) pertain to the Anticipation stage. We recommend beginning with Chapter 5 and then reading the remaining chapters in whatever order seems appropriate, but doing so before moving on to the next section of the book. The Essentials chapter in this section is *The Learning Contract and Supervision* (6), and it may be read as early as pre-internship. Chapter 7 focuses on helping and service work, and for some interns it may not be relevant. Similarly, there are chapters later in the book on the internship site (10) and the community (11) that may be more appropriate to be read during this stage of the internship.

Section Three, *Rhythms* (8–11), looks at the challenges that await once the initial concerns are resolved. In Chapter 8, we introduce the next stage of the internship: the Exploration stage. Chapter 9 is this section's Essentials chapter and offers a variety of tools and approaches for staying engaged and focused on progress as exploration of the internship unfolds (*Advanced Tools for Staying Engaged and Moving Forward*). Chapters 10 and 11 take the intern beyond the initial focus on colleagues and, in some cases on clientele, and ask the intern to consider issues and challenges related to the placement site itself and the community context of the work. For some interns, such as those in a community-organizing agency, a campaign's headquarter, or a politicians' field office, the community is a more immediate focus; those interns may wish to read this chapter earlier.

The final section of the book, *Crescendos* (Chapters 12–14), examines issues and concerns that are common in the latter stages of the internship: *Competence* and *Culmination*. Chapter 12 deals with issues of capability that typically arise as the intern does well in the internship experience. Chapter 13 is an Essentials chapter and deals with professional, legal, and ethical issues that the intern may face or learn about during placement. The final chapter, Chapter 14, guides the intern to end the internship on a number of levels and in productive, meaningful ways.

The Fourth Edition

Each chapter of the book has been updated and augmented, but some particular changes are worth emphasizing:

- Engaged learning is a theme of the book, and in each chapter we discuss ways to be more engaged with the tasks of the stages and the internship.
- The possible experience of disillusionment is discussed at each stage, along with strategies for recovering momentum.
- The opportunities for growth as an engaged citizen and civic professional are emphasized throughout the book, with a designated section of exercises at the end of many chapters (*Civically Speaking*).
- Each of the stage chapters is organized around a chart that serves as a visual guide to the concerns and challenges of the stage.
- Special reflection exercises for interns who come to the internship with considerable life and professional experience have been added in every chapter (*for the EXperienced Intern*).
- There are four "Essentials" chapters in the book. These tend to be loaded with information and strategies and are intended as resources to be consulted throughout the experience—before as well as beyond.
  - Chapter 3 focuses on attitudes, values, skills, knowledge, personal resources, and pieces of information that interns need as they begin.
  - Chapter 6 focuses on the critical components of the Learning Contract and Supervision Plan.
  - Chapter 9 offers a set of advanced techniques and tools for dealing with issues and challenges that tend to arise a bit further in the internship.
  - Finally, Chapter 13 is a comprehensive treatment of legal, ethical, and professional issues.

With Special Appreciation

What keeps us coming back to this book, and to its revisions, is not only the evolution of our ideas which take place in the context of reading and reflecting on the work of the scholar practitioners in the field, but the many conversations we have had with students, colleagues, and site supervisors. Our professional lives are enriched by our contact with faculty and staff in human service and counselor education, civic and democratic engagement, and experiential education. Those spheres overlap, of course, and so rather than try to list individuals by their affiliations we simply extend our thanks and appreciation to those scholars, some of whom we have yet to meet, and colleagues who continue to inform our thinking: Gene Alpert, Richard Battistoni, John S. Duley, Janet Eyler, Andy Furco, Dwight E. Giles, Jr., Georgianna Glose, Jackie Griswold, Garry Hesser, Jeff Howard, Mark Homan, Allen Ivey, Jenny Keyser, Susan Kincaid, Pam Kiser, Ron Kovach, George Kuh, Heather Lagace, Ed Neukrug, Tricia McClam, Lynn McKinney, Diane McMillen, Lynne Montrose, David Thornton Moore, Trula Nicholas, Roseanna Ross, John Saltmarsh, Marianna Savoca, Rob Shumer, Nancy Thomas, Vicky Totten, Mike True, Jim Walters, Marianne Woodside, Ed Zlotkowski, Karen Zuckerman, and colleagues over the years in the human service education programs on our campuses.

We recognize the contributions of national organizations such as the National Society for Experiential Education (NSEE), the National Organization for Human Services (NOHS), and the Association of American Colleges and Universities (AACU). They have been inspiring professional homes for us and have provided us with forums to learn as well as present and discuss our ideas. We thank our many colleagues in these organizations and those who have attended our workshops, shared ideas with us, and reviewed and published our work.

Many reviewers over the years contributed their time and expertise to making this a better book, providing thorough reviews, astute insights, and valuable suggestions at various stages in the writing process. We appreciate the contributions of Eugene Alpert, The Washington Center for Internships and Academic Seminars; Deborah Altus, Washburn University; Peggy Anderson, Western Washington University; Montserrat Casado, University of Central Florida; David Cessna, past president of the Cooperative Education and Internship Association (CEIA); Robert Fried, Northeastern University; Janet Hagen, University of Wisconsin–Oshkosh; Marcia Harrigan, Virginia Commonwealth University; Jill Jurgens, Old Dominion University; Heather Lagace, University of Hartford; Linda Long, SUNY Corning Community College; Roberta Magarrell, Brigham Young University; Tricia McClam, University of Tennessee; Lynn McKinney, University of Rhode Island; Susan Membrino, currently with Bank of America; Punky Pletan-Cross, currently with Hale Kipa, Inc.; Vicki Christopher Rybak, Bradley University; Tom Struzick, University of Alabama; Stephen Waster, University of Wisconsin–Milwaukee; Gardine Williams, Ferris State University; Iris Wilkinson, Washburn University; and Laura Woliver, University of South Carolina–Columbia. Special recognition and appreciation are extended to Dwight E. Giles, Jr., University of Massachusetts–Boston, and Garry Hesser, Augsburg College, for the ways in which their reviews have shaped our thinking and the direction of our work in important ways.

We thank the following reviewers for their contributions of expertise and time to this edition: Connie Boyd, Genesee Community College; Melinda Blackman, California State Fullerton; Pat Lamanna, Dutchess

Community College; Lorraine Barber, Community College of Philadelphia; Joseph Adamo, Cazenovia College; Sharon Jones, University of Georgia; and Nancy Anderson, Warner University.

In addition, our heartfelt thanks go to Professor Steve Eisenstat of Suffolk University School of Law for giving generously of his time and expertise to the chapter on legal issues and to Professor Mark Homan of Pima Community College for his continuing contributions to our understanding of communities.

Our appreciation goes to the staff at Cengage. Seth Dobrin stayed with us during times of transition for us and for the publishing industry. As this book goes into production, Seth has moved on to new professional challenges and we wish him the very best. Amelia Blevins, worked with us through the final editing and production process. In addition, we thank Matt Ballantyne, Don Schlotman, Brenda Carmichael, Ronald D'Souza, Bob Kauser, Judy Inouye, and Charoma Blyden. Our families and friends who have patiently waited for the completion of this edition, give meaning and depth to our lives and inspiration to our work, so thank you and love to Skip and Betty Sweitzer, Sally Sweitzer and Britt Howe, and Phyllis and Dave Agurkis, Deb Allen, Jeff and Judy Bauman, Stephanie Chiha, Cynthia Crosson, Angela Romijn Mazur, Mary Ann Hanley, Anita Hotchkiss, Margot Kempers, Regina Miller, Peter Oliver, Ken Pollak, Bev Roder, Kathy Callan Rondeau, Gin Sgan, Nancy Thomas, and Mary Jean Zuttermeister. Our partners in marriage, Martha Sandefer and Peter Zimmermann; Fred's nephew, Freddy Sweitzer-Howe; and Mary's son, Patrick Zimmermann, as well as Max, Ravi, and Misty, are among the most special blessings in our lives.

About the Authors

H. Frederick Sweitzer is Associate Provost and Professor of Educational Leadership at the University of Hartford in Connecticut. Fred has over 30 years' experience in human services as a social worker, administrator, teacher, and consultant. He has placed and supervised undergraduate interns for 20 years and developed the internship seminar at the University of Hartford. Fred brings to his work a strong background in self-understanding, human development, experiential education, service learning, civic engagement, professional education, and group dynamics. He is on the editorial boards for the journals *Human Service Education* and *Human Services Today*, and has published widely in the field.

Mary A. King is Professor Emerita, Fitchburg State University, where she was faculty and coordinated field placements in Behavioral Sciences, supervised graduate and undergraduate internships in professional studies and liberal arts programs, and instructed service–learning. She has over 30 years' experience working with interns in the field and on campus and brings to her academic work backgrounds in teaching, criminal justice, consultation and counseling psychology, holding several professional licenses. Mary has published in the fields of experiential and human service education and has held positions on national and regional boards, most recently on that of the National Society of Experiential Education, where she currently oversees the NSEE Experiential Educational Academy.

FOUNDATIONS

 CHAPTER *1*

The Lay of the Land

Education is revelation that affects the individual.
GOTTHOLD EPHRAIM LESSING, 1780

I've never learned as much as I did in this internship,"
and, "I learned more in one semester that I did in all my
years of classes."
STUDENT REFLECTIONS

Welcome to Your Internship

You are beginning what is, for most students, the most exciting experience of your education. Chances are you have looked forward to an internship for a long time. You've probably heard your share of stories—both good and bad—from other, more experienced students. And while you may be in the minority on your campus in conducting an internship, you join virtually thousands of interns all over the country. An internship is an intensive field experience, and often a critical component of many academic programs. Internships are conducted in social service and corporate settings, government offices, high tech industry, and research laboratories, to name a few. It's also important to know that there are other kinds of field-based learning experiences, including co-op education, service-learning and course-related practica; and there are other terms by which internships are known, such as field work and field education. We will use the term *internship* in this book to refer to those *learning* experiences that involve receiving academic credit for *intentional learning* at an approved site, under approved supervision, for at least eight hours per week over the course of a semester. Internships are growing in popularity on college campuses and have been recognized as a "high impact practice (HIP)," something that, when done well, promotes high levels of engagement, learning, and development (Kuh, 2008).

THINK About It

Make It High Impact!

George Kuh (2008) and his colleagues have made an extensive study of student engagement and the practices that promote it. They have identified a number of practices that can lead to student engagement and success, but emphasize that they *must* have six key characteristics to be high impact practices (O'Neill, 2010). As you read them below and as you make your way through this book focusing on each of them, you've got yourself a high-impact internship in the making!

- Effortful with purposeful tasks requiring daily decisions
- Including quality feedback
- Opportunities to apply your learning
- Opportunities for reflection
- Building substantive relationships
- Engaging across differences

A Few Basic Terms

Although internships exist at many colleges and universities, different language is often used to describe the various aspects of the experience and the people associated with it. For example, the term *supervisor* sometimes refers to a person employed by the placement site and sometimes refers to a faculty or professional staff member on campus. So, at the risk of boring those of you who are very clear about these terms, we take a moment now to be clear about what we mean by them.

- *Intern*   This is the term that refers to you, the student who is at the site to learn through an internship, even though you may not be called an intern on your campus.
- *EXperienced Intern*   This term refers to those of you who bring considerable life experience or prior internship experience to this internship.
- *Placement or Site*   This term refers to the place where you are conducting your internship. Sites can vary quite a bit, ranging from art museums, K-12 schools, universities, social service agencies, large or small businesses, or court houses, just to name a few. Through the process of finding a placement, you probably are aware of the incredible variety of opportunities that exist in the community.
- *Campus Supervisor(s) or Instructors*   These terms refer to the faculty or professional staff member on your campus who oversees your field

placement. These are the people who may have helped you find the placement, who may meet with you individually during the semester, visit you at the site, hold conferences with you and your supervisor, conduct a seminar class for you and your peers, evaluate your performance, or do all of the above. It is possible for more than one person to fill these roles. Even though they may go by different titles on various campuses (internship coordinator, seminar leader, supervising professor, facilitator, and so on), for simplicity's sake we will use campus instructor or campus supervisor to refer to all those roles.

- *Site Supervisor* Your site supervisor is the person assigned by the placement site to help ensure your learning. This person meets regularly with you, answers your questions, guides you in your work, and gives you feedback on your progress. Most placements assign one site supervisor to one student, although in some cases there may be more than one person fulfilling these functions. Some academic programs use the term *field instructor* to describe this person in order to emphasize the educational (as opposed to managerial) nature of the role.

- *Co-Worker* This term refers to the people who work at your placement, regardless of their title, status, or how much you interact with them. If there are other students at the site, from your school or some other school, they are functioning in the role of co-worker when you are at the placement site.

- *Clients, Population, and Clientele* These terms refer to the people who are served by your placement site or with whom the site does business. Given the wide variety of internships, it is not possible to use one term that works in all settings. For example, in human or social service settings, the term *clients* is very common, but the people served are also called *customers, consumers, residents, students,* or *patients,* depending in part on the philosophy of the site and the nature of the work. Other organizations, such as advertising agencies or public relations firms, have clients as well, although of a different nature and with different needs. Still other settings, such as business or retail, use the terms *customer* or *consumer* more commonly. We will use the terms *clients, population,* and *clientele* to refer to these individuals and groups.

What Can You Learn from an Internship?

It gives meaning to everything you have learned and makes practical sense of something you've known as theoretical.
STUDENT REFLECTION

In our experience, most interns approach their experience excited about what they are going to do and what they hope to learn. And yet, paradoxically, they also typically underestimate the learning potential of the internship. They may be excited about honing professional skills, developing career opportunities, or trying out theories they have learned. An internship can do any or all of those things, and so much more. As is suggested by the title of this book, we view a successful internship as one that facilitates three significant dimensions of your development: personal, professional, and civic. You enter the internship at different points in your development in these three dimensions, based on life and work experience; with care and attention you can (and we hope you will) enhance your knowledge, skills, attitudes, and values in all three dimensions.

Personal Development

The internship is an opportunity for intellectual and emotional development that may be important for your internship but will also be important in your life, whatever path you choose. For one thing, the internship offers an opportunity to develop qualities such as flexibility, sensitivity, and openness to diversity that are critical to your success as a professional, a family member, and a citizen. For another, if you give yourself a chance, you can learn a tremendous amount about yourself during this internship. The experience can be a powerful catalyst for personal growth, providing opportunities to develop a sense of your potential through work under the supervision of experienced and qualified supervisors. There will be opportunities to accomplish tasks independently and test your creative capacities while doing so. In Chapter 4 we will elaborate on the dimensions of self-understanding that are available to you in an internship, and you will learn more in Chapter 12 about what makes an internship fulfilling.

Professional Development

Some students enter an internship primarily for career exploration. They may be studying a traditional liberal arts discipline such as sociology, history, political science, or psychology and want to see some ways in which those disciplines are put into practice. For other students, the internship is the culminating academic experience in a highly structured and sequenced set of courses and field experiences and can be a chance to pull together and apply much of what they have learned. And, of course, there are internships whose purpose is some combination of these two. For everyone, though, the internship is a chance to take the next step: to acquire more of the knowledge, skills, attitudes, and values of a profession or an academic discipline and to explore how well they fit with personal interests and strengths.

The internship also affords you the opportunity to understand the world of work in a more complete way than you do now. It is an opportunity to become socialized into the norms and values of a profession (Royse, Dhooper, & Rompf, 2011). Even if you have had full-time jobs or careers, presumably your internship is taking you into an area in which you have little professional experience at this level. Internships are often described as a time when theory is applied to real-life settings; we believe that the relationship between theory and practice is more complex than that. The internship is a chance to *develop* the relationship between theory and practice, for each should inform the other (Sgroi & Ryniker, 2002). According to Sullivan (2005), professionals excel at the art of what he refers to as *practical reasoning*, which literally means reasoning in and about practice. Professionals must move with fluidity between their understanding of theory and the real, human situations that they face in their work (which do not always quite conform to the predictions of theories). This movement is not easy; theory is abstract and relatively objective, whereas the human context of your work is entirely subjective and concrete. Your experience will help you see where the theories do not quite apply or where you need to search for a new theoretical model to help you. Thus, theories are transformed through their application, and you will be actively involved in that process as an intern. Try not to worry about how complicated this might sound! It happens every day in the workplace and interns do it all the time without realizing it.

Many internship programs also emphasize academic learning, that is, the applied learning of a particular academic discipline. Internships are a wonderful opportunity for this sort of learning, and in some internships it is the primary purpose. Whether your primary goal is to enter a profession or to explore a discipline as deeply as you can, there is an academic component to your learning. There are also important essential abilities that can be strengthened in an internship that go beyond or cut across professions and academic disciplines. The ability to look critically at information, to think creatively, and to look at issues from multiple viewpoints are essential abilities, as is the ability to communicate clearly both verbally and in writing. Solving problems and working in teams are abilities that will serve you at home, at work, and in the community. Many of these abilities are traditional outcomes of what is referred to as a liberal education (Crutcher, Corrigan, O'Brien, & Schneider 2007); they are also critical components of many professions (Lemann, 2004). You may have studied some of them in your undergraduate general education courses; indeed, this is often where important foundations are laid. But if they are not also encouraged and developed in the context of your major area of study, they will have far less effect on you (Crutcher et al., 2007).

Civic Development

You may have been wondering for a while now what this term means and why it joins personal and professional in the title of this book. Well, in our experience, students approach the internship with a wide range of exposure to and understanding of the term *civic*. For some, the term *civics* conjures up topics, such as the branches of government and the legislative process, and seem largely irrelevant to college, not to mention an internship. For others, depending on their life experience, choice of major, or the college they attend, the notion of civic is more robust. Even so, this aspect of an internship may be overshadowed by the expected personal and professional dimensions. Also, placement sites vary in their explicit emphasis on the civic domain of their work and yours.

We invite you to consider civic development as the development of the personal and professional capacity for participation in a healthy democracy (Colby, Erlich, Beaumont, & Stephens, 2003; Howard, 2001). The need for college students to acquire knowledge, skills, attitudes, and values that will allow them to enter the workforce and to function as productive citizens in a democratic society has drawn a good deal of attention on college campuses across this country (National Taskforce, 2012). Importantly, these essential abilities need to be cultivated early and often, and in active engagement with communities.

THINK About It

Re-Imagining What It Means to "Be Civic"

Maybe our discussion of the term *civic* so far seems vague, abstract, or unfamiliar to you. If so, we encourage you to take a moment and think about some of your experiences which are good examples of the term as we are using it. Remember that "participation" is an active word. Participating may require knowledge and skill, but it also requires *action*. Here are some examples of civic participation:

- Signing a petition to lower the speed limit on your street
- Working for a political campaign because you believe the candidate can really make a difference
- Voting in an election in which you understand the issues
- Serving on a committee in your town, school, or office
- Advocating on behalf of local farmers during a drought
- Making a sincere effort to understand the views of people who disagree with you
- Community organizing to collect donations for victims of natural disasters

Your campus, whether it is a physical or virtual one, probably offers opportunities for civic participation as well. For example:

- An office of volunteer services
- A *service-learning* center or service-learning courses
- Student government associations
- Green campus initiatives

Aspects of Civic Development in an Internship

An internship can be a vehicle for civic development in two ways: The internship can help you develop knowledge, skills, attitudes, and values that will make you become a more responsible and contributing member of your community and society regardless of where you live and what you choose to do for your life's work; and the internship can help you become what William Sullivan calls a "civic professional" (Sullivan, 2005). Because this is an aspect of the internship that in our experience is the least well understood by many interns, we spend some time elaborating on these two aspects of civic development.

Regardless of what profession you enter, or even whether you enter a profession at all, the internship is an opportunity for you to learn some of what will help you participate fully and productively in your community. Several authors have written about the various aspects of civic learning (Battistoni, 2006; Colby et al., 2003; Howard, 2001; National Taskforce, 2012), and we will discuss them in more detail when we discuss your Learning Contract, but here are a few examples. Civic *knowledge* might mean learning not just about the challenges faced by the people your profession serves, but about some of the historical and current social forces that bring about those challenges. It might mean, for example, learning that people who are hungry, poor, or underemployed are not necessarily lazy or unintelligent, but that their condition results at least in part from social conditions over which they have no control (Godfrey, 2000). Civic *skills* might mean learning how to advocate successfully for change in a workplace, a neighborhood, or a community to make conditions there more equitable. And civic *attitudes and values* might include the belief that understanding social issues is an obligation for everyone, not just those in politics or journalism.

A *civic professional* is someone who embraces and intentionally attempts to understand the *human* context of the work. For some professions, such as the helping professions, this context begins with the individual. For all professions, however, it includes a broad and complex social context of families, diverse cultures, communities, and political dynamics. A civic professional is also someone who understands the *public relevance* of the profession. To quote William Sullivan, "*To neglect formation in the meaning of community, and the larger public purposes for which the profession stands, is to risk educating mere technicians for hire in place of genuine professionals*" (2005, p. 254). Each profession has an implicit contract with society. Some professions exist only to serve society, and they are funded largely by society because of the public value placed on that service. Even those professions, however, must grapple with the nature of their social mission or contract. For example, there is a history of debate within the field of criminal justice about whether its primary purpose is to protect society in the short term by incarcerating, monitoring, and/or punishing those who have committed crimes or to try and rehabilitate those who have committed crimes so that they may become productive, contributing citizens in the long term. Regardless of what you believe about that issue, an internship in criminal justice is an opportunity to explore it.

All professions also have ethical and moral obligations to the society in which they function, and the work of each professional is by definition connected to a larger social purpose. Journalism should be about more than entertainment; a free press should be an anchor of a healthy democracy. Even the intensely private domain of business can be seen as a public good as well as a private benefit (Colby, Erlich, Sullivan, & Dolle, 2011; Waddock & Post, 2000). Business educators have argued

---

THINK About It

**Civic Responsibility, Civic Professionalism:
How Are You Measuring Up?**

If the concepts of civic learning and civic professional still seem foreign to you, or just an abstraction, or if you'd like to pursue some ways to assess your own civic learning and professionalism, here are two tools to help you.

The *Personal and Social Responsibility Inventory* (PSRI) was developed by the Association of American Colleges and Universities (AACU, 2012) and measures five dimensions of personal and social responsibility: (http://www.aacu.org/core_commitments/documents/PSRI-IowaStateFlier.pdf.)

- Striving for Excellence
- Cultivating Personal and Academic Integrity
- Contributing to a Larger Community
- Taking Seriously the Perspectives of Others
- Developing Competence in Ethical and Moral Reasoning and Action

The *Civic Minded Professional Scale* (CMP) was developed by Julie Hatcher, who describes civic minded professionals as those with a commitment and capacity to work with others in a democratic way to achieve public goods (Hatcher, 2008). The CMP measures five aspects of civic professionalism:

- Voluntary Action
- Identity and Calling
- Citizenship
- Social Trustee
- Consensus Building

that those entering the business world need to understand that business is about more than maximizing profits and creating wealth; that corporations ought to contribute to community issues such as social justice and ecological stability (Godfrey, 2000). The internship, then, is a chance for you to learn about the public relevance and social obligations of a profession—perhaps one that you plan to be part of in a short time—and about how those obligations are (or are not) carried out at your internship site.

So Why Do You Need a Book?

It's as if this was written for me. After reading this, I feel better because now I know what I am experiencing is not an individual thing but part of a natural experience that all interns go through.
STUDENT REFLECTION

The internship is a learning experience like no other. In a classroom learning experience, you learn through readings, lectures, discussions, and exercises; these are the raw materials you are given. You bring your ability to memorize, analyze, synthesize, and evaluate to these materials—these are your learning tools. The arena for learning is the classroom, and classrooms vary in the amount and quality of interaction between students and the instructor, and among students. Some can be very interactive, engaged, and engaging places, with mutual dialogue among students as well as between students and teachers. Other classrooms are interactive, but only between teacher and student; it is almost as if there is a multitude of individual relationships being carried on in isolation. The internship experience is different in every way from those just described; you will be involved in the work and thus will learn experientially. This book is designed to help you and your instructor with this different approach to learning.

Experiencing a Different Kind of Learning

Experience, both intellectual and emotional, is the raw material of the internship. You will be learning mostly through experience, although you may engage in some traditional academic activities as well. However, one noted theorist in the field of experiential education, David Kolb, suggests that experience alone does not automatically lead to learning or growth. Rather, the experience must be processed and organized in some way (Kolb, 1984). More specifically, Kolb contends that you must *think* about your experience, sometimes in structured ways, and perhaps discuss it with others. Reflective dialogue with yourself and your peers is the primary tool for learning in the internship, as it is in any high impact practice. This book is designed to help you structure that reflection and dialogue. It will invite you to think about your internship in a variety of ways, some of which may be new to you. It may also help you anticipate some of the challenges that await you and move successfully through them.

Furthermore, the internship is not just an intellectual experience. It is a *human* experience, full of all the wonderful and less-than-wonderful feelings that people bring to their interactions and struggles. This emotional, human side of the internship is more than a backdrop to the real work and the real learning; it is every bit as real and important.

Relationships are the medium of most internships; they are the context in which most of your learning and growth occurs. Many of you will work directly with clientele, but even if you do not, you will be involved in relationships with a site supervisor, a campus supervisor, other interns on site and on campus, and co-workers at your placement. These relationships offer rich and varied opportunities for learning and growth. This book will help you think about these relationships, capitalize on the opportunities they present, and address problems that may arise.

Facing a Different Kind of Challenge

In addition to excitement and satisfaction, most interns also experience some real difficulties—moments when they question themselves, their career choices, their placements, or all of the above. We like to think of these moments as "crises," but not in the way you are probably familiar with the term. We prefer to think of crises as the Chinese do. The symbol for crisis in Chinese is a combination of two symbols: danger and opportunity (Figure 1.1).

So, while there is some risk and certainly some discomfort in the challenges ahead, there is also tremendous opportunity for growth. We hope this book will help you see both the dangers and the opportunities inherent in an internship and to grow from both. We will encourage you not to run from these crises, but to meet them head on, with your mind and your heart open to the experience.

FIGURE 1.1
Chinese Symbol for Crisis

© Cengage Learning

Learning for High Impact

The truth is that some of you won't need this book; you will learn and survive the challenges just fine without it. But, you will have to decide whether you want to invest in just surviving the internship or thriving in it. If you choose the latter, then this book may be very useful to your learning and experience. Some students use this book as a self-guiding tool; this is often the case when there is no seminar class to accompany the field experience (either face to face or in a distance learning format). Some interns have used the book as a resource to draw upon as needed, especially if they are in the field for only one day a week, the class is one hour, and the focus of the class is the integration of what is being learned in the field with the theories of the academic major. In such a situation, there is little time to discuss the issues in this book in a meaningful way. We have tried to write the book so that you can do just that. However, we believe that for most students, this book, in combination with a skilled campus supervisor, supportive peers, and your intentional effort, can enhance your learning in meaningful and substantive ways.

We first conceived of this book for the helping professionals audience, which has an incredible variety of academic programs and types of placements under its umbrella. When you add the audiences in other professions, such as communications, sociology, nursing, political science, and business, then both the variety of students who use this book and the nature and depth of preparation that they bring to these experiences become even wider. This book is meant as a guide to the *phenomenological* experience of the internship, to help you anticipate and make sense of the full experience of the internship—the emotional as well as the intellectual aspects of your work. But because of the variety of interns just mentioned, it is very important that you consult your campus supervisor(s) early and often as needed, especially if there is no seminar class that accompanies your experience. Decisions about what theories to explore and what to emphasize warrant at the least a consultation with the person overseeing your placement in the field.

The Four Concepts Underlying This Book

Engaged Learning

The term *engagement* is heard quite a bit on college campuses these days. In general, it refers to methods of learning where students are active partners in learning as opposed to passive recipients of knowledge. Hodge, Baxter Magolda, and Haynes (2009) refer to engaged learning as an approach that encourages students to seek and discover new knowledge by exploring authentic questions and problems. When learning is a passive process, teachers are the centers of energy and tell you the information that they think you need to know. But when learning is an *active* process, students are the centers of energy and the teacher's role is to guide or facilitate your learning by taking an interest in your work and coaching you through the experience (Garvin, 1991). As an active participant in the learning process, you play the central role in shaping the *content*, *direction*, and *pace* of your learning.

While being engaged is important in any learning process, it is especially important in an internship. Engagement as a stance toward or way of interacting with the internship experience is the lever for getting the most from an internship by turning experience into learning, empowering yourself as a learner, and being successful in your internship.

Experiential Education

An internship, like other kinds of field instruction, is a form of experiential education (and a potentially excellent example of engaged learning). Although this approach to learning may not be well understood in many places on your campus, it comes out of a long theoretical and practical tradition, as discussed in the following Focus on THEORY box.

A key component to experiential education is *reflection*. Dwight Giles, who has written extensively about service-learning and internships, makes a point about service-learning that we think applies to all the experiences covered in this book. He says that reflection is what connects and integrates the service, or the work in the field, to the learning. Otherwise, whatever theory you study can be emphasized in your classes but not necessarily integrated with practical experience. At the other extreme, practical experience is left to stand on its own. Reflection is the connection, and it is a powerful key to your success, growth, learning, and development (Eyler & Giles, 1999; Giles, 1990; Giles, 2002). This intern captures its meaning in this way, *"I believe that in reading, reviewing, and of course, reflecting on something, someone can learn more about themselves and how they really feel about certain issues."*

David Kolb (1984) originally set forth a cycle of four phases that people go through to benefit from experiential learning, as illustrated in Figure 1.2. In the first phase, *concrete experience* (CE), students have a *specific experience* in the classroom, at home, in a field placement, or in some other context. They then *reflect* on that experience from a variety of perspectives (*reflective observation*, or RO). During the *abstract conceptualization* (AC) phase, they try to *form generalizations* or principles based on their experience and reflection. Finally, they *test* that theory or idea in a new situation (*active experimentation*, or AE) and the cycle begins again, since this is another concrete experience. James Zull's research (2002) on the relationship between the brain, learning, and education suggests that Kolb has it right: Education can affect your brain maximally when you have a concrete experience in the field, reflect upon it, connect it to what you already know,

Focus on THEORY

How Do You *Learn* as an Intern?

Experiential learning has philosophical roots dating back to the guild and apprenticeship systems of medieval times through the Industrial Revolution. Toward the end of the 19th century, professional schools required direct and practical experiences as integral components of the academic programs, for example, medical schools and hospital internships; law schools, moot courts, and clerkships; normal schools and practice teaching; forestry/agriculture and field work (Chickering, 1977). The National Society for Experiential Education (www.nsee.org) describes experiential education as *learning activities that involve the learner in the process of active engagement with and critical reflection about phenomena being studied.*

Perhaps the best-known proponent of experiential education was the educational philosopher John Dewey (1916/44, 1933, 1938, 1940). Dewey believed strongly that "an ounce of experience is better than a ton of theory simply because it is only in experience that any theory has vital and verifiable significance" (1916/44, p. 144). However, he was convinced that even though all real education comes through experience, not all experience is necessarily *educative.* This idea was reiterated by David Kolb (1984, 1985; Kolb & Fry, 1975), who emphasized along with Dewey the need for experience to be organized and processed in some way to facilitate learning. Dewey also felt strongly that the educational environment needs to actively stimulate the student's development, and it does so through genuine and resolvable problems or conflicts that the student must confront with active thinking in order to grow and learn through the experience.

then create your own abstract hypotheses about what you've experienced, and test them actively, which in turn will produce a new CE for you.

You may recognize this cycle from a previous field or work experience or if you already have begun this one. For example, suppose you observe a customer arguing with one of your co-workers. You could then draw on several theories or ideas you have studied to try to understand what was happening, or you might seek out some new information from staff. You then begin to form your own ideas about what happened and why, and you might use this knowledge to guide your own interactions with that or another customer. Once you do that, the interaction is itself a new concrete experience, and the cycle begins again.

Predictable Stages

Over the years, as we have supervised interns and listened to their concerns and read their journals, talked with their site supervisors, and discussed similar experiences with colleagues and students at other

---

FIGURE 1.2
Kolb's Learning Cycle

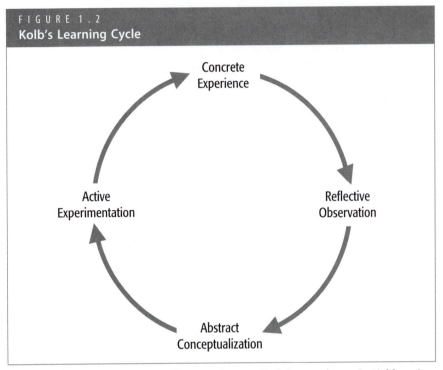

Source: Kolb, D. A., & Fry, R. (1975). Toward an applied theory of experiential learning. In C. Cooper (Ed.), Theories of Group Process. New York: John Wiley & Sons.

institutions, a predictable progression of concerns and challenges has emerged. We have organized these concerns into four stages—Anticipation, Exploration, Competence, and Culmination (Sweitzer & King, 1994; 2012). Understanding this progression of concerns will help you, your campus supervisor, and your site supervisor predict and make sense of some of the things that may happen during your internship and think in advance about how to respond. It will also help you view many of your thoughts, feelings, and reactions as normal, and even necessary. The experience then becomes a bit less mysterious, and for some people that makes it more comfortable.

For example, if you are feeling excited but also pretty anxious as you begin the placement, you may wonder whether that anxiety is a sign of trouble or what may have caused it. Knowing that it is a common and predictable experience in an internship will help you stop worrying about being anxious and let you direct your energy toward moving beyond that anxiety. As one of our students said, *"Now that I know I am not the only one that is concerned with these feelings, I am better able to share them with others without feeling embarrassed."*

Self-Understanding

To make sense of your internship, you need to understand more than a stage theory and the experiential nature of learning in the field; you need to understand *yourself*. No two students have the same experience even if they are working at the same site; that is because any internship experience is the result of a complex interaction between the individuals and groups that comprise the placement site and each individual intern. You are a unique individual, and that uniqueness influences both how people react to you and how you react to people and situations. You view the world through a set of lenses that are yours alone. Therefore, each of you will go through these stages at your own pace and in your own way. Events that trouble you may not trouble your peers and vice versa. Some of you will be very visible and dramatic in both your trials and your tribulations. Others will experience changes more subtly and express them more quietly.

We want to help you think about yourself throughout your internship in ways that we believe will lead you to important insights about you in the workplace and to a smoother journey on your path to personal, professional, and civic development.

In summary, appreciating the experiential nature of the learning that takes place will help you to understand how different learning-by-doing is from traditional classroom learning. The stages of an internship will help you understand internships in general and some of the experiences that you are apt to have during your internship. Understanding yourself will help you recognize the particular style in which you will experience the internship. Combining these pieces of knowledge will give you a powerful tool to understand what is happening to you, to meet and deal successfully with the challenges you face, and to take an engaged stance in making your internship high impact learning and the most rewarding experience it can be.

The Internship Seminar

Many of you will be meeting with an instructor and other interns on campus during the semester. We refer to these meetings as seminars. The word *seminar* comes from the Italian *seminare*, which means to sow or seed. When done well, the seminar is a fine example of engaged learning, providing an opportunity for you to share your experiences and what you've learned in a mutually supportive place; discuss problems you are facing and concerns that you have; and, seek guidance from your campus instructor for the journey you are taking (Sweitzer & King, 1995; Williams, 1975). The class sessions are a medium that is most helpful in integrating intellectual and affective learning, encouraging new understanding and creative responses, and strengthening the effectiveness of interpersonal relationships (Williams, 1975).

A seminar may be a bit different from other classes you have taken. For example, one basic assumption of a seminar is that each person has something to contribute (Royse et al., 2011), unlike many classroom experiences where the assumption often is that only the instructor has something to contribute. If everyone has something to contribute, everyone shares the responsibility for the success of the experience. You have additional responsibilities in this type of class, but then, you also reap additional benefits.

Traditional Model	Seminar Model
• The instructor has the most to contribute	• Everyone has something to contribute
• Many students; one instructor	• All are instructors; all are learners
• Students and instructor blame one another when things go wrong	• Students and instructors take collective responsibility
• The group is a collection of individuals	• The group is necessary for the accomplishment of learning goals

Sweitzer & King, 2003

An effective seminar, whether it is held in a face-to-face classroom or a technology-enabled distance-learning format, affords opportunities for reflective dialogue, support, the development of important relationships, and a variety of new learning experiences. A seminar class is one in which an exchange of ideas takes place, information is shared, and mutual problems are discussed. It is also a forum for problem-centered learning. You will have the chance to hear about and perhaps learn new professional skills, strengthen analytic and problem-solving skills, and develop knowledge of other placement sites as well as the communities in which they are located. For ideas about how seminar classes can be structured, see the Focus on THEORY box below.

It is important to remember that a seminar is not a therapy group and you are not therapists (nor are your instructors in therapist roles, although they may have those skills). However, there may be times when you or someone else encounters a challenge in the internship or has an experience that evokes feelings that need the attention of a counselor or therapist. Your campus supervisor can help you recognize those instances and locate appropriate resources to deal with them.

Finally, if you are seeking to build a community of support, then it is important that there is time in the seminar to reflect upon how that endeavor is going. Early in the seminar, you may want to discuss issues such as listening, effective feedback, and the overall goal of learning to function as a supportive group. As the semester progresses, take time to celebrate your successes and growth in this endeavor, to discuss and try to solve problems, and to give one another feedback about the achievement of the goals.

Focus on THEORY

Barns, Houses, Beauty Contests, or a Free-for-All

Although many courses carry the title of "seminar," they are not all run the same way. Michael Kahn (2003) uses clever metaphors to distinguish different kinds of seminars: the *Free-for-All*, the *Beauty Contest*, the *Distinguished House Tour*, and the *Barn Raising*. In the *Free-for-All* approach, students compete for the teacher's attention and approval. The group itself is not necessary; its only function is efficiency, since a group of twenty-five costs the same to run as a group of five. From the individual student's point of view, however, the other students are only seen as competitors. In the *Beauty Contest* approach, students show off their wonderful ideas and then spend their time thinking of the next wonderful idea, rather than listening to anyone else's ideas. Once again, the size or even the presence of the group serves no educational purpose. In the *Distinguished House Tours* approach, each student might present a case, a piece of writing, or an idea, and the group *does* pay careful attention, perhaps with skilled guidance from the instructor. Then they turn their attention to the next student. In a *Barn Raising* approach, everyone has a role and the group is essential. The group must be of a certain size, and the task can only be accomplished through cooperation and collaboration.

As you think about your internship seminar and its goals, ask yourself, *Where do they fit?* Is the group really necessary, or is it only there for efficiency? We believe that the last two types, the *Distinguished House Tour* and the *Barn Raising*, are the most appropriate models for an internship seminar. Using the first of these two models, each intern can be given a chance to talk about his or her experience, or perhaps present a case. The intern practices the skill of reflection by trying to interpret the story through various theoretical lenses. The group is there to give feedback and support, and perhaps to assist in the reflection and analysis.

If we use the *Barn Raising Metaphor*, then what is the barn? It is what the group is trying to accomplish together. One possibility is that it is trying to build a collective understanding of the profession, using the experiences of different people at different sites. Another possibility is that it is trying to build a collective understanding of the community or communities, of the assets and challenges, and the role that each of the sites and professions might play in maximizing those assets and addressing those challenges.

A third and very common possibility is that the group is trying to create a *community of support*, where everyone provides support, not simply the professor. It is important that interns have a place where they can talk about their experiences, their feelings and reactions, and their struggles and achievements. Although your friends and family can do some of this for you, it is often helpful to have this exchange with others

(continued)

who are undergoing a similar experience. Support groups exist for almost every purpose; perhaps you have participated in some. While the seminar is not a support group in the formal sense, one of its principal benefits is the quality of connections that you develop with your peers. Through these relationships, you give and receive support. In fact, you receive a double benefit—not only do you give and receive support but you also become more skilled at each of these functions. We will talk more about these skills in the next chapter.

In trying to build a community of support, it is also important to remember that the seminar class is a group, and it goes through developmental stages and group dynamics like any other group. It is not hard to describe the atmosphere that you would want in a group where people are sharing ideas, joys, and fears. But the atmosphere of trust, openness, safety, honesty, and feedback that characterize successful groups (Baird, 2011) does not just happen. It happens in stages, over time, with all members investing in it. Because of the sensitivity of some of the information that is shared in the seminar class (about clients or customers, supervisors, co-workers, the site, or yourself), it is very important to be clear early on about what is expected in terms of confidentiality, disclosure of information, and how interns are to conduct themselves in the class. The obligation to keep that information "in the room" is a common guideline for this course.

Overview of the Text

Chapter Organization

This book is organized into four sections. In Section One, *Foundations* (Chapters 1–4), we present the conceptual framework that underlies the book and give you some basic tools to proceed. Chapter 2 introduces you to the developmental stages of an internship, and Chapter 3 introduces you to some critical knowledge, skills, and attitudes, and values in the form of tools that we think you will benefit from right away. Chapter 4 focuses on self-understanding; we ask you to look at and think about yourself in relation to being an intern through a variety of lenses and offer you a range of tools for doing so. We suggest that these chapters be read before the internship begins *or* as a packet of readings during the first week of internship to benefit the most from them.

In Section Two, *Beginnings* (Chapters 5–7), we deal with the issues and concerns associated with getting started in an internship; we call this the Anticipation stage. Some of our colleagues and students have told us that they prefer to know and think about some of these issues and concerns before the actual placement begins, and this section certainly could also be used in that way. Chapter 5 discusses the Anticipation stage itself and outlines the issues you will face as well as our recommendations for facing them in an engaged manner. Chapter 6 emphasizes two critical dimensions of your internship: the establishment of an effective Learning Contract and the beginning of a productive supervision plan. Chapter 7, which will be of special interest to those of you in the helping and serving professions, focuses on clients, clientele, and other populations served by your site. This group of chapters serves you best if you begin reading them the second week in the field, as your work begins to take shape.

Section Three, *Rhythms* (Chapters 8–11), looks at the challenges that await you once the initial concerns are resolved. In Chapter 8, we introduce you to the next stage of your internship: the Exploration stage. Chapter 9 offers you a variety of tools and approaches for staying engaged and focused on progress as your exploration of the internship unfolds. Chapters 10 and 11 take you beyond your initial focus on your colleagues and, in some cases your clients, and ask you to consider issues and challenges related to the placement site itself and the community context of your work. For some interns, such as those in a community-organizing agency, the community is a more immediate focus; those interns may wish to read this chapter earlier.

The final section of the book, *Crescendos* (Chapters 12–14), examines issues and concerns that are common in the latter stages of the internship: *Competence* and *Culmination*. Chapter 12 deals with professional issues that typically arise as you are doing quite well in the internship. Chapter 13 deals with professional, legal, and ethical issues. Of course, some such issues can arise throughout the internship, but it has been our experience that interns often do not *notice* them in a conscientious way, regardless of whether and how often they are covered in class, until later. The final chapter, Chapter 14, guides you to end the internship on a number of levels and in productive, meaningful ways.

Chapter Exercises

At the end of each chapter, we offer you several ways to extend and enhance your learning in the "For Review & Reflection" section. The exercises fall into five categories:

- *Checking In*, which has an exercise to help keep you on target with the purpose of the given chapter;
- *Experience Matters*, which is intended for "EXperienced" interns, that is, those who bring much life or prior internship experience to this internship;
- *Personal Ponderings*, from the Latin *ponderare*, which means to consider deeply and thoughtfully; in this exercise, questions are posed for you to think about and perhaps to write about in your journal in substantive ways;

- *Civically Speaking* focuses specifically on the civic dimension of the internship; and

- *Seminar Springboards,* which are intended for either the individual intern or the whole class, regardless of whether it is held on campus or online, and offer questions and exercises for discussion or skill building.

Not all of these exercises follow each chapter and in many instances, we encourage interns to select those that are most meaningful to them at the time. In the "For Further Exploration" section that follows the "For Review & Reflection," there are additional and annotated resources in case there are areas you want to consider further.

Let the Journey Begin!

The past 8 months have been a journey within a journey…. (Internship is) where everything I have been learning comes to life…really makes a person think long and hard.
STUDENT REFLECTION

You've come a long way already. Just think about it…you are familiar with the terms used in the book; you have an understanding of the dimensions of personal, professional, and civic development that will guide your journey; you are aware of the importance of engagement and self-understanding to the quality of the internship; you have a sense of how learning occurs experientially (Kolb's experiential learning cycle); you know how the book is organized and what to expect at the end of the chapters; you've been introduced to the seminar class; and you know what it will take to make your internship high impact learning. You are ready to begin.

We wish you good fortune, many interesting times, and the best of learning during your journey. Before heading further down the path, take some time to *Get HIP!* Use the reflection exercises that follow for high impact learning and then you'll be ready to move forward and experience this journey.

For Review & Reflection

Checking In

On a Scale of 1-10.   Think about your internship up to this point. If you are yet to begin, think about how you've been spending your time waiting for your internship to start and how you felt about the delay. On a scale of 1 to 10, rate your experience with 1 at the low end and 10 at the high end. After you give a rating, think about why you chose the rating you did. Then consider why you didn't choose the rating below yours *and* what has to happen for your internship to earn the next rating level.

Experience Matters (For the EXperienced Intern)

What has been the most difficult aspect of beginning an internship given your life experience or experience with previous internships? What strengths have you gained from your previous internship or life experiences that can help you meet the challenge?

Personal Ponderings

The beginning of your internship is a good time to review your academic program's expectations of you, your supervisor, and your instructor during the internship. This is particularly important in terms of knowing what you can expect from others and what others, including the staff at the placement site, can expect from you. Take time now to locate any written documents from your program that specify these responsibilities. You will want to make copies and keep them with your other internship paperwork. Before doing so, review them closely and then highlight for each person—your site supervisor, your campus instructor, and yourself—what you weren't aware of before reviewing these documents, what you need to pay attention to, and what may not be on track at this time. It's the last piece of information that you will want to discuss with your campus instructor.

Civically Speaking

What's It All About? Some of you may be quite surprised and perhaps perplexed to see civic development an important part of your internship, where others are not at all surprised, and perhaps even relieved. Those interns tend to major in political science, environmental/green studies, sociology, public service, legal studies, the helping professions (counseling, human services, psychology, social work), and so on. Their work often-times is embedded in such civic issues as leadership, conflict management, diversity training, community assessments, health and medical issues, or the role of the arts in community development. Or, they may attend a college that emphasizes education for citizenship and personal/social responsibility across the curriculum. Regardless of your academic background or your future career plans, civic awareness, readiness, and competence are part of your internship.

Take a moment to think about *What* you now know that you didn't know before you read the introductory civic piece in this chapter. Then ask yourself this question: *So What?* Give yourself time to think about the power of those two words and do your best to respond to them. Then ask yourself this final question which also comes packaged in two powerful words: *Now What?*

Seminar Springboards

Fieldspeak, Agencyspeak, & Seminar Possibilities Internship students often use a language of their own. Your supervisor or co-workers may appear puzzled when you use certain internship terms, even though they are commonly understood on your campus. We call this language *fieldspeak*. There is also *agencyspeak*, which you may use without even thinking about it after a few days at the placement, but it will puzzle your seminar classmates or even your instructor. Review your program's definitions of terms and compare them to the ones in this chapter. Be ready to explain them to people at your placement site. When discussing your internship in class, be sure to translate the important terms and slang that make up your field site's *agencyspeak* so your peers will understand what you are talking about!

Seminar class can be an important part of the internship experience. Now is a good time to think about what you want from the class, especially from your peers and instructor. If there is a syllabus, is there room in that syllabus to incorporate interns' suggestions for topics covered and types of activities that you want included? What role do you see yourself as having in this class? Your peers? The instructor? Share these ideas in class and see just what kind of seminar you can create.

For Further Exploration

Alverno College Productions. (1985). *Critical thinking: The Alverno model.* Milwaukee, WI: Author.

A classic reference for identifying critical thinking skills, operating principles, analytical ability within disciplinary frameworks, and developmental levels of analysis and communication.

As You Sow. (2011). "Corporate Social Responsibility." Retrieved from http://asyousow.org/.

An example of an organization that works with corporations to promote environmental and social corporate responsibility through shareholder advocacy, coalition building and innovative legal strategies ... founded on the belief that that many environmental and human rights issues can be resolved by increased corporate responsibility.

Battistoni, R. (2006). Civic engagement: A broad perspective. In K. Kecskes (Ed.), *Engaging departments: Moving faculty culture from private to public, individual to collective focus for the common good* (pp. 11–26). Bolton, MA: Anker Publications.

An excellent summary of how academic majors can be vehicles for civic development.

Colby, A., Erlich, T., Beaumont, E., & Stephens, J. (2003). *Educating citizens: Preparing America's undergraduates for lives of moral and civic responsibility.* San Francisco: Jossey-Bass.

Thorough discussion of the theoretical and philosophical underpinnings of the education for citizenship movement, as well as sections focusing on examples of best practice at a range of higher education settings.

Coles, R. (1993). *The call of service: A witness to idealism*. New York: Houghton Mifflin.

This book draws on the author's direct experience with service, as well as his experience with countless volunteers, to examine the individual urge toward idealistic action. Strategies for using literature to illuminate service concepts are also discussed.

Collison, G., Elbaum, B., Haavind, S., & Tinker, R. (2000). *Facilitating on-line learning: Effective strategies for moderators*. Madison, WI: Atwood Publishing.

Informative, useful, and classic guide for netcourse instructors for moderating a web-based course such as the internship seminar.

Duley, J. S. (1991). The growth of experiential education in American secondary and post-secondary education: The Role of NSEE. *NSEE Foundation Paper*. Raleigh: NSIEE.

A paper based on the premise that experiential education is a movement; groundwork is traced to a period in the 1920s, followed by the formative years of the 1960s, through a proliferative period of literature, expertise, and resource development in the 1980s, including professional associations reflecting when credibility and acceptance were established.

Duley, J. S. (Winter, 1981). Nurturing service-learners. *Synergist, 9*, (3): 15. National Center for Service Learning, ACTION.

Offers the "Continuum of Pedagogical Styles in Experiential Learning," which identifies the learner's role as shifting from a passive, dependent role to an active, independent role and the corresponding faculty roles in making that transition happen.

Eyler, J. (2009, March). Effective practice and experiential education. Paper presented at the National Conference on Liberal Education and Effective Practice. Clark University.

This commissioned paper examines how classic forms of experiential learning (internships, co-op, service-learning) are effective in helping students solving "real-world" problems and transforming their ideas and values into effective action in field settings.

Eyler, J. (2009, Fall). The power of experiential education. *Liberal Education, 95*, (4): 24.

The author argues that experiential education can not only improve the quality of learning for those in the liberal arts, it can "lead to more powerful academic learning and help students achieve intellectual goals commonly associated with liberal education."

Eyler, J., & Giles, Jr., D. E. (1999). *Where's the learning in service-learning?* San Francisco: Jossey-Bass.

A thorough discussion of service-learning concepts and the results of the first large-scale, systematic study of the impact of service on learning.

Inkster, R. P., & Ross, R. G. (1995). *The internship as partnership: A handbook for campus-based coordinators and advisors*. National Society for Experiential Education. Retrieved from www.nsee.org.

and

Inkster, R. P., & Ross, R. G. (1998). *The internship as partnership: A handbook for businesses, nonprofits, and government agencies*. National Society for Experiential Education. Retrieved from www.nsee.org.

A series of two comprehensive and invaluable handbooks for internship programs and field sites. The first edition focuses on the responsibilities and challenges of the work of the field coordinator; the second edition does the same for the work of the site supervisors. Both editions have significant value for faculty and other campus supervisors in developing their effectiveness in working with placement sites.

Jackson, R. (1997). Alive in the world: The transformative power of experience. *National Society for Experiential Education Quarterly*, 22(3), 1; 24–26.

Excerpts from a keynote address in which the author explores the experiences needed to be considered an educated citizen.

McKenzie, R. H. (1996). Experiential education and civic learning. *National Society for Experiential Education Quarterly*, 22(2), 1; 20–23.

Based on a deliberative democracy seminar sponsored by the Kettering Foundation; participants spanned middle school through four-year colleges and included an international contingent, drawn from a network of those who use National Issues Forums in their classes. The role of deliberation in learning is underscored.

Moore, D. T. (1981). Discovering the pedagogy of experience. *Harvard Educational Review 51(2): 286–300.*

The author argues that the how learning occurs must be understood by the effect of the context in which it occurs and proposes a framework for analyzing the social organization of education in environments beyond the classroom.

Moore, D. T. (1990). Education as critical discourse. In Kendall, J. C. & Associates (Ed.), *Combining service and learning: A resource book for community and public service*. Vol. I. Raleigh, NC: NSIEE (NSEE).

The author argues that experiential education may stand apart as a way to create critical pedagogy, a form of discourse in which faculty and students "conduct an unfettered investigation of social institutions, power relations and value commitments."

Moore, D. T. (2013). *Engaged learning in the academy: Challenges and possibilities*. New York: Palgrave Macmillan.

A critique of whether "first-hand experience is a legitimate and effective source of learning in higher education; pitfalls are identified as are potential uses of engaged pedagogies. Conditions are suggested under which such pedagogies might work best."

NSEE Foundations Document Committee. (1998). Foundations of experiential education, December 1997. *National Society for Experiential Education Quarterly, 23*(3), 1; 18–22.

Describes the common ground of the members of NSEE at the time, reflecting the thinking of the membership and operating assumptions about describing experiential education; intended to initiate discussion about experiential learning and education.

Northwestern Mutual Life Insurance and Financial Services. (2012). Community Service Award. Retrieved from www.northwesternmutual.com.

Northwestern Mutual financial representative interns don't just get to test-drive the full-time career; they can also get engaged in the company's commitment to volunteerism. Through its annual Community Service Award, (http://www.northwesternmutual.com/about-northwestern-mutual/our-company/northwestern-mutual-foundation.aspx#Employees-and-Field) Northwestern Mutual interns like Chad Monheim can get recognition (http://www.youtube.com/watch?v=1ne4qft8gy4&feature=plcp) and charitable funding—for the philanthropic causes they're passionate about. Go to http://www.northwesternmutual.com/career-opportunities/financial-representative-internship/default.aspx to learn more about Northewestern Mutual financial representative internships.

Parilla, P., & Hesser, G. (1998, October). Internships and the sociological perspective: Applying principles of experiential learning. *Teaching Sociology, 26*(4), 310–329.

The authors argue that internships do achieve educational goals by providing opportunities to apply sociological principles, improve analytical skills, and use Mills's concept of "sociological imagination."

Schon, D. A. (1987). *Educating the reflective practitioner.* San Francisco: Jossey-Bass.

and

Schon, D. A. (1995). *The reflective practitioner: How professionals think in action* (2nd ed.). Aldershott, UK: Ashgate Publishing.

Schon's work on reflectivity is a classic and forms a foundation for subsequent work on practical reasoning.

Siemens. (2011). " The Siemens Caring Hands Program." Siemens Corporation. Retrieved from http://www.usa.siemens.com/answers/en/.

A not-for-profit organization of this global company established to receive, maintain, and disburse funds for sponsoring and encouraging charitable activities in which employees donate their time and talents to worthy causes to carry out the company's commitment to impact the communities in which they live and work through volunteerism in a variety of community service activities, including giving campaigns, blood drives, holiday food and gift collections, and disaster relief fund raising when their employees, operations, or business partners are affected.

Sullivan, W. M. (2005). *Work and integrity: The crisis and promise of professionalism in America* (2nd ed.). San Francisco: Jossey-Bass.

Excellent discussion of civic professionalism.

The Washington Center for Internships and Academic Seminars. (2012). Retrieved from www.twc.edu.

Students at the Washington Center (www.twc.edu) intern at government agencies, non-profits, think tanks and businesses such as the Department of Justice, Amnesty International, Merrill Lynch, No Labels and Voice of America. They also have the opportunity to become positive change agents by participating in civic engagement projects on important domestic and international issues including domestic violence, homelessness, veterans, torture abolition and the environment. Students learn about these issues by interacting with national and local community leaders, then become involved in a direct service and/or advocacy project. They can choose to join a TWC-guided project or design their own individual one. For more information about civic engagement at the Washington Center, please see http://www .twc.edu/internships/washington-dc-programs/internship-experience/leadership-forum/civic-engagement-projects. Articles about recent civic engagement activities can be found at TWCNOW http://www.twc.edu/twcnow/news/term/civic% 20engagement.

References

AACU. (2012). *The personal and social responsibility inventory.* Retrieved from http://www.aacu.org/core_commitments/PSRI.cfm.

Baird, B. N. (2011). *The internship, practicum and field placement handbook: A guide for the helping professions* (6th ed.). Upper Saddle River, New Jersey: Pearson Prentice Hall.

Battistoni, R. (2006). Civic engagement: A broad perspective. In K. Kecskes (Ed.), *Engaging departments: Moving faculty culture from private to public, individual to collective focus for the common good* (pp. 11–26). Bolton, MA: Anker Publications.

Chickering, A. W. (1977). *Experience and learning: An introduction to experiential learning.* Rochelle, NY: Change Magazine Press.

Colby, A., Erlich, T., Beaumont, E., & Stephens, J. (2003). *Educating citizens: Preparing America's undergraduates for lives of moral and civic responsibility.* San Francisco: Jossey-Bass.

Colby, A., Erlich, T., Sullivan, W. M., & Dolle, J. R. (2011). *Rethinking undergraduate business education: Liberal learning for the profession.* San Francisco: Jossey-Bass.

Crutcher, R. A., Corrigan, R, O'Brien, P, & Schneider, C. G. (2007). *College learning for the new global century: A report from the national leadership council for liberal education and America's promise.* Washington, DC: Association of American Colleges and Universities.

Dewey, J. (1916/1944). *Democracy and education.* New York: MacMillan Publishing Co.

Dewey, J. (1933). *How we think.* Lexington, MA: D.C. Heath and Co.

Dewey, J. (1938). *Experience and education.* New York: MacMillan Publishers.

Dewey, J. (1940). *Education today.* New York: Greenwook Press.

Eyler, J., & Giles, Jr., D. E. (1999). *Where's the learning in service-learning?* San Francisco: Jossey-Bass.

Garvin, D. A. (1991). Barriers and gateways to learning. In C. R. Christensen, D. A. Garvin & A. Sweet (Eds.), *Education for judgment.* Boston: Harvard Business School Press.

Giles, Jr., D. E. (1990). Dewey's theory of experience: Implications for service-learning. In J. C. Kendall Associates (Ed.), *Combining service and learning* (Vol. I, pp. 257–260). Raleigh, NC: National Society for Internships and Experiential Education.

Giles, Jr., D. E. (2002). *Assessing service-learning.* Paper presented at the Annual Conference of the National Organization for Human Service Education, Providence, RI.

Godfrey, P. C. (2000). A moral argument for service-learning in management education. In P. C. Godfrey & E. T. Grasso (Eds.), *Working for the common good: Concepts and models for service-learning in management* (pp. 21–42). Sterling, VA: Stylus Publications.

Hatcher, J. A. (2008). *The Public role of professionals: Developing and evaluating the civic minded professional scale.* Dissertation for Doctor of Philosophy, Indiana University, Indianaoplis, IN. Retrieved from http://hdl.handle.net/1805/1703.

Hodge, D. C., Baxter Magolda, M. B., & Haynes, C. A. (2009). Engaged learning: enabling self authorship and effective practice. *Liberal Education, 95*(4), 16–22.

Howard, J. (Ed.). (2001). *Service-learning course design workbook.* Ann Arbor, MI: OCSL Press.

Kahn, M. (2003). The seminar. Rereived from www.sonoma.edu.

Kolb, D. A. (1984). *Experiential learning: Experience as the source of learning and development.* New York: Prentice Hall.

Kolb, D. A. (1985). *Learning style inventory.* Boston: McBer & Co.

Kolb, D. A., & Fry, R. (1975). Toward an applied theory of experiential learning. In C. Cooper (Ed.), *Theories of group process.* New York: John Wiley & Sons.

Kuh, G. D. (2008). *High-impact educational practices: What they are, who has access, and why they matter.* Washington, DC: Association of American Colleges and Universities.

Lemann, N. (2004). Liberal education and the professions. *Liberal Education, Spring,* 12–17.

National Task Force on Civic Learning and Democratic Engagement (2012), *A crucible moment: College learning and democracy's future.* Washington, DC: Association of American Colleges and Universities.

O'Neill, N. (2010). Internships as a high impact practice: Some reflections on quality. *Peer Review, 12*(4), 4–8.

Royse, D., Dhooper, S. S., & Rompf, E. L. (2011). *Field instruction: A guide for social work students* (5th ed.). Upper Saddle River, NJ: Prentice Hall.

Sgroi, C. A., & Ryniker, M. (2002). Preparing for the real world: A prelude to a fieldwork experience. *Journal of Criminal Justice Education, 13*(1), 187–200.

Sullivan, W. M. (2005). *Work and integrity: The crisis and promise of professionalism in America* (2nd ed.). San Francisco: Jossey-Bass.

Sweitzer, H. F., & King, M. A. (1995). The internship seminar: A developmental approach. *National Society for Experiential Education Quarterly, 21*(1), 1; 22–25.

Sweitzer, H. F., & King, M. A. (2003). The Internship seminar: Approaches and best practice. Annual Conference, National Organization for Human Service Education, Nashville, TN, October, 2003.

Sweitzer, H. F., & King, M. A. (2012). *Stages of an internship revisited.* National Organization for Human Services. Milwaukee, WI.

Waddock, S., & Post, J. (2000). Transforming management education: The role of service learning. In P. C. Godfrey & E. T. Grasso (Eds.), *Working for the common good: Concepts and models for service-learning in management* (pp. 43–54). Sterling, VA: Stylus Publications.

Williams, M. (1975). The practice seminar in social work education. In M. Williams (Ed.), *The dynamics of field instruction.* New York: Council for Social Work Education.

Zull, J. E. (2002). *The art of changing the brain: Enriching the practice of teaching by exploring the biology of learning.* Sterling, VA: Stylus.

CHAPTER 2

Framing the Experience: The Developmental Stages of an Internship

As I read through the stages, I was comforted in knowing that I was not the only one experiencing all the various emotions. Knowing that others have felt the same way calmed my fears a bit to know that (what) I was experiencing was "normal."

STUDENT REFLECTION

E ach intern's experience is unique, and yours will be, too. You may have a different experience from other interns at the same placement or from any previous field experiences you have had. Placement sites differ, too, and you may be in a seminar with peers who are doing very different work with very different groups of people. We continue to be amazed and enriched by the diversity of experiences that interns have as well as the diversity of their personal, professional, and civic development; it is one of the factors that makes working with interns gratifying, even after many years. Over time, we have noticed some similarities that cut across various experiences. Many of the concerns and challenges that interns face seem to occur in a predictable order. Our experience, our review of years of student reflections, plus our study of other stage theories have yielded a developmental theory of internship stages that helps to guide the thinking behind this book. For this edition of the book, we have shifted from five stages to four. If you are interested in knowing more about this change, we refer you to the Preface. But for those of you familiar with the former version, we will refer to this one as DSI-2.

We have identified four developmental stages that students tend to experience in an internship: *Anticipation, Exploration, Competence,* and *Culmination.* Learning in each stage is driven by concerns that you experience; the concerns reflect what is most meaningful to you at that time in your internship. We use "concerns" here in two ways: one to signal interest and one to signal worry. So you may be concerned with (as in interested in) extending your learning as the internship progresses, but you may also be concerned (as in worried) that you will not be able to meet all your commitments. The stages are not completely separate from each other; rather, concerns from earlier and subsequent stages often can be seen, in less prominent ways, during the current stage. Certain concerns and issues are apt to be particularly prominent during a designated stage, along with related feelings. We cannot predict how quickly you will move through the stages; we can only predict the order in which the stages will occur. Your rate of progress through the stages is affected by many factors, including the number of hours spent at the agency, previous internships or field experiences, your personality, the personal issues and levels of support you bring into the experience, the style of supervision, and the nature of the work.

Each of the stages has its own obstacles for you to deal with and its own opportunities for you to grow through. Again, engagement is key. When you *engage* the challenges of each stage, you meet them head-on in the most productive ways and take the necessary actions to move through them—not around them. We will try to help you see what those actions are in general terms and give you tools to figure out what the best actions are for you in particular. *Disengagement,* on the other hand, means being *passively* involved in your learning and in your experience at the site. When this is your approach to learning—to the work of the internship— then you are letting things happen to you by letting them take their course when action is needed; hunkering down or waiting it out when changes are needed; and withdrawing emotionally and even physically, when dealing intentionally with the situation is what's needed.

Table 2.1 gives an overview of the four stages. Before we move into a description of each stage, though, it is important to recognize that there are times when the going can get tough, and extreme levels of disengagement can occur. This experience can result from a series of disengaged responses to smaller challenges, or from some combination of external and internal circumstances that challenge the limits of your knowledge, skills, and internal resources. Some of these circumstances are unfair, or even tragic, such as a loved one getting a serious diagnosis or losing their employment, the loss of a loved one or pet, or a traumatic event that happens to you or someone very close to you. When such circumstances affect your ability to function in the internship, a leave of absence from the

TABLE 2.1		
Developmental Stages of an Internship (DSI-2)		
Stage	**Associated Concerns**	**Critical Tasks**
Anticipation	Getting off to a good start	Examining and critiquing assumptions
	Positive expectations	Acknowledging concerns
	Acceptance	Clarifying role and purpose
	Anxieties	Developing key relationships
	Capability	Making an informed commitment
	Relationship with supervisor	
	Relationship with co-workers	
	Relationships with clientele	
	Life context	
Exploration	Building on progress	Increasing capability
	Heightened learning curve	Approaching assessment and evaluation
	Finding new opportunities	of progress
	Adjusting expectations	Building supervisory relationships
	Adequacy of skills and knowledge	Encountering challenges
	Real or anticipated problems	
Competence	High accomplishment	Raising the bar:
	Seeking quality	accomplishment and quality
	Emerging view of self	Having feelings of achievement
	Feeling empowered	and success
	Exploring professionalism	Maintaining balances
	Doing it all	Professionalism
	Ethical issues	
	Worthwhile tasks	
Culmination	Saying goodbye	Endings and closure
	Transfer of responsibilities	Redefining relationships
	Completion of tasks	Planning for the future
	Multiple endings	
	Closing rituals	
	Next steps	

Sweitzer & King, 2012

internship may be the most appropriate response. In the vast majority of situations, though, the challenges are yours to work out in the internship with the help of your supervisor and your support systems. Regardless of how you get to extreme levels of disengagement, the experience can leave you with deep feelings of discomfort. We call this experience *disillusionment*.

Although *disillusionment* is a common term, used to apply to anything from mild disappointment to profound discouragement, here we use the term to describe a fundamental shift in how you *feel* about the internship. It has been described by many as being in a whirlpool, moving downward in a spinning motion. The Italians have a word for this phenomenon; they call it a *vortice*. In English, it is called a vortex. There is no predicting when or even whether disillusionment will occur. It can occur early in placement, when anticipatory concerns are highest, or later in placement when competency concerns emerge. It can even occur during the final stage of your internship. Most of the time, though, if it occurs, it is during the Exploration stage. That is the time during the internship when the workload increases, a shift occurs to more responsible work, and there are greater demands on the intern for developing and demonstrating placement-specific skills.

We cannot say whether you will experience this pervasive feeling of losing control over your internship in any or all of the stages; that depends on the circumstances that arise as well as your ability to navigate them. However, once you are in disillusionment, you will know it; you will feel it as a crisis. When it is a personal crisis, such as those precipitated when the tragic situations described earlier occur, you will need to take an engaged approach and connect with both your campus and site supervisors to discuss whether taking a leave of absence from the internship makes sense. If it doesn't, then, like those who struggle with internship-based problems, you will need to develop a plan of support and structure to keep you on track so you can get yourself through a very difficult time.

In each of the stage chapters in the book, we discuss what the experience of disillusionment is like during that stage. Regardless of the causes, there is a way out of disillusionment. It does require concerted effort and extra support to get back on track; provided your supervision plan and support networks are in place, you'll find that the suggestions we offer in Chapter 9 to manage this crisis will let you do just that.

The Developmental Stage Model (DSI-2)

Internship is like a diamond, in that it is multifaceted; it is also like a roller coaster with its highs and lows.

The focus (of the week) has been for me to normalize my feelings and allow the process to happen.

Allowing the stages to happen allows the intern to learn and have positive learning experiences.
STUDENT REFLECTIONS

Stage 1: Anticipation

As you look forward to and begin your internship, there is usually a lot to be excited about. Interns often look forward to the internship for several semesters, and it is your best chance to actually "get out there," do what you want to do and, for many of you, take a step toward the profession you want to enter. For most interns, however, along with the eagerness and hope there is inevitably some anxiety, which is good and necessary

| In Their Own Words | Voices of Anticipation |

I had a lot of anticipation going into my internship, even at training I asked if it was normal to feel as nervous as I did. The trainer reassured me that if I didn't feel nervous, something would be wrong.

This stage reminded me of my first year in high school and college. I just wanted to be accepted and didn't know how to do that. Although I am starting to gain a sense of what is expected from me in the internship, I am still wondering what staff members and clients think of me. Do they think I am stupid, lazy, ignorant, etc.?

for learning. It may not be very visible, even to you, but there are enough unknowns in the experience to cause some concern and anxiety for anyone.

When it comes to interns, this anxiety generates the first set of concerns and challenges, which generally center on the self and key relationships in the internship, such as the supervisors, co-workers, and often the clientele (customers, patients, contractors, citizen groups, or clients). We often refer to this as the *"What if?"* stage because interns wonder about things like: *What if I can't handle it? What if they won't listen to me? What if they don't like me?* or *What if my supervisor thinks I know more than I really do?* You will probably be concerned about what you will get from the experience and what it is really like to work at this particular site. Many interns wonder whether they "really can do this" and what will be expected of them.

Some interns report fears that they are not really competent, that they have gotten this far only by lots of good luck, and that in their internship they will surely be found out. You may also wonder about your role: You are not in a student role while in the placement, but you are not a full-fledged employee either. Depending on your personal situation, you may also be concerned about your family and the effect such a demanding experience will have on them.

You are going to interact with a number of people during your internship, and it is natural to wonder what to expect from them and whether they will accept you in your role as an intern. It won't surprise you to read that most interns are concerned about how they are received and treated by staff in the role of an intern. You may, for example, be unsure of the role and responsibilities of the site supervisor and you will be unusual if you don't wonder what your supervisor will think of you. Those working directly with clientele inevitably wonder about how they will be perceived and accepted by them and the kinds of demands, behaviors, or problems that go along with the work. And, by now, you probably already are wondering how you will manage all the other responsibilities in your life while you conduct your internship and who will be there to support you.

In this initial stage, you may not yet be learning the specific things you set out to learn, and that can be frustrating. If you look back at the last few paragraphs, though, you will see that there is plenty here to challenge you, and it is important that you face and *engage* those challenges, rather than ignore or chafe at them. What is most important at this stage is that you learn to define your goals clearly and specifically, begin considering what skills you will need to reach them, and develop a realistic and mutually agreed on set of expectations for the experience. You have probably read, thought about, or maybe heard a lot about the site where you will be working for your internship. It is inevitable that you will make assumptions, correctly or incorrectly, about many aspects of the internship. Some of these assumptions come from stereotypical portrayals in the media of certain client groups (such as those with mental illness) or organizations (such as Wall Street firms, detention centers, corporations, law offices, green businesses, etc.); others may come from your own experience with certain issues or problems. As much as possible, you need to recognize and acknowledge these assumptions and then critique and examine them because that is your responsibility as an intern. A final key challenge to moving beyond this stage is feeling accepted by and developing good key relationships. As you meet these challenges, your commitment to your internship seems to become more informed and renewed.

If you have started your internship, you might be thinking to yourself that these matters don't really apply to you, that you have had some uncomfortable feelings, but nothing you would label as "anxiety," at least not yet. We have found that students who have the pressure of a graded internship that is essential for graduation can feel much more stressed and experience more intense anxiety than those who have an optional internship selected for personal interest or career exploration.

On the other hand, we have worked with interns who feel somewhat incapacitated at this stage; this is the shape of *disillusionment* at this time. The "what ifs" can be overwhelming, and in some cases they are exacerbated by negative experiences early in the internship that leave the intern feeling helpless and disempowered. If this happens to you, it is even more important to take the time, with support from on and off campus support systems, to actively change these circumstances. Your internship is too important to let negative experiences set the tone for your journey. Whenever you engage and emerge from *experiencing disillusionment*, you not only move forward, you feel empowered as a learner.

Stage 2: Exploration

After the Anticipation stage comes a stage where you grow toward feelings of competence, but you know you are not quite there. What you are

| In Their Own Words | Voices of Exploration |

I am so excited to be given a few responsibilities—real responsibilities—while I am shadowing my supervisor and co-workers and reading manuals about the site. It makes me feel like they think I can do the work.

Here I am in a site I always dreamed of and talking with real professionals as if I were one of them. They asked me what I thought this week about 2 situations and how the customers might respond. And, they listened to me. I felt SO good!

I was a manager for 20 years. I didn't know how I would handle being a beginner again. But, this week, both my supervisor and her supervisor asked me how I would handle a situation if it occurred when I was the manager. I felt really good about them recognizing expertise that I have that may be put to use in this field I am preparing for.

It feels good to be learning something new every day and talking about it with my supervisor. It feels like I have traveled so far in the first 4 weeks—I am even using language I didn't know before I started my placement.

doing is finding your rhythm; you are *exploring*. This is an interesting and affirming time in the internship because you are beginning to branch out into the work, taking on responsibilities that are important to the work of the site. There is an apprenticeship quality to this stage because of how closely you work with your supervisor and/or others at the site. You may also be interacting with the broader professional community, meeting colleagues at sites affiliated with yours. And, very importantly, you will learn throughout this stage how to do the work you went there to do, initially under close supervision and gradually more independently until you are ready to "fly" on your own with oversight supervision.

Interns face two challenges at this stage: assessing and improving, and growing through the challenges that inevitably come with learning. Assessing and improving means moving forward with development on all the dimensions: personal, professional, and civic. It means staying in touch with learning goals and seeking feedback on progress toward them. It also means taking frequent inventories of successes and failures on a variety of fronts. Growing through the challenges means recognizing them as part of the learning process and using effective tools to smooth out and learn from these inescapable ups and downs.

Exploration does not occur on a straight line. Some paths are fruitful and others are not; sometimes you have to back up and try a new path. There are bumps, potholes, and dead ends along the way. For one thing, there is almost always a difference between what you anticipated about your internship and what you really experience. If the concerns of the Anticipation stage have been adequately addressed, you will be less likely to encounter a wildly different reality in this Exploration stage from what you expected. Most likely, though, there will be some discrepancies as that is normal; some of them will be troubling and that is normal also. Furthermore, issues may arise that you simply never considered. Challenges can come at any time, from any direction, and in a variety of sizes. There are, though, some common aspects of the internship that often present challenges for many interns: the work itself, issues with the people, and issues with the larger community or political context. This is also a time when interns are being formally evaluated, and that can be upsetting on a number of fronts. Feelings associated with these challenges often include frustration, anger, sadness, disappointment, and discouragement. You may find yourself directing any or all of these feelings in any number of directions, including toward yourself.

As the saying goes, "The only way around is through," and the way to get through the Exploration stage is to engage it: Face and study what is happening to you. Some interns resist acknowledging any problems, even when their level of performance or satisfaction is dropping. You may fear that any problems must somehow be your fault or that you will be blamed for them. You may think that "really good" interns would never have these problems. Paradoxically, though, our experience with interns has shown us that taking a disengagement approach, namely, failing to acknowledge and discuss problems, can diminish your learning experience, your performance, and your evaluations both on site and on campus. If you refuse to acknowledge the problems you encounter, or if you are not able to approach them in a proactive, solution-oriented manner, you will likely feel increasingly disconnected from your experience. It is important to keep in mind that *having problems* is a necessary and essential element of an internship; otherwise, growth and development will not occur because you grow through the challenge of handling the problems.

If you let smaller problems go unaddressed, or if you encounter major problems that overwhelm you, you may slide toward disillusionment, which in the Exploration stage is marked by a pervasive and sometimes debilitating sense that something has gone very wrong with your internship, or even with you.[1] If you do not decide to do something productive

---

[1] In some cases, the intern's primary focus is not the work of the internship, but the alluring location of the internship site. For example, some students use an internship in part as a way to be in a major city, like New York, Chicago, or Washington, DC; to be part of a cultural experience, such as Hollywood, Wall Street, or Disney Productions; or to work or study internationally. In the case of such geospecific internships, the experience of disillusionment, if it occurs, is likely to focus on some of the unexpected realities of life in this new and intriguing place. Of course, some interns are focused equally on the work and the location. For them, there may be two possible sources of feelings of disillusionment in the internship experience.

In Their Own Words	Disillusionment—and Empowerment!

At a certain point, the internship was not what I expected it to be.

I was still unsure of what I would be doing next, and getting up early was very difficult for me. I found myself wanting to go back to bed rather than getting up and going to my internship.

There has been a loss of focus for 3 weeks in regards to my not doing what I went there to do. I should have discussed the issue sooner with my supervisor, but I kept telling myself that next week will be busier. I am feeling frustrated going in each week and still there was no real work for me—and it is written right there in the contract that I will be doing it.

I recognized, sensed and felt the great struggle and satisfaction exploring change can initiate within others as well (as me).

I chose to confront this situation and figure out just what was happening to me that made me feel unsatisfied.

Once I confronted these anxieties, though, everything worked out. I am beginning to confront this situation by making something good of this.

about the situation, then at best learning and growth will be limited; at worst, the placement may have to be renegotiated or even terminated. On the other hand, we remind you again that each time you engage with and overcome an obstacle, you will feel more independent, more effective, and more empowered as a learner. You will have a sense of confidence that comes not just from what you have accomplished, and not from denying problems, but from the knowledge that you can grapple with problems effectively.

What you may need in order to stay engaged is a good tool kit to ride out these bumps. We will give you a beginning set of tools in the next chapter, Essentials for the Journey (Chapter 3); and, we will give you a more advanced set in Chapter 9.

Stage 3: Competence

As your confidence grows, you will forge ahead into a period of excitement and accomplishment. This is the stage that every intern looks forward to—the reason for the internship. Morale is usually high, as is your sense of investment in your work. Your trust in yourself, your site supervisor, and your co-workers often increases as well. This is the stage where interns often begin to identify themselves as professionals, rather than exclusively as interns. As an emerging professional, you have a solid platform from which to expect, or even demand, more from yourself and your placement; that is what engagement looks like at this stage. You

---

The Competence stage consists of being confident in myself. I am looking forward to that aspect of my internship. It's not that I have never been competent in a task before. It is just that the sense of professionalism will be much greater.

(The site supervisor) gave me some good guidance and lots of space to create what I wanted. This is usually a good thing for me. To be left alone to do what I wish. I felt intimidated by my audience so I want to be perfect in what I present and not look stupid. So, I have had some difficulty getting it all together.

I feel that all my years of schooling have been for this exact purpose. The minute I walked in the door, I felt at home...and in my element.... It's like nothing I have ever felt before.

While I was aware that I had experienced a difficult day (at work), my focus at the internship at night was to be more attentive. It meant putting into perspective all that occurred during the day in order to be effective at night. Balancing the work during the day continues to be the hard work of negotiating the intellect and the affect.

---

may find that you want more than you are getting from your assignments, your instructor, or your supervisor. Many interns also report that during this time, they are better able to appreciate the professional and ethical issues that arise in their placements and are more willing to discuss them with their supervisors.

This stage is not without its predictable issues, and you will read more about them in Chapter 12. This is a point in the internship where interns can begin to feel the press of trying to meet all their commitments in and outside the internship. At the same time, interns find that they are leveling off, reaching a plateau in the learning curves of the internship. If no one is pushing them to go further, and they are not pushing themselves, they can feel somewhat disengaged from their internship—almost like just "going through the motions." To prevent a slide into further disengagement or even disillusionment, the intern needs to deal with the need for greater challenge by bringing that need to the attention of the supervisor. Waiting for the supervisor to figure this out won't work. Supervisors expect interns to take responsibility to recognize their needs and seek to have them met in proactive ways.

If this dynamic goes on long enough, engagement in the internship changes and the intern can disengage to a point of apathy. The high levels of morale and investment that characterized the internship up to this point are replaced with complacency and boredom. The aspirations that once steered the intern with high levels of achievement are replaced by a need to "just get by." As in the other stages, if disillusionment occurs, the intern's responsibility, as painful as it may be, is to take the time, with support from your on and off campus support systems, to be proactive in bringing things under control. You have done it before, which may make it easier this time, and you certainly can do it again.

Stage 4: Culmination

The approaching end of the internship, coupled with the end of the semester and in some cases with the end of the college or university experience, can raise some significant issues for you. You may experience a variety of feelings as this time approaches. Typically, there is both pride in your achievements and some sadness over the ending of the experience. If you are working in the helping or service professions, you even may feel guilty about not having done enough for your clientele.

Those of you who are ending your college years may be concerned with continuing your education, finding employment, surviving the economy, or perhaps all of these. Relationships with friends, family members, lovers, and spouses that have been organized around your role as a student have to be reorganized around your new role in life. In any case, there are many good-byes to be said and good-byes are never easy. For some people, they are very difficult.

Often, interns find ways to avoid facing and expressing these feelings, particularly the negative ones; this is the shape of disengagement in the Culmination stage. Avoidance behaviors may include joking, lateness, or absence. Some interns may devalue the experience—they begin saying it hasn't been all that great or find increasing fault with the placement site, the community, the supervisors, or the clientele. Many interns find themselves having a variety of feelings and reactions, some of them conflicting and changing by the hour. This can be very confusing and upsetting, especially if you are not aware that this is more typical than not.

To engage the concerns of this stage, you need to focus on your feelings (whatever they may be) and have appropriate ways to express them; find satisfying ways to say good-bye to staff, supervisors, community groups, customers, patients, clients, and in some cases, other interns, both at the site and in an internship seminar on campus; and make

---

In Their Own Words	Voices of Culmination

I looked forward to entering the Competence stage, but I am not looking forward to (the) stage—when placement ends.

The end of this experience will be sad...everything is ending...but the reflections will be great.

plans for the future on a number of fronts. Of course, if you do not pay attention to the concerns of the Culmination stage, the internship will end just the same. However, you could be left with an empty and unfinished or unfulfilled feeling if the concerns are not addressed in productive ways. In some cases, interns struggling with the Culmination stage actually sabotage their internship by allowing their discomfort about ending it to color their perceptions of the entire experience and affect others' perceptions of their work. This sabotage is one shape of disillusionment in this stage.

Connecting the Stages and Civic Development

As you read about the stages, you should be able to see opportunities for personal, professional, and civic development inherent in the journey. Again, we find that the civic dimension can be elusive for some interns, and so we offer this chart to help reinforce the civic possibilities of each stage.

Stage	Opportunities for Civic Development
Anticipation	• Develop awareness of concepts, vocabulary, and definitions of civic learning and civic professionalism in an internship setting • Include the civic dimension in goals, activities, and assessments in the learning contract • Cultivate openness to diversity • Understand the concept of cultural competence • Clarify own cultural profile
Exploration	• Clarify values and differences in values between what the profession expects and your personal as well as emerging professional values • Clarify the meaning of the concept of "public good" and what responsibility you see yourself as having to it as an intern • Clarify the responsibility to resolve differences through discussion and reasoning while demonstrating effective communication skills • Learn organization's civic mission and public relevance • Determine degree of site's involvement in community • Conduct community assessment
Competence	• Demonstrate knowledge of the public relevance of one's work • Demonstrate continued refinement of *practical reasoning* skills • Demonstrate knowledge of the civic responsibility of the profession • Demonstrate civic professionalism in the human context of the work • Demonstrate knowledge of agency's social obligation to the community
Culmination	• Assume identity of emerging civic professional • Demonstrate commitment to civic effectiveness as a person and a professional

Sweitzer & King, 2012

Conclusion

Reading about (the) stages, I was amazed to see how much in common I had with the student reflections.

Reflecting on a recent confrontational situation at my site ... I began to read (about the stage), but after a few pages, I stopped. I had two immediate reactions: one, I wish I had this material a few weeks ago, it would have helped me greatly ... and two, I would prefer to read (this) text from start to finish ... (even now that my internship has ended) ... to follow the internship process "postmortem."

I am aware that my stages and concerns are not unique to me ... One strategy was to respond to the concerns of each of the stages. The vocabulary (of the stages) gave me a way to express myself...to work to normalize my feelings and behaviors ... this week I acknowledged feelings, reassessed goals, expectations and support systems, and began to develop strategies. and I shared concerns and started to identify unfinished business. These experiences spurred me to make decisions, (to) choose behaviors I would not have in the past.

STUDENT REFLECTIONS

Now you have a better sense of what is ahead in your internship and in this book. Keep in mind that even though these stages are predictable, especially when viewed by those observing you do your work, both the pace with which you move through the stages and the phenomenological experience of being in each of them will vary a great deal from intern to intern. As you *experience* your internship, you will find that the chapters that follow consider each stage in greater depth. We encourage you to remain focused, as well, on aspects of yourself, which you will explore in depth in Chapter 4. The chapter that follows provides you with essential tools to begin and engage your internship. And, you are ready to do just that.

For Review & Reflection

Checking In

Metaphorically Speaking. So, now you know. You will be having an *experience*, not only an internship. The use of a metaphor can be a powerful way to think about the experience. The metaphor can be sensual, visual, as well as emotive. Think about one that captures your internship. It's important to choose one that can expand as the internship matures. There are some examples from our interns below. Then think about the next step in your metaphor—What does it look like? How does it feel? How does it sound? Smell? Many interns choose to write in their journals about their metaphor and how it develops over time.

Internship is:

- The frame of the house that will be my life.
- A thousand-piece puzzle with many colors.
- A blossoming flower starting to develop new and different petals.
- An extremely fast roller coaster.
- A sailing trip plotted with hopes and wrought with uncertainties and challenges.
- A skydiving adventure; once I am out the door there is no turning back.
- A caterpillar that will eventually cocoon and fly as a beautiful butterfly.
- A new relationship, filled with hope, curiosity, excitement, mystery, anxiety, and awkwardness.

Experience Matters (For the EXperienced Intern)

As you read the description of the developmental stages of an internship, did the stages remind of you of any other experiences you have had? If so, in what ways? What strengths do you draw from the similarities? What challenges do expect to face knowing what you know?

Personal Ponderings

What is your reaction knowing that there are stages of an experience you already are experiencing? How can knowing about the stages being helpful to your experience? How could it compromise your experience?

Seminar Springboards

Think about the issues raised for you by your understanding of the stages and discuss them with your peers. Are there issues that the stages do not obviously address? Create a plan for managing those issues with your peers and campus instructor.

For Further Exploration

Birkenmaier, J., & Berg-Weger, M. (2010). *The practicum companion for social work: Integrating class and field work* (2nd ed.). Boston: Allyn & Bacon.

Thorough, contributive fieldwork text for the social work student.

Baird, B. N. (2011). *The internship, practicum, and field placement handbook* (6th ed.). Upper Saddle River, NJ: Prentice Hall.

Comprehensive and especially useful to graduate students in the helping professions.

Boylan, J. C., Malley, P. B., & Reilly, E. P. (2001). *Practicum & internship: Textbook and resource guide for counseling and psychotherapy* (3rd ed.). Philadelphia: Bruner-Routledge.

Takes a comprehensive approach to many aspects of graduate counseling internships.

Chiaferi, R., & Griffin, M. (1997). *Developing fieldwork skills: A guide for human services, counseling and social work students.* Pacific Grove, CA: Brooks/Cole.

Offers a developmental framework of stages for the intern-supervisor relationship.

Chisholm, L. A. (2000). *Charting a hero's journey.* New York: The International Partnership for Service-Learning.

Based on Joseph Campbell's tale of the hero's journey, this is a guide for reflective journal writing in a cross-cultural context, for students involved with study abroad and service.

Cochrane, S. F., & Hanley, M. M. (1999). *Learning through field: A developmental approach.* Boston: Allyn & Bacon.

Takes a developmental approach to the fieldwork experience in social work education.

Faiver, C., Eisengart, S., & Colona, R. (2004). *The counselor intern's handbook* (3rd ed.). Belmont, CA: Wadsworth Thomson Learning.

Takes a focused, pragmatic approach to counseling field experiences.

Gordon, G. R., & McBride, R. B. (2011). *Criminal justice internships: Theory into practice* (7th ed.). Cincinnati, OH: Anderson Publishing.

Offers the undergraduate criminal justice intern a comprehensive guide to issues specific to a criminal justice internship.

Grobman, L. M., Ed. (2002). *The field placement survival guide: What you need to know to get the most from your social work practicum.* Harrisburg, PA: White Hat Communications.

A collection of previously published articles from The New Social Worker that identify and address important issues across the duration of the fieldwork experience.

Jowdy, B. & McDonald, M. (2008). *The Impact of experienced-based learning on students' emotional competency.* VDM Verlag Dr. Muller Aktiengesellschaft & Company: Saabrucken, Germany. 978-3—8364-3766.

The study examines the impact of a sport event management course on students' emotional competency and considers whether students' emotional competency can be increased when the students are not introduced to emotional intelligence theory.

Kiser, P. M. (2011). *Getting the most from your human service internship: Learning from experience* (3rd ed.). Belmont, CA: Brooks/Cole.

Offers a framework for processing and integrating previous learning while engaging in field experiences.

Lacoursiere, R. (1980). *The life cycle of groups: Group developmental stage theory.* New York: Human Sciences Press.

Explains Lacoursiere's theory in detail and discusses its application to many kinds of groups.

Milnes, J. A. (2003). *Field work savvy: A handbook for students in internship, co-operative education, service-learning and other forms of experiential education.* Enumclaw, WA: Pleasant Word/ Division of Winepress Publishing.

A nuts-and-bolts guide filled with advice that leads the student through the many tasks of a structured experience in field-based learning.

Schutz, W. (1967). *Joy.* New York: Grove Press.

A group development theory that has had an effect on our view of internships and talks a great deal about acceptance, inclusion, and control issues.

Sheehan, B., McDonald, M., & Spence, K. (2009). Developing students' emotional competency using the classroom-as-organization (C-A-O) approach. *Journal of Management Education, 33*(1), 77–98.

The paper describes the C-A-O pedagogy and complementary experiential learning activities used to develop students' emotional competency (e.g., self-confidence, self and social awareness); empirical support found for continued and increased use of this pedagogy in management education.

Sweitzer, H. F., & King, M. A. (1994). Stages of an internship: An organizing framework. *Human Service Education, 14*(1), 25–38.

Gives more details of how Lacoursiere's and Schutz's works informed our thinking about the developmental stages of an internship model.

Sweitzer, H. F., & King, M. A. (1995). The internship seminar: A developmental approach. *National Society for Experiential Education Quarterly, 21*(1), 1, 22–25.

Discusses our general approach to working with interns in a seminar class from a developmental perspective.

Sweitzer, H. F., & King, M. A. (2012). *Stages of an internship revisited.* National Organization for Human Services. Milwaukee, WI.

Described the revised Developmental Stage Model (DSI-2), the stages of Anticipation, Exploration, Competence, and Culmination, and the experience of disillusionment.

Models That Frame Field Experiences

Chiaferi, R., & Griffin, M. (1997). *Developing field work skills.* Pacific Grove, CA: Brooks/Cole.

Chisholm, L. A. (2000). *Charting a hero's journey.* New York: The International Partnership for Service-Learning.

Cochrane, S. F., & Hanley, M. M. (1999). *Learning through field: A developmental approach.* Boston: Allyn & Bacon.

Gordon, G., & R., McBride. (2011). *Criminal justice internships: Theory into practice* (5th ed.) Cincinnati, OH: Anderson Publishing.

Grant, R., & MacCarthy, B. (1990). Emotional stages in the music therapy internship. *Journal of Music Therapy, 27*(3), 102–118.

Grossman, B., Levine-Jordan, N., & Shearer, P. (1991). Working with students' emotional reaction in the field: An educational framework. *The Clinical Supervisor,* (8), 23–39.

Inkster, R., & Ross, R. (1998, Summer). Monitoring and supervising the internship. *National Society for Experiential Education Quarterly,* (23), 4, 10–11, 23–26.

Inkster, R., & Ross, R. (1998). *The internship as partnership: A handbook for businesses, nonprofits, and government agencies.* National Society for Experiential Education. Retrieved from www.nsee.org.

Kiser, P. M. (1998). The integrative processing model: A framework for learning in the field experience. *Human Service Education, 18*(1), 3–13.

Lamb, D., Barker, J., Jennings, M., & Yarris, E. (1982). Passages of an internship in professional psychology. *Professional Psychology,* (13), 661–669.

Kerson, T. (1994). Field instruction in social work settings: A framework for teaching. In T. Kerson (Ed.), *Field instruction in social work settings* (pp. 1–32). New York: Haworth Press.

Michelsen, R. (1994). Social work practice with the elderly: A multifaceted placement experience. In T. Kerson (Ed.), *Field instruction in social work settings* (pp. 191–198). New York: Haworth Press.

Rushton, S. P. (2001). Cultural assimilation: A narrative case study of student-teaching in an inner-city school. *Teaching and Teacher Education,* (7), 147–160.

Siporin, M. (1982).The process of field instruction. In B. Sheafor & L. Jenkins (Eds.), *Quality field instruction in social work* (pp. 175–198). New York: Longman.

Skovholt, T. M., & Ronnestad, M. H. (1995). *The evolving professional self: Stages and themes in therapist and counselor development.* New York: John Wiley & Sons.

Sweitzer, H. F., & M. A. King. (2008). Dimensions of an Internship: Personal, Professional and Civic Development. Annual Conference, New England Organization for Human Service Education, Boston, MA, April, 2008.

Sweitzer, H. F., & M. A. King. (2012). *Stages of an internship revisited.* National Organization for Human Services. Milwaukee, WI.

Wentz, E. A., & Trapido-Lurie, B. (2001). Structured college internships in geographic education. *Journal of Geography,* 100, 140–144.

CHAPTER 3

Internship Essentials
Tools for Staying Engaged

Throughout this book we use the metaphor of a journey to describe your internship experience. As we discussed in Chapter 1, active engagement with that journey is key for getting the most from your internship. Maintaining this active stance and regaining it when you lose it (and you will sometimes), will make a substantial difference in your experience. In this chapter, we discuss more thoroughly the features of engaged and disengaged learning as they apply to internships, and then present our ideas about what you will need at the outset of your journey to remain engaged. These essentials include an awareness of the assets and challenges you bring to the journey, information you can easily get if you don't have it, skills you need to start developing, attitudes you need to cultivate, and resources you need in order to take an empowered stance. You may find that you already have some of what you need. You may also decide that your particular internship does not require some of the attitudes, skills, and knowledge that we describe. We encourage you to make those decisions in collaboration with your campus instructor or supervisor.

As you read through this chapter you are going to encounter a good deal of information, and we encourage you to refer back to this "tool kit" often as you progress through the initial weeks of your internship, even beyond that. To make that reference a little easier for you, we have created a chart that shows the major resources in this chapter and where you can find them (see Chart 3.1).

As much as we hope these tools will be all you need, we know that may not be so. While most interns do not face substantial or deeply upsetting challenges in the first few weeks, it can happen. We have included a more advanced tool kit in Chapter 9, which provides you with resources for such situations. You can consult as needed at any time.

Internship Essentials

Chart 3.1 A Guide to the Tools for Staying Engaged

The Engaged Learner

A successful internship depends on the intern taking responsibility for his or her learning in the field; we have referred to this approach as being an *engaged* intern and learner. In the Think About It box that follows, are common characteristics we have observed in interns who are engaged learners (Sweitzer & King, 2012)

THINK About It

Profiling the Engaged Intern

The engaged learner has what it takes to be successful in the internship. The skills and qualities of this type of learner that are relevant to the internship experience are listed below. In many ways, this list serves as a useful guide for the intern, ensuring that the intern stays on the path to success—both in the internship and in the career that follows.

Engaged interns invest in learning when they…

- Take responsibility to seek out knowledge and understanding.
- Take charge of pace, direction, and shape of what they learn.
- Desire to be successful in learning.
- Are excited about learning and welcome learning opportunities.
- Ask insightful and challenging questions.
- Self-motivate.
- Think critically and question the givens.
- Self-direct through tasks.
- Are intellectually curious.
- Commit themselves to meaningful work.
- Solve problems strategically and effectively.
- Have a sense of personal agency.
- Self-initiate direction and tasks.
- Are proactive in the face of challenges.
- Seek connections between what they learn from various sources.
- Seek to understand the wider context of their learning.
- Take pride and delight in knowledge and in the work.
- Persist in the face of adversity and challenge.
- Work independently as well as collaboratively.

A Word of Caution… The over-engaged learner may be overdoing it! Burnout may be inevitable if this engaged learner doesn't take stock of all that he or she is committing to and take action to prevent it (Swaner, 2012).

Focus on THEORY

Forms of Engagement

In writing about the various factors that lead to successful, engaged learning, Lynn Swaner discusses three forms of engagement that we consider especially important for internships (2012):

- *Holistic Engagement.* Engaged learning is more than using your brain to think (the cognitive domain). Engaged learning is an affective experience as well; it draws on your "heart" as well as your "head." The balancing and integration of the cognitive and affective domains is one of the challenges and opportunities of an internship.

- *Integrative Engagement.* This form of engagement focuses on combining various learning experiences toward a common goal. Throughout this book we encourage you to reflect on your experience from multiple perspectives, including the ones you bring, the ones you encounter along the way, and ones that others may suggest to you. Completing Kolb's learning cycle (see Chapter 1) means considering the implications of all your reflections, drawing some conclusions from them, and (most importantly) taking some action as a result of your conclusions.

- *Contextual Engagement.* An engaged learner is able to place learning in a broader context. Throughout this book, we will encourage you to consider the social, political, and civic contexts of your work and your learning, as well as the contexts of your personal and professional aspirations in relation to the internship.

As you look over the list, make note of those qualities or skills that you already have. If you do not recognize many of the characteristics as yours, there is a good chance that you are not yet a fully engaged learner. On the other hand, if most of the characteristics do not describe your ways of learning, there is a chance that you are entering the internship as a *disengaged* learner. If so, you have decisions to make. Choosing to become an engaged learner puts you on the path to a successful internship. Choosing not to puts you on the path of disengagement and risks a failed internship. Fortunately, your supervisors are prepared to work with you in developing the skills of the engaged learner.

Next, we move to a discussion of the attitudes, values, skills, and knowledge that will help you remain engaged in your internship now and in the weeks to come. Chart 3.2 provides a visual depiction of this profile of engagement.

CHART 3.2
The Engaged Intern

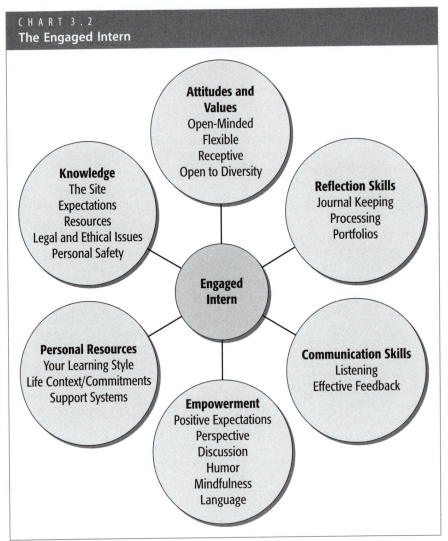

Attitudes and Values
Open-Minded
Flexible
Receptive
Open to Diversity

Knowledge
The Site
Expectations
Resources
Legal and Ethical Issues
Personal Safety

Reflection Skills
Journal Keeping
Processing
Portfolios

Engaged Intern

Personal Resources
Your Learning Style
Life Context/Commitments
Support Systems

Communication Skills
Listening
Effective Feedback

Empowerment
Positive Expectations
Perspective
Discussion
Humor
Mindfulness
Language

© Cengage Learning 2014

Essential Attitudes and Values

As we mentioned in the previous chapters, many professions have lists of essential attitudes and values. What we present to you here are those we believe are essential to success in any internship.

Being Open-Minded

You approach an internship with at least some preconceived ideas about the profession, the work you will be doing, your likes and dislikes, and so on. In some cases, the internship may confirm every one of those preconceptions. In most cases, though, some will turn out not to be true. It helps a great deal if you keep your mind open to the way things *really* are, as opposed to the way you wish they were or thought they might be. If you keep an open mind, you can also learn that there are many ways to do a good job, that many styles work well depending on the situation, and that norms and customs that may seem strange, useless, or even destructive to you in fact serve an important purpose.

Being Flexible

We hope that you and your campus support people have planned your internship carefully; if so, the internship site is clear with you about what will happen, when, and with whom. In most organizations, however, situations change, and sometimes suddenly; emergency shelters are a good example. Other settings, such as corporations, are more stable, but no organization is immune, and when situations change for the organization, sometimes they change for you. The more flexible you can be, the better you will be able to respond. If you can keep focused on the learning opportunities that are there for you and that may arise while you are there, you will be less attached to the specific experiences through which you learn.

Being Receptive

Consistent with the principles of high impact practice, there are multiple opportunities for feedback in an internship and it is imperative that you be receptive to constructive feedback on your work. If you are not, you will surely make a bad impression, but more importantly you will miss out on great learning opportunities. Supervisors, co-workers, other interns, instructors, and even clients (if you have them) can be valuable sources of information about your strengths and weaknesses. We don't mean to say that you should accept all feedback at face value, only that you should consider it from a nondefensive position. If you think of feedback as criticism, you are apt to become defensive. If you react defensively, you will be less likely to get feedback in the future, and even less likely to benefit from it.

Being Open to Diversity

It is very likely that you will meet people whose cultural backgrounds are different from your own, influenced by factors including race, class, gender, ethnicity, and sexual orientation. There is a natural human tendency to react with some suspicion to differences, and for some people their first response is to equate difference with being somehow wrong. Others are tolerant of differences, but assume that their way of doing things is really the best way and attempt to get others to change. Yet others take a "live-and-let-live" approach, appearing to accept diversity but not showing any real interest in it. All of those stances result, again, in lost opportunities for learning and growth. If you are willing to engage with, learn about, and learn from members of other cultural groups, if you look for the strengths and assets in their norms and values, then diversity will enrich your experience, making you a better professional and a better citizen.

THINK About It

Do Others See You as "Just an Intern"?
Is That How You See Yourself?

Many interns find relief in knowing they are not being treated as "just an intern." As a matter of fact, feeling like "just an intern" is one of the most prevalent fears and sources of anxiety that interns report having when they begin their internships. You might want to give some thought to these questions now so you can head off an attitudinal shift before it happens at the placement site, whether it's your attitude or those of site supervisors.

- What exactly does *just as an intern* mean to you?
- How would you know if you were being treated as "just an intern"?
- Have you been feeling this way? If so, what is it about your internship that cultivates this attitude in you? In your supervisors or co-workers?
- If not, what is it about your internship that leaves you feeling more than "just an intern"?

Essential Skills

Your attitudes form a solid base for you to move forward, while your skills are the hands-on tools you will need. Again, many more skills will be presented later, both in this book and at your placement. But here are some that we consider crucial for you to have right away.

The Skills and Habits of Reflection

Reflection to me is the connection that you make between an experience and all of your feelings that surround it. I believe more and more that reflection can only make someone better. Just looking at myself in regard to my internship, school and personal life, if I never stopped to reflect on the emotions that I was feeling, the thoughts that I was having, and the

*knowledge that I was being taught, I would never learn, and essentially
never grow.*
STUDENT REFLECTION

Reflection is a fundamental concept in any kind of experiential
education activity, and the internship is no exception. It is also the hallmark of professional practice (Schon, 1987; Schon, 1995; Sullivan, 2005).
In order to turn your experience into a *learning* experience you need to
stop, recall events, and analyze and process them. Although this may
sound daunting, you actually do it all the time. If you are walking back
from class and find yourself mulling over the remarks of a professor,
or wondering how a classmate came up with a particularly interesting
comment, you are reflecting. If you are in the car and start to think
over an argument you have had with a friend, trying to figure out
what happened and how you could have handled it differently, you are
reflecting.

Make It Deliberate and Consistent

The previous examples are instances of spontaneous reflection; we want
you to make reflection a deliberate and regular habit, which requires
intentionality as well as time. Reflection requires that you look back on
your experience; for your internship the reflection should start at the
beginning of the experience and be integral throughout (NSEE, 1998).

Use Varied Perspectives

Reflecting on experience from multiple perspectives means drawing on
internal and external "lenses" for reflection. Internal lenses include
values, reaction patterns, learning styles, and cultural identity. External
perspectives include theories, the written "wisdom of practice" set down
by others, and the opinions and perspectives of friends, co-workers, and
others. When you use external perspectives, you have the opportunity to
step outside of the ways you normally make sense of the world to consider different possibilities.

Take the Time

Developing the habit of productive reflection takes patience, practice,
and discipline. It means setting aside quiet time to think, because, as
one of our students put it, *"the best answers come from the silence within."*
It also means resisting the temptation to just keep going from one activity to the other, in your internship or in your busy life; another intern
noted, *"the internship proceeded at such a fast pace that I often felt it
was one step ahead of me."* Paradoxically, we have found that one of the
best ways to stop the internship from getting ahead of you is to make
time to think.

Try the Techniques

There are lots of techniques for reflection; we will discuss some specific ones below, but it is not a comprehensive list and you will need to find the one or ones that work best for you, your instructor, and your situation. Remember, too, that there is a difference between reflecting and recording, although the two can overlap. Your campus instructor, your placement site, or both may have specific ways that you are to keep a record of what you have done. These records may be used for documentation and kept in official files. They may be used in supervision as well. The primary purpose of reflection, though, is to promote your growth and learning, and the primary audience for a reflective technique is yourself.

Make it High Quality

Eyler and Giles (1999) offer important guidelines for selecting and assessing potential reflective techniques. These principles are called the "Five Cs": *connection, continuity, context, challenge,* and *coaching.* In keeping with these principles, you need to make reflection a habit; structure and connect your reflection to learning goals; and, make sure you are challenged to reflect more deeply and through a wider range of lenses. You also need to work with your campus contact to ensure that your choice of techniques makes sense for your particular situation and that you get the coaching and guidance you need for those techniques to be used to your best advantage.

Focus on THEORY

Eyler and Giles's Five Cs

The principle of *connection* refers to the importance of connections... between the classroom and the field, the campus and the community, your experience and your analysis of it, your feelings and your thoughts, the present and the future, you and your peers, the community, and your campus and site supervisors (Eyler & Giles, 1999).

The second principle is that of *continuity,* which refers to learning as a lifelong process for you, and the importance of developing a habit of reflecting on your field-based learning experiences, be they service-learning, practicum, internship, co-op education, or study abroad. *Context* is the third principle and refers to the thinking and learning you will be doing in the field with the tools you are given, the concepts you are learning, and the facts of a given situation. This principle also refers to the importance of the style you use and the place where your reflecting takes place.

The fourth principle is that of *challenge,* and it refers to the need for you to be challenged in order to grow. It is the challenge of a new

(continued)

Focus on THEORY *(continued)*

situation and new information, such as that of your internship that creates an ambiguous situation, at odds with your perspectives, and requires that you resolve these differences by developing "more complex and adequate ways of viewing the world" (pp. 184–185).

The fifth principle is that of *coaching*, which refers to the emotional as well as intellectual support you will need during your internship; the sources of these forms of coaching could come from both your campus and site supervisors, your peers, and your personal support system. Emotional support means a safe space where you know that you can express your feelings and share your insights without criticism and with respect. By intellectual support is meant having the opportunities to ask questions, think in new and different ways about situations, and question in retrospect how you think about issues and situations that challenge you (p. 185).

Keep a Journal

I believe that class work and journals are critical to internships because they allow support from peers, feedback from teachers, and reflection on your own work and feelings.

STUDENT REFLECTION

One of the most powerful tools for reflection that we know of is keeping a journal. Your instructor may require you to keep a journal of some kind, but even if it is not required in your setting, we strongly recommend that you keep one. We also suggest that you write an entry following every day that you go to your internship. Although it may occasionally seem like a chore, if you put time into it, journal keeping will give you a way to see yourself growing and changing. It also forces you to take time on a regular basis to reflect on what you are doing. A well-kept journal is a gold mine to be drawn on for years to come. It becomes a portfolio of the experience as well as a record of the journey.

Make Time   Again, perhaps the most important thing you can do for your journal is to allot sufficient time to do it. Doing it over lunch on the due date is not a good approach! For many of you, it is going to take practice and focus to learn to write in your journals in the most effective and productive way. As you plan your days and weeks, leave at least 30 minutes after each day at your internship to write.

Adapt It to Your Needs   If you have learning challenges that make it impossible or difficult for you to write this type of assignment, or if writing does not come easily to you, then you might consider investing in some voice recognition software such as Dragon 10. If that is not feasible, with your instructor's permission, you could tape-record your journal for your supervisor's review. Of course, you will need to negotiate these arrangements with your instructor, but a little time and thought should yield a method that allows you to reflect comfortably on your experience and maintain a dialogue with your instructor.

Work in Cyberspace For those of you who are doing your internship at a great distance from campus, or as part of a distance-learning program, the journal takes on added importance. In addition to the benefits already mentioned, the journal and responses to it are a way for you and your campus instructor to have a continuing conversation about your work and your reactions to it. Advances in web-based technology such as Blackboard make it easy to send journal entries back and forth. If you do not have access to these technologies, e-mail can work just fine.

Keep It Safe If you do decide to keep a journal, make sure you are very clear with your instructor, supervisor, and clients about the intent of the journal and issues of confidentiality. If your journal is for your personal use only, then there is no issue. You have full responsibility for its contents and for ensuring that what you write is for your eyes only. However, if you want or are required to show it to other interns, your instructor, your supervisor, or anyone else, you must be careful not to disclose information about clients, the placement, or even yourself that is supposed to be kept private (see Chapter 13 for a discussion of the privacy issues in the pervasive new HIPAA regulations). Discuss this issue with your instructor and your site supervisor before going too far with your journal. You may also be concerned that you cannot be completely candid in your journal if some of the people you are writing about are going to read it. Some interns keep their journals in loose-leaf format and merely remove any pages they wish to keep private before showing the journal to anyone else.

Choose a Structure There are many different approaches to journal writing, and many different reflective techniques. Your instructor may have forms and techniques that you are required to use, but we would like to discuss just a few of the more common forms here.

Unstructured Journals The simplest form of journal writing is just to take time after each day to think back on what stood out for you that day. Although there is no "right" length for these entries, they should record what you did and saw that day, new ideas and concepts you were exposed to and how you can use them, and your personal thoughts and feelings about what is happening to you. It may be helpful to divide what you learn at an internship into categories: (a) knowledge, (b) skills, (c) values, (d) personal development, (e) professional development, and (f) civic development. Try to include all these categories in your journal.

Focus on SKILLS

When You Don't Know What to Write

Many interns tell us they are afraid that there are going to be days when there is just nothing to say. Well, our experience is that you won't have that happen very often, but there may be some days when writing is difficult. For those days, here are some questions to consider, generated by a community service program a number of years ago that apply as well today as they did then (National Crime Prevention Council, 1988):

- What was the best thing that happened today at your site? How did it make you feel?
- What thing(s) did you like least today about your site?
- What compliments did you receive today, and how did they make you feel?
- What criticisms, if any, did you receive, and how did you react to them?
- How have you changed or grown since you began your work at this site? What have you learned about yourself and the people you work with?
- How does working at this site make you feel? Happy? Proud? Bored? Why do you feel this way?
- Has this experience made you think about possible careers in this field?
- What kind of new skills have you learned since beginning to work at this site, and how might they help you?
- What are some of the advantages or disadvantages of working at this occupation?
- If you were in charge of the site, what changes would you make?
- How has your work changed since you first started? Have you been given more responsibility? Has your daily routine changed at all?
- What do you think is your main contribution to the site?
- How do the people you work with treat you? How does it make you feel?
- What have you done this week that makes you proud?
- Has this experience been a rewarding one for you? Why or why not?

Source: "Reaching Out: School-Based Community Service Programs," by the National Crime Prevention Council, 1988, p. 101.

Other Kinds of Journals (Inkster & Ross, 1998; Compact, 2003; Baird, 2011) Some other forms of journals that have been used with interns include:

- **Key Phrase** journals are those in which you are asked to identify certain key terms or phrases as you see them in your daily experience. A more expanded version of this concept will be discussed later.
- **Double Entry** journals are divided into two columns. In one column you record what is happening and your reactions to it. In the other you record any ideas and concepts from classes or readings that pertain to what you have seen and experienced.
- **Critical Incident** journals identify one incident that stands out over the course of a day, or a week, and analyze it in some depth.

Use Processing Techniques There are a number of processing techniques that you may want to include in your journal. Some techniques are specific to a discipline, such as "verbatims" in pastoral counseling or "process recording" in social work. If you are not in one of these fields, you can use different ones at different times, or you can use one consistently, depending on your preference or that of your instructor. There are merits to both choices. Switching techniques from time to time may let you see things you have been missing. On the other hand, using a consistent technique allows you to look back over several entries and look for patterns. Common techniques include process recording (Sheafor & Horejsi, 2003), SOAP notes and DART notes (Baird, 2011). In two *Focus on SKILLS* boxes in this chapter, you can read about other techniques that are less common but that we think are very useful. Three column processing is particularly useful for those just learning the skill of reflection (Weinstein, Hardin et al., 1975; Weinstein, 1981). It provides a structured way of writing about incidents (both positive and negative) that you recall so that you can look for tendencies and patterns. The integrated processing model (Kiser, 1998) is a particularly useful approach for developing the art and skill of practical reasoning, as it guides you back and forth between theory and practice.

Consider a Portfolio Another use for your journal is as part of a portfolio of your experience. You can use a portfolio as a personal record, but many academic programs require them. A portfolio is often used to document and reflect on your entire journey in your academic major, and your internship experience can make a valuable contribution. There are many types of portfolios, and a discussion of them is beyond the scope of this book; however, there are resources listed at the end of the chapter if you want to explore portfolios further. Portfolios also can be valuable assets in an interview for a job, for graduate school, or even for

Focus on SKILLS

Weinstein's Three Column Processing

Gerald Weinstein (1981; Weinstein, Hardin, & Weinstein, 1975) developed a method of reflecting on events that may be helpful with your reflective journal. Take a moment at the end of the week to recall any events that stand out in your mind. Select one or two (they can be positive or negative). Divide a piece of paper into three columns. In the left-hand column, record each action taken by you or others during the event.

Record only those things that you saw or heard, such as "she frowned," "he said thank you," or "they stomped out of the room." List them one at a time. Now, review the list and try to recall what you were thinking when the different actions occurred. When you recall something, enter it in the middle column, directly across from the event. For example, you may have been thinking "what did I do now?" when the people left the room. Finally, read the list again and try to recall what you were feeling at the time each action and thought occurred. Record what you recall in the right-hand column. For example, you may have felt embarrassed, confused, or angry when they walked out.

another internship. If you are interested in using portions of your journal for this purpose, start planning for and discussing that project with your instructor. There may be items you will want to remove, such as highly sensitive comments about a person or organization. There may also be items you want to include, such as samples of your work, photographs, brochures, and so on. Your instructor can help you think about this now so that you know what sorts of items to seek out and save and so that you avoid violating confidentiality.

Communication Skills

You may have already studied the skills of effective communication in your courses. There is a wide range of such skills, and we cannot cover them all here. Once again, we have chosen two that we believe are essential for an internship.

Active Listening

Listening well is an art, a skill, and a gift to the other person. *Active* listening means giving the other person your full attention and making sure that you understand what the person is trying to tell you. It means avoiding the impulse to rush in with advice or to tell a story about something

Focus on SKILLS

Kiser's Integrative Processing Model

Kiser's model focuses on helping you integrate past learning with present experience. The integrative processing model (1998, 2012) consists of six steps:

1. *Gathering Objective Data from Concrete Experience*   In this step you select an experience that you have seen or been part of. You can use a written, videotaped, or audiotaped account of the experience.

2. *Reflecting*   In this step you record and assess your own reactions to the experience. You may respond to particular questions or you may use a less structured format.

3. *Identifying Relevant Theory and Knowledge*   Here you seek out or recall ideas that can help you make sense of the experience in a variety of ways.

4. *Examining Dissonance*   Now you review all the ways you have looked at the experience to see whether there are any points of conflict. These conflicts may be between or among competing theories; between what the theory says should happen and what actually did; between what you believe and what the agency seems to value; or between any two or more aspects of the experience. Sometimes this dissonance is resolvable, and sometimes it is not.

5. *Articulating Learning*   Here you look back over your writing and thinking and write down the major things you have learned from thinking about this experience.

6. *Developing a Plan*   This comes next, and here is where you consider the next steps in your learning and your work. You may identify areas you need to know more about and places to pursue that knowledge. In addition, you may identify new goals or approaches you plan to use in your work. Taking these next steps can be another new experience, and the cycle can begin again.

similar that happened to you (Johnson, 2008). There are, of course, many other ways to respond to someone, and they all have their strengths and weaknesses. The Focus on THEORY box that follows presents one way to categorize and think about responses.

Listening is the most important component of support. It sounds simple, but think about how rare it is that someone really wants to listen to you, especially when you are struggling (attention wanders, small-talk intrudes, unwanted advice is given). It is a wonderful experience just to have someone listen quietly, attentively, and empathically to whatever you want to say. As one student put it: *"Knowing that others are feeling* similar feelings doesn't make those feelings go away, but it does make me feel better about having them. I feel more comfortable now opening up to my class-mates and feel better equipped in encouraging them." So listen carefully, and listen actively. Use skills such as paraphrasing to be sure that you under-stand what the person is trying to say. You will often find that these tech-niques also help the person to elaborate even more.

Once you have listened, and listened carefully, we suggest that you ask the person what she or he needs at the moment. Advice, analysis, problem solving, or reassurance can be wonderful if they are wanted. Or, it may be enough just to be heard.

Giving Feedback

You may be in situations at the site as well as in an internship seminar where you are asked to give feedback to a peer, a co-worker, or even a supervisor. Effective feedback is specific and concrete, as opposed to vague and general; it should refer to very specific aspects of the situation being discussed. It is descriptive rather than interpretive. You can tell the person how you feel about what they did or did not do, but it is not your job to tell them why they did it. Feedback is usually best delivered using an "I" statement, rather than a "you" statement. *I had a hard time under-standing what was happening when you were describing the situation* is better than *Your story was really confusing.* If you have feelings about what was said, state them directly and attach them to a specific statement or portion of the story. And finally, feedback should always be checked with the receiver to see whether you have been understood and whether you have understood the other person.

Focus on THEORY

Modes of Responding

Theories about and guidelines for effective communication abound in the helping professions. You may very well have read some of them. We offer here, as one way to think about responding, categories of responses sug-gested by David Johnson in his extraordinary book *Reaching Out* (2008). Imagine that you or a peer has just told a story about something at the internship. A patient has relapsed, a co-worker or supervisor has been very harsh in their criticism, a community meeting turned into a shouting match, or whatever other event you want to imagine.

Advising and Evaluating. If you have taken a helping skills class, you probably know that giving advice is often not the best approach, but when a friend or classmate is struggling, it can be awfully tempting to offer your heartfelt suggestions. And it can be a great relief as well. But

(continued)

it is not always helpful, and it certainly does not empower your peer to confront and resolve the situation—not to mention the next one! We have found that it is usually best to hold your advice until and unless it is asked for.

Analyzing and Interpreting.   Here, the listener uses theory to interpret what has happened. It may be an opinion about the underlying psychological dynamics of the people involved, a sociological analysis, or any number of other interpretations. These thoughts can be contagious as other students move into the intellectual realm they know so well. But they can also be very distracting, especially if that is not the direction in which the person telling the story wants to go.

Questioning and Probing.   There is a difference here between asking clarifying questions, so you are sure you understand, and asking questions that take the conversation in a different direction. So, if a student tells a story about a customer and someone asks, "Is this the same customer you spoke about last week?" that is clarifying, but "Who referred this customer to your agency?" is not.

Reassuring and Supporting.   The term *support* in this context may be confusing, because we are talking about building support, but as Johnson uses the terms, this way of responding has as its goal to calm the person and reduce the intensity of feelings. Sometimes that is fine, but sometimes it sends the message "don't feel bad." And sometimes what the person needs is just to experience their feelings, at least for a few minutes. Sometimes the listener will do what we call a "me too," relating a similar incident that happened. It may be helpful for someone to hear whether others have had a similar problem and how they handled it, but let the request come from that person.

Paraphrasing and Understanding.   These are other terms for active listening, meaning that the goal of these responses is to make sure that the person is able to explain and elaborate fully the actions, thoughts, and feelings that occurred. There are a lot of specific techniques for achieving this goal, and there are resources suggested for you at the end of the chapter.

Essential Personal Resources

The most valuable assets you have for engagement in this journey of learning are what you personally bring to it. By this we mean the wealth of your life experiences, aspirations, knowledge, hopes, expectations, attributes, and relationships. These qualities and resources are your personal reserve of vitality and endurance; they have the potential to give you good feelings and get you through difficult moments. We focus on three resources in particular because of their importance to the learning experience: learning style, life context, and support systems. These resources can be powerful assets for you if they are in place and you know how to use them; they can also be sources of challenge for you if they are not in place or you are not sure how to get the most from them.

Your Learning Style

Learning style means the way that you most effectively take in and process information, and knowledge of your learning style is critical for you, your campus instructor, and your site supervisor (Birkenmaier & Berg-Weger, 2010). There are many theories that examine learning style, and they are not necessarily mutually exclusive. Your learning style is influenced by genetics, family norms and values, gender, other cultural factors, and your personal learning history. Take a moment right now to think about the ways that you learn best; these are your strengths as a learner. Perhaps you have already thought about this issue and you can answer the question in general terms. If not, think about learning experiences you have had that were successful and unsuccessful for you. What has been your experience with learning from readings? From group discussions? From hands on, trial-and-error approaches?

If you are involved in an internship that is not well matched to your learning style and strengths, you can still learn, but something needs to give. You may be able to take steps to change or augment the learning experience so that it is better suited to your strengths. On the other hand, you may need to try to stretch your learning repertoire and strengthen a style that does not come easily to you. In this way, a mismatch of styles can be an opportunity for growth.

Your Life Context

My biggest concern is stress, basically from outside the internship. The stress to get all the papers done on time, working many hours a week at my job to pay bills, and still put in 30 hours in the internship.
    STUDENT REFLECTION

Your life context consists of all the other things going on in your life in addition to the internship. That context will vary according to your family situation, your social life, and the configuration of your academic program. You will have some responsibilities outside your placement, some expectations placed on you (or being continued), and possibly some stress. Here are some areas to consider:

Academics

In some programs, interns take very few, if any, other classes during their internship semester. In other programs, the internship can be the

---

THINK About It

How Would You Describe Your Learning Strengths?

Here are some statements from our students about their learning strengths. Try developing one of your own:

- *When learning something new I understand it much better when I'm able to try myself. Instead of asking me questions and quizzing me, I'm much better at just showing you what I can do. I also do not learn when people stand right over me. The best way for me to learn is hands-on experience; please take the time to let me.*

- *I am an active learner. I learn best when observing or doing things hands on. While in the process of learning I may write a few notes so that I can refresh my memory if need be. Being an active learner allows me to process information effectively while experiencing it. I also like to reiterate what I am hearing if I feel I do not fully understand a process or subject. This method allows me to figure out if I understand what was explained or what parts I misunderstood.*

- *I seem to be able to learn in a variety of environments and with a variety of styles. I tend to be an auditory learner. On most occasions I sit and listen and take very few notes but I can recall the highlights of the topic with minimal difficulty. Please keep this in mind as I can attend numerous conferences and trainings that cater to many learning styles.*

---

equivalent of one course, leaving a full-time student with three or four other courses to carry. Some students even need to carry an overload because they must finish college within a certain time frame. The internship tends to be at least as demanding of your time as one course, or even two. Our students also report that internships demand a good deal of emotional energy, much more than a traditional course does.

The work can be exhilarating, and many interns find themselves thinking about their work when they are not at the placement or putting in extra time to read and research issues relevant to their work. But the work can also be emotionally draining, and the situations some interns face with individuals and communities can be heartbreaking and frustrating. Many interns have told us that they find themselves thinking and talking about the internship a great deal, which sometimes makes it hard to focus on other classes.

Employment

Some internships are paid; many are not. If you have been holding down a job in addition to attending classes, you will want to think about whether you can and should give it up during the placement. If you are going to be employed during your internship, think about the schedule you will have and how it will fit into the time demands of your placement and coursework.

Roommates

If you live with other people, they are part of your life context. Schedules for housekeeping, sleep, quiet time, study, and entertainment are areas of negotiation with roommates, and you've probably already managed a way to live with others. However, you may need to do it again depending on the demands of your placement. You may be up earlier, home later, or not available for cooking or other chores at your expected time.

Family

In this category we are including your family of origin (your parents, guardians, siblings, and anyone else who lives with them) and your nuclear family (your spouse or partner and children). You may be living with any, all, or none of these people, but you undoubtedly have responsibilities to them. What are they? How flexible are they? Could they be changed for this semester if the demands of the internship warrant it?

Intimate Partners

If you have a special partner in your life, whether or not you live with that person, this is another area of responsibility. It took time and energy to build that relationship, and it will take time and energy to maintain and nurture it during the internship. How much time have you been devoting to that relationship? What does that person expect of you, especially now that you have begun your internship?

Friends

Although some friends are closer than others, every friendship is a responsibility, too. Think about the amount of time you spend with friends, individually or in groups, and about how much time you hope to spend with them this semester. Think, too, about the demands those friendships put on your time and energy.

Other Commitments

Many interns make the mistake of planning to "just insert" an internship into an already busy life and expecting it all to work out somehow. Some of our busiest students fall prey to this misconception, especially those raising families, working full time, caring for a family member, or seriously involved with athletics. We are going to encourage you to look a little more objectively at your life context. If you list all the people and activities that you will be committed to this semester and estimate the number of hours each week or month each of those commitments is going to take, you may be shocked to discover that you have committed yourself to more hours than exist in a week or a month. If so, just as when you are doing your financial budget, you will have to make some compromises. Often, this compromising involves talking with others to whom you now have commitments about how you need to rearrange things, just for the short run.

Many interns report feeling that they have less time than they ever had. When we ask them to conduct an inventory of their commitments, as just described, they find that they should have enough hours in the day and week, but they just do not feel as if they do. Sometimes that comes from poor estimating. For example, if a class lasts for an hour, you need to remember that you will spend some time getting to and from class. One strategy you can try is to keep careful track for a day or two of everywhere you go, how long it takes to get there, and how long you spend there. However, another reason for this feeling is that interns often underestimate the psychological commitment and emotional demands of the internship. Whether you realize it or not, your emotional energy is likely to be taxed by an internship, and that is going to affect how available you feel for other activities and responsibilities.

Yourself

All of us have activities we engage in for pleasure and as a source of stress reduction. Hiking, playing the piano, painting, lifting weights, meditating, and reading are just a few examples. You will need to take the best care of yourself that you can to meet the demands of the internship. You may be able to cut down some on hobbies if necessary, but you will need to be realistic about the time required to take proper physical and emotional care of yourself.

Your Support Systems

Staying engaged requires that you balance all the demands we were just discussing in a good support system; these are the people who give you what you need to get through life's challenges. Your support system will be an important part of helping you meet the demands of your internship and the other demands in your life. As one of our interns pointed out in a journal entry, *"Support teams are a way to relieve stress and frustration ... they listen to you, give advice, and give support. It is very comforting to know that you have someone to fall on when needed."*

Everyone needs some support, more at some times than others, and you are no different. Sometimes, people in the helping professions have trouble accepting that they need support, so they do not seek it out and often decline it when it is offered. They give to others unstintingly and enjoy being people that others can count on, but they are better at giving than receiving. Most of the time, sooner or later, they become exhausted and are of little use to anyone. It may be that they have an image of the perfect helper—someone who never needs help. "How can I help others," students have asked us, "if I have problems myself?" Others know they need support, but they seem unable to accept it. Accepting your need for support and developing a strong support system are critical skills that could be called into play during the internship.

Regardless of the nature of your internship, your support system is made up of many different people, and you will need different kinds of support at different times. Here is a partial list of the kinds of support you might need. You may be able to add to it (Seashore, 1982).

- *Listening*   Sometimes you just want someone who will listen to you without criticizing or offering advice. The person listening should be someone to whom you can say almost anything and on whom you can count not to grow restless or frightened. Think of these people as your "sounding boards."

- *Advice*   On the other hand, sometimes you need sound advice. You may not always follow it, but you need a source of advice that you can trust. Think of these people as your "personal consultants."

- *Praise*   There are times when what you need most is for someone to tell you how great you are. If they can be specific, all the better. Think of these people as your "fans."

- *Diversion*   Some people are friends you can count on to go out and have fun with. You don't have to talk about work, your problems, or anything else. You just have activities you enjoy together. Think of these people as your "playmates."

- *Comfort*   When we were children and we became ill, there was nothing we wanted more than pure comfort. A comfortable place to rest, good food, music we enjoy, all without having to lift a finger! At times, ill or not, this is still just the kind of support we need. Think of the people in your life who comfort you as your "chicken soup people."

- *Challenge*   There are times when challenge is the last thing you want. At other times, though, someone who will push you to do more, look at things in a different way, and confront problems or inconsistencies in your thinking is the best friend you have. Think of these people as your "personal coaches."

- *Companionship*   It is good to have people in your life with whom you feel so comfortable that you can do anything, or nothing, with them. Sometimes you may not care what you are doing, only that you are not doing it alone. Think of these people as your "buddies."

- *Affirmation*   Another kind of support comes from people who have some of the same struggles that you do. It is a lonely and depressing feeling to believe, or suspect, that you are the only one who is troubled by a particular issue or set of circumstances. Knowing that others feel the way you do, even if they can't change it, can be very helpful.

Sometimes there is no substitute for someone who has been through what you are going through. Think of these people as your "comrades."

As we said earlier, you are not going to need all these kinds of support at the same time. You are also not going to be able to get all the support you need from the same person or group. As you read the list, think of people who are helpful in your life with one aspect of support but not with others. Both of us have friends who are fun to be with and provide wonderful diversion, but they do not listen well at all! You will know soon enough if you have called on the wrong person for support. For example, you may need good advice, but call on the listener. Even if there were one person in your life who could provide all these kinds of support, that person would soon become exhausted if you asked him or her to meet all your support needs. Similarly, some people who can provide a particular kind of support may not be available at any given time. Try not to become frustrated. This is an opportunity to learn about your needs and about how different people can best help you meet them.

Think about these categories and add some of your own. Now think about how well each of those needs is currently met in your life. You may find that there are some gaps that you need to address. Your support system will be stronger if you have more than one person in each category. Remember that the internship may, by itself or in combination with other things occurring in your life, create a greater need for support than you now have. The internship may also strain your existing support system. If most of your friends are not involved in internships, for example, they are not going to know how you feel, and they may not understand what you do or how difficult it is. "That's work?" they may say, or "You get credit for that?" They may not understand your need to get to sleep early, work on weekends, or cut back on your social life. Or they may have negative attitudes about the population you are working with, like those who are homeless, have addictions, or have HIV/AIDS.

If you have discovered gaps in your support system, we urge you to take steps to fill them. Cultivate new friends; discuss your needs with friends and family. Like investing in self-understanding, it will pay off for you now and later.

Essential Knowledge

In addition to essential personal resources, attitudes, and skills, there is also knowledge that is essential to have as you begin this journey in learning. In this section, we will introduce the sorts of knowledge you may need as well as some issues you could encounter in your role as an intern that could have professional, ethical, and/or legal implications.

Cultural Competence

Chances are that in your internship and your life you are going to interact with people whose cultures are different from your own. The term *culture* is defined in many different ways, but for our purposes, we use the definition offered by Donna Gollnick and Philip Chinn (2005) that culture is a shared and commonly accepted set of beliefs, practices, and behaviors. As such, culture does not simply refer to race, ethnicity, and nationality, but also to characteristics such as gender, social class, and sexual orientation. *Being culturally competent means being able to work and communicate effectively with people from a range of cultural backgrounds.* Such competence is not something you achieve quickly, easily, or completely; it is a journey rather than a destination. Components of cultural competence include awareness of your own cultural influences (we will help you explore this component extensively in Chapter 4), knowledge of the important aspects of various other cultures and the ways they affect worldview, and adapting your skill set to include approaches that achieve the desired effect with a variety of cultures (Diller, 2011). For example, we just discussed communication skills and the importance of listening. In some cultures, eye contact is a critical component of listening; in others, it is a sign of disrespect. In some cultures, asking for help is a sign of weakness. What you can do now is try to find out the cultural backgrounds of the people you will be working with (and/or serving) and try to learn something about them.

Information About the Site

You will learn a lot about your internship site as the internship progresses, but there are some things you should know before you begin. Depending on how the placement process on your campus works, you may already know most or even all of this information. If not, it is important that you find out.

You need to know where the site is located and your options for getting back and forth. If you are planning on using public transportation, be sure to find out how far you will have to walk to and from the bus or subway and what sort of neighborhood(s) you will be walking through. You also need to know the norms of dress and behavior at your placement site. If you arrive either over- or underdressed, that will be a momentary embarrassment that is easily corrected, but with a little research you can avoid it. If you start out addressing the others who work with you in an informal way (using their first name, using slang when you talk) and the norm of the workplace is to be more formal, it can be more than a brief embarrassment. Finally, you need to understand the rules and conventions about confidentiality. In medical and social service settings, this is a huge issue, but it can be an issue in other settings as well. You may want to come home on the first day and tell your friends and family all about something that occurred or you may want to take a file home with you. And that may all be fine, but it may not be, and crossing those lines can be considered a serious ethical transgression. It could cost you your internship and result in sanctions by your academic program.

Knowing What Is Expected

It is important that there is a clear, preferably written, and mutual understanding among all parties—yourself, the site, and the academic program— about what your start and end dates will be, your hours, those days (if any) that you will not be there (for example, during semester breaks) or that the site will be closed, and your basic responsibilities. There should also be a clear understanding of any other conditions or requirements of you, the internship site, or the academic program. In many programs, this document takes the form of an agreement or contract, which is signed by all parties and kept on file. If your program does not use such a document, you should ask how you can get some sort of written understanding of these expectations and issues.

On-Site Resources

Most interns have assigned supervisors on-site, and we believe that supervision is a critical component of a successful internship. We will discuss supervision in detail in Chapter 6, but as you begin you work, you should know who will be supervising your work, providing you with feedback, and answering your questions. It may be a single person,

---

THINK About It

You Have Rights!

As an intern, you may feel like you don't have any special rights, but you do. What's very important is for you to be aware of them and determine whether they are being respected in your internship.

- The right to a field instructor who knows how to supervise, i.e., has been adequately trained and skilled in the art of supervision
- The right to a supervisor who supervises consistently at regularly designated times
- The right to clear criteria when being evaluated
- The right to growth-oriented, technical, and theoretical learning that is consistent in its expectations.
- The right to an adequate amount and variety of cases to ensure learning.

(Munson, 1987, pp. 105–106)

a team, or a rotating group of individuals. You also need to know which supervisors you will meet with, how often, and what the general format of those meetings will be. In some cases, in addition to a supervisor, you will be assigned a co-worker to support and assist you with direction. It is especially important that you know who to go to in a crisis situation if your supervisor is not available.

Campus Resources

If you are attending an internship seminar, the instructor is one of your campus resources. However, that person may or may not have arranged the placement and may or may not be able to troubleshoot with you if problems arise. You may also have a campus staff or faculty member who visits you at your site; reads your journal, papers, and assignments; and provides you with feedback. Other campus resources might include faculty in your academic program, an internship or co-op office, service-learning office, or career services office, and, of course, your peers.

Liability Insurance

We should note that there is considerable risk and liability involved in working in a direct service internship, be it with patients, employees, students, customers, or clients. Many campuses require liability insurance coverage for all students who are involved in off-site learning, such as internships, service-learning, course-based practica, and co-op education. If your campus has a blanket liability insurance policy that covers your work in the field, it is important to ask your campus instructor for a copy of the verification of coverage so you are fully informed about it before you begin your work in the field.

Professional, Ethical, and Legal Issues

There some aspects of your internship that you may encounter early on and could necessitate discussion, usually with your supervisors or your co-workers. For your needs at this point, we find it's important to become aware of the breadth of the issues as well as how they are named. If you are particularly interested in knowing more about these issues than we cover here, as well as information about issues with professional, ethical, and legal dimensions, you will want to peruse Chapter 13, where we discuss these issues in substantial detail. The following list lays the groundwork for knowledge about your role as an intern and the three related issues of academic integrity, competence, and supervision.

- *Academic integrity* issues include a quality field site, responsible acceptance and learning contracts, and a seminar class that ensures a "safe place" for reflective discussions (Rothman, 2002).
- *Competence* issues include knowing your limitations and finding a balance between challenging work and a realization that you have

exceeded your level of competency. It is important that you know the limits of your skills and seek help as needed (Gordon & McBride, 2011; Taylor, 1999, p. 99).

- *Supervision* issues include the assignment of an appropriate supervisor who knows how to supervise interns in a particular setting and can appropriately deal with such complex issues as client abandonment, the dynamic of attraction in the supervisory relationship, and quality evaluations of the intern.

As you look over the list of issues common to interns in Table 3.1 you might feel a bit overwhelmed by it. But, there is no need to. You do not need to remember the issues, nor know what each one means. As you

TABLE 3.1
Internship Issues Across Campus Programs and Academic Disciplines

Intern Role Issues (All Interns)

- Right to quality supervision
- Responsibility to confront situations in which educational instruction is of poor scholarship and nonobjective
- Disclosure of risk factors to all potentially affected parties (to site supervisor about intern; to intern about site supervisor)
- Behaving consistently with community standards and expectations
- Awareness of risk status of agency
- Active involvement in the placement process and consideration of more than one placement site
- A clearly articulated learning contract that identifies mutual rights, responsibilities, and expectations
- A service contract with the agency that defines the limitations of the intern's role
- Liability insurance
- The prior knowledge clause
- Assurance of work and field site safety
- Assumption of risk as limited to ordinary risk
- Employer-employee-independent contractor relationship

- Compensation: stipend, scholarship, taxable/tax-free
- Deportment
- Negligence
- Malpractice
- Implication of federal funds and related statutory and regulatory requirements
- Use of college work-study funds for interns
- Grievance processes
- Informed consent in accepting the internship
- Respecting the prerogative and obligations of the institution
- Responsibility to confront unethical/illegal behaviors
- Public representation of self and work
- Disclosure of status as intern
- Boundary awareness
- Boundary management
- Personal disclosures
- Criminal activities
- Political influences/corruption
- Subversion of service system
- Office politics

experience your internship over time, you will develop an awareness of them and know what they mean in practice. We will revisit this list in Chapter 13, where we have a more in-depth discussion about them and cite examples of dilemmas that interns have faced. For the time being, you are responsible only for reading over the list and taking time to consider those issues that have meaning for you.

Information About Personal Safety

Depending on where you are interning, you may be concerned about your safety. This is not a pleasant topic to discuss at a time when we are trying to buoy your enthusiasm, but it is important. In our experience, there are interns who worry when there is no realistic cause for concern, and there are interns who perceive no risk—and hence take no precautions—when the risk, while manageable, is very real. Whether you have these concerns or not, it is important that you make sure that you assess your level of risk and develop a plan to minimize it, and that you seek help with this crucial task.

Assessing Your Levels of Risk

You should be made aware of any and all potential risks to your safety before you start your internship, which can come from the organization, the community, and the client population. If you are not made aware of these risks, then it is important to ask about them.

Client Risk Levels   When it comes to clients, three factors should be considered: the client's developmental stage, motivations, and immediate situational factors and conditions (Baird, 2011, pp. 157–159). Details on these and other client issues related to personal safety can be found in Chapter 7.

Site and Community Risk Levels   Location and hours of operation are another set of variables. Some sites are located in risky neighborhoods (i.e., at risk for violence or crime), especially for someone who does not live in that community. Some placement sites have conducted as complete an assessment of the risk factors involved in their work as they can; others have been less attentive. Some have thorough procedures for staying safe and for handling risky situations and others don't or might not be nearly as thorough. Agencies have the responsibility to be thorough in their assessment of risk factors and levels and to respond accordingly with policies and procedures for staying safe. Your responsibility is to determine how well your site meets these criteria for safety.

Birkenmaier and Berg-Weger (2010) offer excellent suggestions to assist you in your fact-finding efforts, including finding out the number and nature of violent or abusive incidents that have occurred at your placement in the past, arranging a tour of the surrounding neighborhood, and gathering information about other neighborhoods you may be visiting or traveling through.

Minimizing Risk Levels   If you feel in any way unsafe or otherwise uncomfortable traveling to and from your site due to location, hours of operation, or security measures in or near the building, consult with the site supervisor and co-workers immediately to determine how they ensure their personal safety working at the site.

If you suspect the possibility of violence by clients or their family/ friends, it is critical that you consult with others immediately, starting with the campus placement coordinator who arranged your placement, your campus supervisor, your site supervisor, and co-workers as to the history of violence at the field site. If there is a history of violence, it is time to meet with both of your supervisors and develop a safety plan.

If you have not already been informed about the relevant policies and procedures for safety, crisis management protocol, and reporting policies and procedures, then it is time to ask your supervisor for them. When you get them, ask to copy them so that you can spend some time studying and understanding them. That way, you know ahead of time just what to expect of your co-workers, your supervisors, and yourself should a safety emergency develop. If there is *any* possibility of client violence against you, you need to be trained in physical restraint that is safe, nonviolent, and respectful of the client. The staff at the site already has the training and your supervisor can refer you to the next class. If you find out from your peers that there are better choices for the training, this is not the time to be shy. Ask about them and ask if and how you can attend. You want the best training possible for yourself.

Again, if there is any possibility of client violence against you, be sure to request training in universal precautions for contact with bodily fluids such as blood or vomit. In addition, you want to ask if there is special training you can or should receive, such as cardiopulmonary resuscitation (CPR), basic first aid, and hygiene management for communicable diseases (Baird, 2011).

The Developmental Stages of an Internship (DSI-2)

As we noted earlier, how you learn in the field differs considerably from how you learn in a traditional classroom on campus. The four developmental stages of an internship—*Anticipation, Exploration, Competence,* and *Culmination*—reflect your phenomenological experience while you are learning. These stages were described in Chapter 2 and will be discussed in detail in subsequent chapters of the book. Each stage of an internship brings with it certain challenges. When interns *engage* those challenges they meet them head on, continually empowering themselves as they resolve challenges in productive and proactive ways. When interns avoid, shrug off, procrastinate about, or otherwise fail to engage those challenges, they become disengaged by not taking responsibility and in turn disempower themselves. If you skipped Chapter 2, or were not asked to read it, we strongly suggest you go back and read it now.

Essentials for Empowerment

Next, we identify five sources of personal power, combining knowledge, skills, attitudes, and values, that will ensure the vitality and endurance you need to remain engaged and make your way through the challenges of learning experientially in your internship. They are the powers of positive expectations, perspective, discussion, humor, language, and mindfulness—which, if used effectively, will empower you in the ways you go about your work.

The Power of Positive Expectations

In a famous study, Rosenthal and Jacobson, an educator in San Francisco (1968), found that children whose teachers believed they were intelligent in fact did "improve in their school work to a significant degree," whereas children who were not believed to be as smart achieved significantly less. Their research on the accuracy of interpersonal predictions found that one's expectations of another's behavior can become an accurate prophecy simply for its having been made (1968, p. vii). This *self-fulfilling prophecy*, as it's come to be known, occurs when people live up to the labels and expectations of others. What we expect of others is in fact what we get because of how we treat others. In your internship, what you expect of your supervisor, co-workers, community leaders, clientele, and others will affect how you interact with them and, in turn, how they work with you. So, too, will their expectations of you.

The Power of Perspective

Have you noticed that when you look at a glass that is half-filled with liquid and half-empty of liquid, you respond to this classic exercise in perspective-taking based on whether you are more optimistic or more pessimistic in your thinking that day? That is because your perspective is affected by your experiences, your needs and values, and your feelings. Do you recall the story of *The Wise Men and the Elephant*? In that story, six blind men each argued about what an elephant looked like. They based their perceptions of the elephant on their "blindness," hence the concept of blind spots affecting one's perceptions and thoughts.

We all are prone to blind spots because we bring our self in its entirety to bear on our perspectives. In our seminar classes, we use exercises that demonstrate that not everyone perceives illustrations similarly—case in point the glass that is half-filled or half-empty. In your internship, using the power of your perspectives to question your possible blind spots and how they can contribute to biases, prejudices, or oppressive actions will make the difference between an empowering experience and a disempowering one. What should you be striving for in terms of your perspective taking? Bradley (1995) identified the capacity to view situations from multiple perspectives and place them in context as a high level of perspective taking. So, too, is having the capacity to perceive conflicting goals within and among individuals while recognizing that the differences can be evaluated.

The Power of Discussion

One of the most challenging aspects of the internship for many students is the necessity to talk at times with supervisors and other colleagues about frustrations with their work or supervision. Many interns tend to believe that having such a conversation must be confrontational in nature, so the conversation rarely happens. However, the issues that frustrate you are real issues, such as lack of supervision or quality supervision, sexual approaches by a co-worker, ethical issues that suddenly surface, perhaps even legal infractions, and so on. Think about it for a moment. You probably have experienced some frustrations in your employment or in volunteer work. How did you deal with them at that time? Would that way be the best way to handle them in this instance? Because of the negative connotations associated with the term *confrontation*, we ask our interns to think in terms of a *discussion* instead. Discussions allow perspectives of situations to be brought into the open and talked about. Confrontations can set up turf and battle lines that are difficult to rise above and even more difficult to move beyond. Discussions, on the other hand, allow you to work *with* your supervisor to address the issues and develop realistic resolutions. If you think these sorts of sensitive discussions will be difficult for you, ask your campus instructor or supervisor to create scenarios for you and your peers to role play so that you can move beyond the confrontation trap into the power of discussion.

The Power of Humor

Some people just love to laugh, but others don't. But if you are one of those who enjoy a good laugh and has a good sense of humor, then you have an invaluable tool at your disposal to help you through tough times during your internship. Humor used appropriately has been shown to lift us above feelings of fear, despair, and discouragement. It helps us cope, especially with anxiety, and gives us strength to get through adversity; it keeps us "in balance" when all seems to be falling apart. Its physiological benefits have been well documented: It's good medicine for the heart and for the mind. It heightens and brightens mood, releases tensions, and leaves us feeling uplifted, encouraged, and empowered. Very importantly, it allows us to transcend predicaments, be flexible, and see alternative ways of looking at situations. It allows us to deal with high stress that we can't escape from by making fun of it, removing us from our pain and providing us the strength we need to get through difficult times. Its therapeutic values have been recognized for some time. If we can laugh at our setbacks, we can't feel sorry for ourselves. That is a very empowering way to move beyond crises (Fanger, 1999).

The Power of Language

The language we use to describe our experiences is often a reflection of how we feel about the experience. However, some theorists see this dynamic in reverse, maintaining that the language we use to describe an experience actually affects how we feel about it, which can in turn affect our behavior. You may recall that we referred to the word *crisis* as implying both danger and opportunity. Suppose you are having an especially difficult time with a supervisor. You might say to yourself—and to others—"I just can't take this anymore. I shouldn't have to put up with this kind of treatment, but I don't know what to do; I am stuck there!" Reframing that statement as an opportunity might sounds like this: "I am going to be supervised by lots of different kinds of people and while this may not be the most fun I have ever had, it is a real opportunity for me to learn how to deal effectively with someone like that." Psychologists Robert Kegan and Lisa Lahey have argued that the way we speak about work actually has an impact not just on our experience but our performance (2001). We will return to this theme later in the book.

The Power of Mindfulness

By now you are well familiar with the media's attention to the issue of multitasking and the research findings that our brains cannot do two independent things at the same time that both require conscious thought. We know that when we are mindful, our attention is focused on doing one thing well at a time; being mindful fits well with mono-tasking.

Mindfulness is a centuries-old practice central to contemplative traditions, poets, and religions such as Buddhism. Mindfulness meditation is described as the capacity to be fully present in the moment, that is, to pay attention to what is going on in the actual moment in time, without allowing our thoughts to drift to the past, to the future, or to other aspects of the present. Mindfulness is being used more and more in mainstream health care in this country because of the growing evidence that it can increase our enjoyment in life, expand our capacity to cope, and possibly improve our health, emotionally as well as physically (Schatz, 2004).

Practicing mindfulness doesn't mean that you won't be able to take on a number of tasks at the same time, push through a project with intensity, or allow yourself to experience many different emotions. Practicing mindfulness does mean that you can make the multitasking expected of you a conscious choice and do it deliberatively; that you'll be aware that you are pushing through the project while paying attention to salient details; and, that you will have insight into your feelings and be more aware of your choices in managing as well as responding to them (Schatz, 2004). Mindfulness meditation can be learned on your own or by taking a brief course on your campus, through your local hospital, or at the local community center.

Conclusion

Just as you probably feel like there is so much to learn, we feel like there is so much more we want to tell you, so much we want to call to your attention. As the book and your internship progress, you will use a wider array of tools and look at your experience through a wider and more sophisticated set of lenses. For now, though, these are the basics.

For Review & Reflection

Checking In (Select the Most Meaningful)

1. *Safety First.* As a group, put aside some time to talk about the safety issues that each of you faces. Frame your discussion by focusing on risk factors, risk levels, and safeguards that need to be in place. Remember...all risk levels that warrant safeguards need to be brought to the attention of your campus supervisor. An important piece of such a discussion is acknowledging the feelings that the risk levels evoke and how you will manage them so you *thrive* instead of *survive* each day in placement.

2. *The Journal.* Review the kinds of journals and reflective techniques discussed in this chapter. Which ones seem interesting and useful to you? Before you use the technique of your choice, you must decide the larger purpose of your journal besides receiving credit for doing it! That includes acknowledging who if anyone will read it. Once that decision is made, you'll need to decide how it will be done (handwritten, taped, typed, or computer generated, etc.) and what materials you'll need to have in place.

3. *The Portfolio.* It's important to know this early in the internship whether or not a portfolio of the experience is required or advised. If neither, you should decide now if you want to create a portfolio on

your own. If so and you consider using your journal as a portfolio of your internship, think about the sorts of things you might include. If you want to create a portfolio, you do want to think about what your ideal portfolio would look like so that you can start to identify and collect the necessary documentation.

Experience Matters (For the EXperienced Intern)

Your Life Context.   Because you are an EXperienced intern, by definition you have had more experience than other interns. Consequently, there is a good chance that your life is more complicated in terms of responsibilities and history than a student who hasn't "lived as much." Think about your life context as described in this chapter. Which components of your life context seem to be assets for you as you begin your internship and which may be liabilities? How might your life, career, or job history work and not work to your advantage? What is it that you most want from your supervisors and your seminar peers when it comes to the experience you bring to the internship?

Personal Ponderings (Select the More Meaningful)

1. Creating a concise statement about your strengths as a learner will serve you well in a variety of contexts. Practice developing one at this early point in your internship; consult the examples in this chapter for guidance.

2. Think about your strengths and weaknesses when you consider the essential attitudes described in this chapter. Give a specific example of times when you have and have not displayed these attitudes. What differences did you notice in the responses you got when you used these engaged ways of approaching situations.

Seminar Springboards

1. *Response Skills.*   Choose a real or hypothetical (written-out) experience at an internship and practice the categories of responding. Have one intern tell the story and then have others take turns responding, each in a different category. One gives advice, one analyzes, and so on. Then ask the person who told the story how each response felt. Talk about which modes of responding were easier and which ones you may need to work on.

2. *Feedback Skills.*   Devote a class, or a portion of the class, to practicing effective feedback. Again, you may use a real or scripted situation. Have each person practice giving feedback and have others use a checklist or rating scale to point out strengths and areas for development. This exercise could also be done in triads or small groups.

For Further Exploration

Baird, B. N. (2010). *The internship, practicum and field placement handbook: A guide for the helping professions* (6th ed.). Upper Saddle River, NJ: Pearson Prentice Hall.

Excellent coverage of safety issues as well as a variety of reflective techniques.

Birkenmaier, J., & Berg-Weger, M. (2010). *The practical companion for social work: Integrating class and field work* (3rd ed.). Boston: Pearson Allyn & Bacon.

Particularly thoughtful discussion of safety issues and how to advocate for yourself. Coverage of intrapersonal and interpersonal issues, including communication skills.

Compact, C. (2003). *Introduction to service-learning toolkit: Readings and resources for faculty* (2nd ed.). Providence, RI: Campus Compact.

Comprehensive and impressively useful sourcebook.

Collison, G., Elbaum, B., Haavind, S., & Tinker, R. (1999). *Facilitating online learning: Effective strategies for moderators.* Madison, WI: Atwood.

Informative, useful guide for netcourse instructors for moderating a web-based course such as the internship seminar.

Eyler, J., & Giles, Jr., D. E. (1999). *Where's the learning in service-learning?* San Francisco: Jossey-Bass.

Thorough coverage of the theory and practice of reflection.

Johnson, D. W. (2006). *Reaching out: Interpersonal effectiveness and self-actualization* (9th ed.). Boston: Allyn & Bacon.

Excellent discussion of feedback and modes of responding, with useful exercises.

Kabat-Zinn, J. (2005). *Full catastrophe living: Using the wisdom of your body and mind to face stress, pain, and illness.* NY: Random House Inc.

Anniversary edition of this classic work on mindfulness meditation and healing; tapes available to accompany book which can be used for self-teaching.

Kiser, P. M. (2012). *Getting the most from your human service internship: Learning from experience* (3rd ed.). Belmont, CA: Brooks/Cole.

A more complete explanation of the integrated processing model.

Kuh, G. D. (2008). *High-Impact educational practices: What they are, who has access, and why they matter.* Washington, DC: Association of American Colleges and Universities.

Describes educational practices that have a significant impact on student success, which benefit all students and especially seem to benefit underserved students even more than their more advantaged peers.

Shafir, R. (2006). *The zen of listening: Mindful communication in the age of distraction.* Wheaton, IL: Quest Books.

Inspirational work on the practice of mindfulness for listening more fully.

Zubizarreta, J. (2004). *The learning portfolio: Reflective practice for improving student learning.* San Francisco: Jossey-Bass.

A useful collection of articles on portfolios.

References

Baird, B. N. (2011). *The internship, practicum and field placement handbook: A guide for the helping professions.* Upper Saddle River, New Jersey: Pearson Prentice Hall.

Birkenmaier, J., & M. Berg-Weger. (2010). *The practical companion for social work: Integrating class and field work.* Boston: Pearson Allyn & Bacon.

Bradley, J. (1995). A model for evaluating student learning in academically based service. In M. Troppe (Ed.), *Connection cognition and action: Evaluation of student performance in service learning courses,* Denver: Education Commission of the States/Campus Compact.

Compact, C. (2003). *Introduction to service-learning toolkit: Readings and resources for faculty.* Providence, RI: Campus Compact.

Diller, J. V. (2011). *Cultural diversity: A primer for human services.* Pacific Grove: CA, Thomson.

Fanger, M. T. (1999). *Humor and social work: Are you serious?* Boston, Massachusetts Chapter: National Association of Social Workers.

Gollnick, D. M., & P. C. Chinn. (2005). *Multicultural education in a pluralistic society.* Upper Saddle River, NJ: Prentice Hall.

Gordon, G. & McBride, R. B. (2011). *Criminal justice internships: Theory into practice* (7th ed.). Cincinnati, OH: Anderson Publishing.

Inkster, R., & R. Ross (1998). Monitoring and supervising the internship. *NSEE Quarterly Summer:* 10–11; 23–26.

Johnson, D. W. (2008). Reaching out: Interpersonal effectiveness and self-actualization. Boston: Allyn & Bacon.

Kegan, R., & L. L. Lahey. (2001). *How the way we talk can change the way we work.* San Francisco, CA: Jossey-Bass.

Kiser, P. M. (1998). The integrative processing model: A framework for learning in the field experience. *Human Service Education 18*(1): 3–13.

Munson, C. E. (1987). Field instruction in social work education. *Journal of Teaching in Social Work, 1*(1), 91–109.

National Crime Prevention Council. (1988). Reaching out: School-based community service programs. Washington, DC.

NSEE (1998). Foundations of experiential education, December 1997. *NSEE Quarterly 23*(3): 1; 18–22.

Rosenthal, R., & L. Jacobson. (1968). *Pygmalion in the classroom.* New York: Holt: Rinehart and Winston, Inc.

Rothman, J. C. (2002). *Stepping out into the field: A field work manual for social work students.* Boston: Allyn & Bacon.

Schatz, C. (2004). The benefits of mindfulness. *Harvard Women's Health Watch* 11(6): 1–2.

Schon, D. A. (1987). *Educating the reflective practitioner.* San Francisco, CA: Jossey-Bass.

Schon, D. A. (1995). *The reflective practitioner: How professionals think in action.* Ashgate Publishing.

Seashore, C. (1982). Developing and using a personal support sytem. In L. Proter & B. Mohr (Eds.), *Reading book for human realtions training.* Arlington, VA: National Training Laboratories., 49–51.

Sheafor, B. W., & C. R. Horejsi. (2003). *Techniques and guidelines for social work practice.* Boston: Allyn & Bacon.

Sullivan, W. M. (2005). *Work and integrity: The crisis and promise of professionalism in America.* San Francisco: Jossey-Bass.

Swaner, L. E. (2012). The theories, contexts, and multiple pedagogies of engaged learning: What succeeeds and why? In D. W. Harward (Ed.), *Tranforming undergraduate education: Theories that complete and practices that succeed* (pp. 73–90). New York: Rowman and Littlefield.

Taylor, D. (1999). *Jumpstarting your career: An internship guide for criminal justice.* Upper Saddle River, NJ: Prentice Hall.

Weinstein, G. (1981). Self science education. In J. Fried (Ed.), *New Directions for Student Services: Education for Student Development.* San Francisco: Jossey-Bass.

Weinstein, G., Hardin, J. & Weinstein, M. (1975). *Education of the self: A trainers manual.* Amherst, MA: Mandella.

CHAPTER 4

Starting with You: Understanding Yourself

*I have been in my placement for several weeks and have
challenged my own philosophies many times. It frightens
me to think that the very foundation on which I have
based my life is being challenged by clients who I
believed were going to be textbook cases. Not that
I assumed that I was entering a vacuum, but I didn't
think that my own beliefs could be shaken in such a short
period of time. Maybe I am making no sense at all.*

Maybe I am trying to make too much sense.
        STUDENT REFLECTION

Introduction

As you become more involved in your placement, what would you like to
know? What knowledge, if you had it, would make you feel more prepared? When we have asked interns these questions over the years, we
have gotten all kinds of answers, including those focused on co-workers,
clientele, intervention techniques, community dynamics, and the organization's rules. Those are all good responses and important aspects of the
internship to think about. What we hear less of, though, and what is just
as important, is that you need to know about yourself, especially if you
have chosen to work in the helping or serving professions. In Chapter 3,
we talked about the personal resources and qualities that prepare you for
your journey and can leave you "feeling good" throughout your internship. Understanding yourself, that is, being aware of *who* you are as a
person and as an intern and *why* you are the way you are, will allow
you to realize the value of your strengths and the resources you have at

your disposal so you can use them effectively when the time comes. It will also allow you to anticipate and prepare for challenges that could arise as a result of your ways of interacting.

Meaning Making

All of us are constantly engaged in a process of trying to understand what is happening to us; we struggle to make sense of the world in which we live, the people in it, and the experiences we have. Robert Kegan, a developmental psychologist, believes that the most fundamental human activity is "meaning making"—that is, making meaning or making sense of our lives (Kegan, 1982, 1994). David Kolb, about whose work you read in Chapter 1 also refers to learning as making meaning (A. Y. Kolb & Kolb, 2005; D. A. Kolb, 1984). Perhaps nowhere is this activity more important than in the helping and service professions. If you are preparing for a career as a nurse, a counselor, or an advocate, you have been encouraged to consider the complex factors that lead to human problems and their solutions. Now that you have begun your internship, you are aware that it stops being an abstract exercise; you aren't thinking about human problems in general anymore, but about the ones that confront you daily. Even if you are not entering the service professions, problem solving will be an important part of your job; it is an important part of *every* job. Very importantly, you must consider an additional factor in thinking about problems, and that is *you*. You must pay attention not only to the sense you are making of what you encounter but of *how* you are making sense of it. Doing so will make you a more effective intern, a more effective practitioner in the future, and a more sensitive individual as you journey through the stages of an internship.

Dealing with Difference

You will deal with many people whose experiences and values are very different from your own. In those cases, self-understanding will help you avoid two major pitfalls: professional myopia and the tendency to regard difference as deviance (Sweitzer & Jones, 1990). At the heart of combating these tendencies is the ability to see your own reactions and views as only one among many possibilities. For this reason, understanding some of the sources of your personal characteristics and responses is critical. For example, understanding how your family influenced your beliefs and personality, coupled with knowledge of the diversity in family patterns and dynamics, will let you be more objective in your assessment and opinion of others.

Because everyone has their own system of meaning making, not only will two people see the same thing differently, but they will often see two different things altogether. If you are not aware that your point of view is only one of many, you may fail to consider other possibilities before

---

THINK About It

Making Sense of Daily Encounters

Suppose you are confronted with the following situations at your internship and must make sense of them:

- Your supervisor, in correcting your work, makes some insensitive statements.
- One of your co-workers is too aggressive in advancing some ideas.
- One of your clients is backsliding.

What factors do you consider? What is the problem and who or what needs to change? The very words we just used to describe the situations imply that the problems, and hence the solutions, lie in other people. However, you are part of the problem too, and not just in what you may have done to evoke another person's response. For example, you think your co-worker is being too aggressive. Too aggressive for whom? Some people would not find that behavior too aggressive or even describe it as aggressive at all. Why did you? What needs to happen now? Does your co-worker need to tone it down? Do you need to learn to be more tolerant? Do you need to see things from other perspectives? It is not a simple matter to make sense of situations in the internship, as you are realizing.

---

making decisions about causes and solutions; we call this *professional myopia*. For example, suppose you are an intern in a predominantly White high school and you hear a Black student complain of feeling out of place and uncomfortable, perhaps even unable to work to capacity. Part of the problem may be that certain features of the school itself, such as staffing patterns, menus, and recreational opportunities, are suited to middle-class White students. If you are not sensitive to these dynamics, you may describe the problem as one of adjustment or even cultural deprivation. Your efforts, then, will be aimed at helping the Black student feel better about a situation that may in fact be unfair or discriminatory.

When people are aware of differences between themselves and others, they sometimes confuse *difference* with *deviance*. They do not describe it as something different; they describe it as something wrong. You are not going to like every person you encounter, every client assigned to you, nor everyone you work with, but making the distinction between difference and deviance can help you accept and empathize with a wider variety of people. More importantly, it can help you think carefully about what needs to change. For example, a family may choose to sacrifice some degree of their children's comfort and education to allow elderly grandparents to live in relative comfort. The family then becomes the target of a complaint by the school system. Their decision may reflect a cultural value about the relative responsibility families have to children and elders. If you do not see the decision in this light, you might condemn the family for their treatment of the children, thereby damaging your relationship with them. Furthermore, you might then choose to concentrate on "changing" the family, never considering the option of urging the community to accommodate a wider variety of family customs and structures.

Recognizing the Stages

Finally, the more you understand about yourself, the better you will be able to recognize the stages of the internship as they happen. It will not be enough to know, or suspect, that you are in stage 1 or stage 2. You need to consider some questions across the stages. For example, What does that stage mean for you as a unique individual? What strengths can you draw on, and what personal traps should you avoid? To move through the stage you are in, are there aspects of yourself, as well as aspects of the internship, that you must face and try to change? If you can discuss these issues in your journal and talk about them with peers, instructors, and supervisors, you will recognize the challenges sooner and move through them more smoothly.

It is important to realize that self-knowledge is not like other knowledge. You are changing all the time, and your system of making meaning is changing with you. In addition, just as there is always something new to see and appreciate in a work of art, there is always more to know about yourself. So, self-understanding is a process to which you must become accustomed and committed. It takes work, but it will pay you big dividends.

The Special Relevance of Self-Understanding to the Helping Professions

Self-understanding also plays an important role in helping human service workers fulfill their responsibilities effectively and responsibly. As an intern in a helping profession, regardless of your specific responsibilities, you are usually trying to form relationships with clients. In the context of those helping relationships, you attempt to respond, or help clients respond, to a variety of human problems. Although you may never have thought of it this way, in fulfilling these functions, you occupy a position of power and influence over clients (Brammer & MacDonald, 2002). Not only do you form opinions and make judgments but you also make those opinions known to others in individual and group sessions, reports, and staff meetings.

Furthermore, you may be asked to recommend for or against various kinds of intervention. Sometimes the internship site controls vital resources, in some cases as basic as food and shelter. Increased understanding of the feelings, beliefs, tendencies, and styles that make up your system of meaning making will help you be more effective in all of your functions. It will also help you be sure you are using the power inherent in your role for the welfare of your clients.

A number of authors have commented on the role of self-understanding in forming effective relationships (G. Corey, Corey, & Callanan, 2011; M. S. Corey & Corey, 2011; Levine, 2012; Schram & Mandell, 2006). If you can clarify and discuss your feelings and personal patterns, you can serve as examples for clients who are struggling to do the same. Furthermore, if you have confronted and modified aspects of yourself that you didn't like, you are more likely to communicate a belief in the capacity for change (M. S. Corey & Corey, 2011). Finally, awareness of ways in which your experiences are similar to those of clients can help establish empathy and trust.

Self-understanding will also help you avoid *projection*, the unconscious tendency to believe that you see in others feelings and beliefs that are actually your own, of which you are unaware because of their unconscious nature. This tendency can affect your ability to understand, accept, and empathize with another person. If a client brings up an issue with which you are uncomfortable in your own life, you may avoid the subject or decide that the client is being difficult (G. Corey et al., 2011). If you are angry at a client and unaware of that anger, you may decide that the client is angry at the world. If you have trouble with assertiveness, you may react negatively to an assertive client, a supervisor, or a co-worker. Projection can also lead to poor decisions about client needs. You may fall into the trap of inadvertently using clients to meet your own needs (G. Corey et al., 2011). For example, if you have trouble with assertiveness, you may push clients to be assertive in part because it vicariously fulfills your need to be assertive. Increased self-awareness will help you make more conscious choices in all of these situations.

Components of Self-Understanding

It goes without saying that self-understanding is an enormous area to consider, and one that requires a continuous commitment on a personal level. We have chosen several topics within the realm of self-understanding to concentrate on in this chapter. Our goal is to introduce you to these areas, to give you some things to think about, and to provide you with resources to further explore these areas. In so doing, we hope that this chapter becomes a resource for you throughout your internship. Toward that end, we have summarized complex theoretical information, some of which may be new to you. You and your instructor will need to decide which of these areas merit your time and energy, as a class and individually.

Knowing Your Values

A value is an idea or way of being that you believe in strongly, something you hold dear and that is visible in your actions. In the previous section, we argued that self-examination is important; that is a value for us. You may believe strongly in taking care of your family, in serving your country in some way, or in the existence of a deity. You are probably very aware of some of your values, especially ones that have been challenged, debated, or highlighted in the media (e.g., values about abortion, euthanasia, gun ownership, or freedom of speech).

Others, though, are so much a part of you and are shared by so many people in your life that they don't seem like values—they seem like truths. For example, both of us always placed a high value on punctuality (especially professionally) and even assumed at one time that people who were habitually late for professional appointments were irresponsible. We have come to understand, however, that not all cultures share the same view of time. It is important that you take the time to clarify your values and give some thought to how you will respond when faced with co-workers, clients, or members of the public who do not share those values.

You will almost certainly encounter someone at your internship who does not share some of your values, and you should think about how you might respond if the issues arise. It might be a co-worker, a customer, a client, or even a supervisor whose value system is at odds with yours. Discussions about values can be lively and interesting if both parties are open to the discussion, but discussion is not always advisable or appropriate, especially in an internship. If there are some values that you cannot or will not accept in a client, a co-worker, or a supervisor, that is something you want to know as soon as possible and perhaps to discuss with your supervisor or campus instructor.

THINK About It

A Values Check

Values permeate your life; we could never list all the important areas of value. However, here are some that may be especially important in your internship. To help you clarify your values in these areas, we ask you to think about select questions and answers to these questions, especially those that describe how you believe things "should" be, as these are important clues to your values. Think also in terms of how these values affect the work of your internship.

(continued)

THINK About It *(continued)*

- *Sexuality*   How do you feel about homosexuality? Bisexuality? Heterosexuality? Transgendered individuals? Teenage sex? Premarital sex? Monogamy? Extramarital sex? Various sexual practices?

- *Family*   How should a family be structured? Are single-parent families okay? Should one parent stay home with the children? Should there be a "boss"? If so, who should it be? How should decisions be reached? Should grandparents or other relatives live with the family?

- *Religion*   How important is religion to you? How do you feel about other religions? Would you ever do something that a leader of your religion said was prohibited?

- *Abortion*   How do you feel about abortion for yourself? For others? Are there circumstances under which you believe it is morally wrong? Morally justified? Should teens be able to choose on their own? Should the father of the fetus have a say?

- *Euthanasia*   Should a person be able to choose to end his or her own life? If so, under what circumstances?

- *Self-disclosure*   What kind of things is it okay to tell someone you hardly know? For example, would you tell that person your financial problems? Your family difficulties? Your income? Which of these things would you tell a close friend? Are there things you would never tell anyone?

- *Honesty*   How do you define this word? Are there different kinds of lies? Is it ever okay to lie? If so, when and to whom?

- *Autonomy*   Do you think people should normally make their own choices and accept responsibility for their lives? Or do you tend to see people as more responsible for one another? How large a role do you think fate plays in a person's life?

- *Work*   How hard do you think a person should work? What do you think about people who work only enough to get by, and no more? What about people who seem to have no desire to find a better job or make more money? Those who always push themselves to work harder, no matter what? How about people who don't want to work at all?

- *Acceptance*   How important is it to you that people from different cultural traditions be able to do things in a way that is consistent with that culture, even though it may be unusual by mainstream standards?

- *Hygiene*   How often do you think a person should bathe? Should a person use deodorant? What would you think of a person who showed up at your office with dirty hands? A dirty face? Dirty feet? Dirty clothes?

(continued)

THINK About It *(continued)*

- *Freedom of Speech*   Do you think people's speech should be limited or restricted if it is hurtful to members of a cultural group, such as women, or Muslims? What about statements disagreeing with U.S. military policy? With democracy as basic form of government?

- *Time*   How important is it for you to be on time? That others are punctual?

- *Alcohol*   Is moderate use of alcohol okay? Is it ever okay to get drunk? If so, how often? What about binge drinking—is it a dangerous practice? Harmless fun? Is it up to the individual? Do you think alcoholism is a disease?

- *Drugs*   Are there some illegal drugs that you think are okay to use in moderation? If so, is it also okay to sell them? What are your thoughts about the use or abuse of over-the-counter drugs or prescription medications? Marijuana for medical purposes?

It is important to ask yourself how strongly you feel about these values in your own life and about them in general, and whether you are open to changing them. You might also want to think about how you came to have these values. Did you choose them consciously, after careful thought? Did they come to you from your family, your friends, or from a cultural group to which you belong? Did you accept them more or less uncritically? Perhaps you are not really sure where they came from and why you hold them. That is fine, but you may want to think critically about them before they are challenged.

Recognizing Reaction Patterns

Reaction patterns are ways that you respond—your thoughts, feelings, and actions—to particular kinds of situations. Some patterns are helpful and work well. Others are distinctly unhelpful; they do not appear to get you what you want, and yet you repeat them in spite of yourself. One of the most frustrating, stressful experiences people can have is one in which they find themselves doing something they don't want to do or reacting in a way they don't want to react. Here are some examples:

- A friend has seen your in-class presentation and offers some constructive feedback. As the conversation goes on, you find that you are getting angrier and angrier and are having a hard time listening. You keep coming up, mentally or verbally, with defenses for every perceived criticism, and you imagine telling your friend off.

- You are at a party and people are talking about some hot topic. You have something to say but can't seem to say it. Since others are very vocal, it is easy, although frustrating, for you to just sit there. Later, someone says exactly what you were going to say and everyone seems impressed.

- You are struggling with an intimate relationship. A good friend asks you how it's going with that person, and to your surprise, you hear yourself saying that things are fine.
- A friend calls late at night and asks that you meet him right away. The matter does not seem like an emergency, but you leave your assignments undone and go off to meet your friend.

These are not situations in which you later find, after much reflection, that you made a mistake. They are situations in which you know immediately afterward, or even during the situation, that you are not responding the way you want to. In fact, almost as soon as it is over, you can think of several ways to handle the situation that would have been better.

Think about situations like this that have happened to you and jot down a few of them. You may find that these are isolated incidents or that they occur with just one person. You may also find, however, that these responses are part of a pattern for you. You may find that in general you are defensive about criticism, unable to speak in groups, unable to say "no" even when you want to or to ask for help. Gerald Weinstein (1981) describes these tendencies as dysfunctional patterns. Please note that, in this case, dysfunctional does not mean useless, nor does having a dysfunctional pattern make you a dysfunctional person. On the contrary, these patterns are clues to some important aspects of yourself. We both have them, and so does everyone else. The reason we are asking you to think about these patterns is that you may "bump into" them during your internship. We will talk more about what patterns you might encounter and what to do about them later in the text, in Chapter 8 and 9.

THINK About It

Everyone Has Dysfunctional Patterns—Even You!

See if you can identify any such dysfunctional patterns in your life by using the format suggested by Weinstein (1981):

> *Whenever I'm in a situation where _____, I usually experience feelings of _____. The things I tell myself are _____, and what I typically do is _____. Afterward I feel _____. What I wish I could do instead is _____.*

Here is an example to help you:

- *Whenever I am in a situation where I feel angry at a friend, I usually experience feelings of anxiety and self-doubt. The things I tell myself are, "Take it easy.*

(continued)

THINK About It *(continued)*

It's not that bad. There's probably a good explanation, and besides, you don't want to upset him." What I typically do is smile, joke, or protest very weakly. Afterward I feel as if I let us both down. What I'd like to do is find a clear, respectful way to tell my friend what is upsetting me.

And here are two from student journals:

- I am trying to build my self-esteem and confidence. It is hard for me to hold back my emotions when something upsets me, especially given my fear of doing something wrong or failing. I appear to be happy and cheery, but it takes so much out of me that I become negative and sometimes grumpy. This gets in the way of me liking myself.
- I feel guilty that I need to rest and refuel my batteries, but this is my life. I dislike that I want to do everything perfectly.

Recognizing Family Patterns

For many people, their family of origin is the group with which they spend the most time until they form their own family or leave home for some other reason. Your experience in your family of origin is a powerful influence on who you are. Each family has its own way of doing things. Often, we are so accustomed to the way our family is that we assume that everyone's family is that way. For example, while both of us come from families where eating dinner together was very important, they differed considerably in how that time was spent. For one of us, conversation at the table was quiet and polite, and interruptions were frowned upon. For the other, the dinner table was lively, and often loud, with many conversations taking place at the same time. Imagine the shock for one of us visiting the other's family for dinner or for either of us visiting a family whose members came and went from the table or ate in different parts of the house.

We are going to encourage you to think about two common features of family life: rules and roles. Although very few families have a list of rules posted on the wall, they all have unwritten rules that tell everyone what they can and cannot do. As a child, you usually learn the rules by breaking them; someone reprimands or disciplines you, and you gradually figure out what is and is not acceptable. Sometimes families have rules that they are not even aware of. For example, one of us once listened to a family over the course of a weekend as they talked about all their relatives, living and deceased. There was one who was never mentioned, though. The family was quite surprised when this observation was shared, but upon reflection agreed that they almost never talk about that person. In observing this informal tradition, they were obeying a rule, even though they would never have called it that. Some family rules are a combination of individual family patterns and cultural norms (which we will explore more a little later on).

Here are some examples of family rules. Use this list to stimulate your thinking about the rules in your family:

- Keep family business in the family.
- Never question a decision or disagree with an opinion from someone older than you.
- If your brother or sister picks on you, handle it yourself. Don't go crying to your parents.
- Guests are always welcome for dinner—you don't have to ask.
- Don't talk about sex.
- Grandparents must always be consulted on important decisions.
- Mom and Dad need 15 minutes to relax after work before you ask them anything.
- No one is entitled to privacy, except in the bathroom.

Family roles tell you who in the family performs which functions. Some of them are pretty formal. One parent, for example, may pay the bills, do the cooking, or handle the discipline. There are other kinds of roles, though. There might be a family jokester, who is counted on to make people laugh, or the mediator, who tries to help settle conflict between family members. Some children have the role of the "good" child, and their mistakes are often overlooked or treated lightly, whereas others have the role of "bad" child and are treated more strictly.

THINK About It

The Rules and Roles in Your Family

Here are some questions to help you think about your family:

- Who has the final say in an argument or dispute?
- Who is in charge when the parent or parents are not home?
- Which child is the smart one? The talented one? The athletic one?
- Who can you count on to help you out of a jam?
- Which child gets the most leeway from the parent(s)?
- What do you think your own role is or was in your family of origin?
- Do you see yourself playing that role in other areas of your life—with your own family, friends, or groups you are in?
- What are or were the most important rules in your family of origin?

How many of them still influence you today?

In your internship, you will undoubtedly meet people whose family patterns and norms are different from your own. That can be pretty confusing as you try to make sense of their behavior and reactions. Understanding the sources of some of your behavior and feelings will, again, help you look more thoroughly and empathically at others.

Also, if you are not careful, you may carry a rule or a role from your family into your internship that is not helpful in that context. For example, your role as the jokester in your family, while it may serve to ease tension at home, may be annoying in a staff meeting, especially one where there is tension. Similarly, your role as the comforter may make it difficult for you to let a client struggle through his or her own problems.

Understanding the Way You Think

We discussed the importance of knowing how you learn in Chapter 3. As you progress in your internship, that knowledge will continue to be important, so we discuss it in a bit more depth here. As we mentioned earlier, knowledge about your learning and thinking styles can help you be an advocate for yourself in designing and modifying learning experiences at the internship. If you are working with people as clients, this knowledge can help you be more effective as you help them learn to do new things, such as change their behavior, locate resources, or improve their skills. If you can learn to adapt interventions to a variety of learning and thinking styles, you will be more successful in this attempt. If you understand diversity in learning and thinking styles, you are less likely to view those different from your own as deviant or deficient.

There are a great many theories of learning and thinking styles that may interest you—far too many to discuss here. You might be categorized as a concrete or abstract learner (D. A. Kolb, 1984), as random or sequential (Gregorc, 2004), connected vs. separate (Belenky, Clinchy, Goldberger, & Tarule, 1986) or dualistic vs. relativistic (Perry, 1970). In the box below, we highlight one theory that draws on some of the theories cited above and that we think has a range of applications to your internship.

Focus on THEORY

Baxter Magolda's Stages of Thinking

In Magolda's model of intellectual development, college students demonstrate thinking in any of four developmental stages and can do so at the same time. For example, on a given day, depending on the situations, you may be thinking in any of the four stages, depending on what you are thinking about. You may think differently when in a favorite class

(continued)

Focus on THEORY *(continued)*

than in a class you don't want to take; you may think differently when discussing certain topics or when working your part time job. As you read the summaries below of the stages, think about how you fit into this particular theory of thinking—as a student taking classes, as an employee, and as an intern. Are there differences? If so, how do you account for them? Focusing on yourself as an intern, are there one or more of these stages that you would like to work on?

Absolute Knowing At this stage, knowledge is believed to be unchanging. Students characterize thinking as "right or wrong" with no middle ground, or black or white with no shades of grey. Professors, experts, and authorities are seen as having "the answers" and their responsibility is to give those answers to the students. The students are responsible for memorizing the material, that is, recalling and absorbing it, and taking it in without questioning or challenging it.

Transitional Knowing Students' thinking at this stage reflects doubts at times about how the certainty of knowledge; some knowledge is certain, but some is no longer seen that way. The uncertainty of reality is accepted at times, as is the reality that professors, experts, and authorities are no longer "all knowing." Students now want to understand instead of absorb and recall knowledge; they want to make their own judgments about knowledge and uncertainties.

Independent Knowing At this stage, students realize that knowledge is uncertain. They have their own beliefs and opinions now, and they realize that others do as well. They value the opinions of their peers and consider those opinions as they make their own judgments. Students look to professors, experts, and authorities to support their independent views by giving them opportunities to explore knowledge in independent ways. Students in this stage take responsibility for their learning—actively producing knowledge by thinking through issues and expressing themselves.

Contextual Knowing This is the stage when students think for themselves and are able to work with uncertainties. They understand that knowledge is contextual and that it is individually constructed from ideas and evidence. They realize that they construct knowledge themselves and that the professor, expert, and authorities are partners in developing knowledge with them. They make judgments when situations are ambiguous and support their opinions by evidence that takes the context into consideration. They are open to new evidence and can have a change of mind.

(Baxter Magolda, 1992; Moon, 2011)

Your Style of Communicating During Conflict

Everyone experiences conflict and there will likely be some level of inter-personal tension or conflict in your internship, when your wants and needs conflict with those of a peer, a co-worker, or even a supervisor. There are, of course, a variety of ways to approach conflict but most of us have an approach that we use more often than not. Often this approach is something we learned as children or adopted as part of the rules and roles in our families. We encourage you to identify your typical way of approaching conflict for two reasons. First, as we have said before, if you do not see your approach as one of many, you will tend to react to other approaches as being somehow "wrong." Second and equally important, you can learn to put a variety of styles into your repertoire, and choose among them according to the situation. Often people use an approach over and over, even when it does not work for them; it becomes a *reaction pattern*.

In his insightful and useful book, *Reaching Out*, David Johnson explains that when you engage in a conflict, there are two set of concerns that you carry with you; concern for accomplishing your goals and concern for your relationship with the other person or people (Johnson, 2008). Each of us weighs these concerns in a conflict, but that does not mean that they are equal weight in our approach. According to Johnson, there are five possible combinations of these two sets of concerns:

- *Accommodating* or *smoothing* reflects high levels of concerns for the relationship and low levels of concern for your goals. People who use this style are likely to simply go along, giving in to the other person's wishes, fearing that not doing so will damage or break the connection.

- *Forcing* is the opposite of accommodating; it is an approach that combines high levels of concern for your goals and low levels of con-cern for the relationship(s). People who use this style are not very concerned about whether they are liked, and seek to "win" the conflict by getting what they want, no matter what.

- *Withdrawing* blends low levels of concern with achieving your goals with low levels of concern for the relationship(s). People who use this style are quick to try to defuse a conflict by avoiding the situation or the people, and even denying to themselves that there is a conflict at all.

- *Confronting or collaborating* is a style that blends high levels of con-cern for your goals and high levels of concern for the relationship. People who use this approach tend to remain engaged with the other person until they can find a solution that meets both of their needs.

- *Compromising* is the seeking of a middle ground, reflecting moderate concern with achieving your goals with moderate concern for the rela-tionship. People who use this style tend to seek or propose solutions in which everyone gets some portion—but not all—of what they want.

THINK About It

Conflict—What's Your Style?

Review the five styles of conflict described earlier. Which one sounds most like you? Which ones did you experience in your family, as a child, or now? Which one seems *least* like you? Can you think of people you like who seem to use that style? Finally, are there kinds of situations in which you tend to use one style—for example, you might be more forceful in a work or class situation but more compromising in your intimate relationships.

Cultivating Cultural Competence

We introduced the concept of cultural competence in Chapter 3, describing culture as a shared and commonly accepted set of beliefs, practices, and behaviors (Gollnick & Chinn, 2005). We also said that one key component of cultural competence is understanding your own culture. Actually, we might say cultures, because as Gollnick and Chinn point out, there are both macro-cultures and micro-cultures. The macro-culture consists of those beliefs and practices shared by the majority of citizens. So, for example, in the United States, a belief in democracy is shared by most (although not all) citizens and would be considered part of the macro-culture. Micro-cultures, on the other hand, are those beliefs and practices that come from membership in a smaller group or subgroup. There are, for example, attitudes and beliefs that are more accepted by women than by men, and vice versa. Your cultural profile consists of these subgroups, the degree of identification you have with them, and the extent to which your membership in these groups has an impact on the way you view the world and react to people and situations.

In your internship—and in your life—you will encounter more people whose cultural profiles are different from yours than those whose cultural profiles are the same as yours. The difference may be slight or substantial, the impact small or profound, but it is important as a person, a professional, and a citizen to cultivate cultural competence—the ability to interact effectively with people with different cultural profiles. There are many components to cultural competence, but understanding your own profile and your own attitudes is the place to begin.

Attitude Toward Difference

Do you see yourself as a prejudiced person? Do you hold any stereotypes about people of a different gender, race, ethnicity, sexual orientation, or religion? If you are like many of our students, you will answer these questions "no" or "not anymore." If you have qualified for an internship, you have probably been exposed to the core values of tolerance, pluralism, and respect for individuals that are at the heart of many professions. And naturally, you want to see yourself as a person who adheres to those values. If a part of you suspects that you do have some lingering prejudices, you may be keeping them hidden for fear that your peers or your professors will think badly of you or block your progress in your program.

Here is what we believe and invite you to consider. Everyone carries some stereotypes and prejudices, including us. If you work at it, you can learn to see your prejudices and make progress in overcoming them. However, at the same time, you will surely discover other, more subtle ones. The first step, though, toward being a nonprejudiced person is to confront and accept the prejudices you have. If you hide or deny them because you are ashamed or think that a good person would never have any, they will never change. Listen to what this intern had to say: *"Throughout my college years I held opinions of people, groups and cultures. I never once stopped to think about how my preconceived notions affected the way I acted towards people. I truly thought I was an open-minded person until I caught myself acting in this way."*

THINK About It

You've Got Stereotypes!

Try this exercise. Pick a subgroup that has been the target of some discrimination and of which you are not a member. Depending on your own subgroup membership, some examples are Blacks, Muslims, lesbians, blue-collar workers, Native Americans, and many others. Now, think about all the stereotypes you know about that group. Remember, we are not asking which ones you believe. Just write down as many as you can think of. You might want to get together with a classmate or two and expand your list. Now, look at your list. Where did the stereotypes come from? If you don't believe them, how did you come to "know" them?

Most students have little trouble coming up with a substantial list. The reason stereotypes are so easy to recall is that they are literally all around us. Think about the group you picked. How many members of this group do you actually know? In some cases the answer will be few or none. So where do your impressions of them come from? Some, of course, come from family and friends. Another part of the answer, though, is the way members of these groups are depicted in the media. Usually, members of various subgroups are portrayed in very limited, stereotyped roles. Think, too, about the books you read in school. How many members of these groups were in those books? How were they portrayed? The point here is that we have all "learned" harmful stereotypes about other groups (and even about our own).

Joanne Levine, in her informative book *Working With People* (Levine, 2012), cautions professionals to be sensitive to their use of "the paranoid 'they'." If you find yourself thinking that "they" are too aggressive, too concerned with money, or too standoffish, that is a red flag for a stereotype or prejudice that you hold. We encourage you to explore and face your prejudices, and we recommend some resources for doing so at the end of this chapter.

Once you have identified some prejudices, you can work on changing them. For example, through an experiential exercise like the one we just encouraged you to do, Fred once discovered that he held some stereotypes about Hispanic men:

> *I had to admit that I felt uncomfortable in groups of Hispanic men, and that I thought of them as being in general more violent and temperamental than me. Some reading about Hispanic cultural and family life helped me to get a more balanced picture of this group, with whom I had very little contact.*

Mary recalls that there was a time when she was somewhat apprehensive about working with same-sex couples in therapy. She attributed the discomfort to being generally uninformed about the homosexual culture:

> *After reading a good deal and attending workshops, it became apparent that my discomfort stemmed from such stereotypical beliefs as gay individuals are more maladjusted, unhappy, and pathological than their heterosexual counterparts. It was only after an exploration of these beliefs that I could better understand the culture, as well as the basis for my own homophobic thinking.*

Your Cultural Profile

We now invite you to think about your own cultural profile, which you will recall consists of the subgroups to which you belong, the degree of identification you have with them, and the impact your membership in those groups has on your worldview.

You are a member of many subgroups. Some of them are temporary; you can move in and out of them as you choose. For example, you are now a college student, but you could stop being one at any time, and eventually you will no longer be a student. Other subgroups are relatively permanent; your membership in them is determined primarily by accident of birth. Your race and ethnicity are examples. Your age will change constantly, of course, but you will always be part of an age group. You are also part of an age cohort, sometimes called a generation. There are pieces of music and historical events that have enormous power for people of your particular age. Some have argued that sexual orientation is determined at birth; others believe it can change. However, your status as heterosexual, homosexual, bisexual, or transgendered is a relatively permanent feature of your life. Social class is another important subgroup, and here you need to consider both the social class in which you were raised and the one to which you now belong (Payne, 2005).

Knowing the subgroups you belong to is only part of the picture. Another important part is how strongly you identify with that group. There are Jews, for instance, who identify very strongly with their Jewish heritage. They observe the traditions and holidays, obey dietary laws, and travel to Israel. There are also Jews who do none of these things. Individuals have varying degrees of identification with subgroups to which they belong. Although both of us are very aware of issues that confront us as a man or woman, we vary considerably in our identification with other subgroups. Fred, for example, knows almost nothing about his ethnic heritage, which is a mixture of Irish, English, and German, and is aware of very little influence from that subgroup. Mary, on the other hand, identifies strongly with her Irish and to a degree with Lithuanian heritage, and does have a good sense of how those traditions affect her as a person; however, she still struggles at times to fully appreciate the effects of her race (Caucasian) in her personal and professional life.

Your Worldview

As you move through your day, you encounter people and events, you make sense of them the best you can, and you react and respond to them. In many ways, the meaning you make and your reactions and responses are a result of the way you look at things—your worldview. Your worldview affects your perceptions, as well as what you learn and remember and how you think and act (Axelson, 1999). And whether you know it or not, you look at the world through cultural lenses. For example, many of our women friends are much more concerned than our men friends about walking across campus alone at night; they "see" a more dangerous situation. As members of the middle class, both of us grew up simply assuming that we would go to college—the question was where.

Some groups are afforded certain privileges by society (Doane, 2003; McIntosh, 1989). You may find that hard to believe, given all the publicity that affirmative action and other programs receive, but many of the privileges are subtle. For example, both of us are White, heterosexual, and Christian. We can go into a department store and wander around as long as we want; friends of ours who are Black or Hispanic have told us that they are frequently followed by store detectives or questioned by clerks after only a few minutes. Wherever we have gone to school or have worked, the major holiday in our religion is at least two different days off for us and with school breaks accommodating the holy days. When we are out in public with our partners, we can hold hands, hug, or even kiss without worrying about who may be watching or disapproving. These two instances are not the case for our Jewish friends or our gay or lesbian friends, respectively.

Thinking about the groups you belong to, how strongly you identify with each one, and how each impacts your worldview can help you construct a cultural profile. Such a profile is a snapshot of your cultural awareness at this point in your life. You will see one that one of us filled out, and there follows a blank for you to consider.

Category	Group	Importance to Worldview
Race	White	5
Ethnicity	Irish	1
Social Class	Middle	4
Gender	Male	4
Health	Good	4
Age Group	Middle Age	4
Geographic Region	New England	2
Sexual Orientation	Heterosexual	4
Religion	Unitarian	1
Language	English	1
Ability/Disability	Able Bodied	4

© Cengage Learning 2014

THINK About It

Diagram Your Cultural Profile

Category	Group	Importance to Worldview
Race		
Ethnicity		
Social Class		
Gender		
Health		
Age Group		
Geographic Region		
Sexual Orientation		
Religion		
Language		
Ability/Disability		

Of Special Relevance to
The Helping and Service Professions

The Power and Peril of Motivation:
Wanting vs. Needing

At some point in your education, someone has probably asked you to think or write about the reasons you are considering a particular profession as a career. Perhaps, too, a placement coordinator or prospective supervisor has asked you why you are interested in a particular kind of internship. Understanding and reminding yourself of these motivating factors can be a source of strength. Yet every one of them can be a liability as well, especially in social service settings or other settings in which you are working closely with people in need. Corey and Corey (2011) discuss several possible reasons for entering the helping professions or needs that workers bring with them, including the need to have an impact, the need to care for others, the need to help others avoid or overcome problems they themselves have struggled with, the need to provide answers, and the need to be needed. They also point out that each one of these motivations can cause problems.

It is important to make a distinction here between wanting and needing. All the reasons just listed are good ones for entering the helping and service professions, provided you substitute "want" for "need." For example, wanting to care for others is fine, but as Corey and Corey (2011) point out, it is not fine if you always place caring for others above caring for yourself. Sometimes what you need conflicts with what the placement wants or needs. For example, let's assume you have worked a very full week and are very tired. Your supervisor explains that someone is out sick and asks if you would mind working over the weekend. If you choose to say "yes" to these requests from time to time, that is fine. However, if you find that you cannot say "no," ever, then you may be someone who needs to put others' concerns first in order to feel good about yourself. That can happen for many reasons, but none of them offset the price you will pay if you don't learn to attend to your own needs as well as those of your clients.

Providing answers is great, too; coming up with a solution for a problem is a wonderful feeling. However, there are going to be problems you cannot solve. Furthermore, there will be times when it is better to let someone struggle to find an answer than to jump in with a lot of advice, even if your advice is on target and would solve the problem faster. If you need to be the one with the answers, these situations are likely to be difficult for you. One intern realized in a journal entry, *"There is some part of me that has a need to rescue these women."*

If you have struggled with a particular issue in your life, such as depression, substance abuse, or an eating disorder, and feel like you have emerged from that struggle, pay special attention to your motivations as they relate to this issue. Often, past struggles are the reason people choose the helping professions. They want to help others deal with or avoid problems that they experienced and feel that their experience will be an asset. In many cases, it is. It can be a powerful motivator for you, and it can be a source of hope for both you and your clients. However, it can also be a trap. You may have learned a lot about your own struggle and your own path, but there is always more to learn about how to help others deal with the same issue. The techniques that worked for you may not work for others, and newer, more effective approaches may have been developed. You need to remain open to that possibility. You also need to remember that your path is not necessarily the best path for others. Just as people have different learning styles, they have different healing styles, and your job is to fit their needs. If you have clients, they may need to struggle, just as you did, and it may be that you cannot and should not "save them" from that struggle.

Finally, some interns try to use the placement to resolve personal issues that they themselves have not yet resolved. If you encourage, or insist, that a client do what you cannot (e.g., be assertive or show love) or rush to protect a client from a critical parent (in a way that you never were), you are falling into this trap.

THINK About It

When Wants Turn to Needs

Take some time to think about your own motivations for entering the sort of placement you have chosen. If those desires turn into needs, what kinds of problems could that create for you?

A Psychosocial Lens on Wanting and Needing

Particularly if you are preparing for a career in the helping professions, you have probably had a course in developmental psychology in which you had some exposure to the ideas of Erik Erikson (1980). In fact, you may have had a lot of exposure to Erikson and are now thinking to yourself, "Erikson? Again?" Well, yes, but just a little bit, and with a different twist. Although some of Erikson's ideas have been questioned and criticized since he first advanced them, some of the basic tenets of his work are valuable tools for self-understanding. There are issues in each stage that are not always obvious from the familiar names of the stages (trust vs. mistrust and so on). Your early experience with these issues often reverberates into your later life, and we find that some of these issues are evoked or otherwise become relevant in internships.

Many of you are familiar with Erikson's first four stages: trust vs. mistrust, autonomy vs. shame and doubt, initiative vs. guilt, and industry vs. inferiority. At the core of each stage, though, is a basic need that is especially important at a moment in a child's life. As an adult, those issues are still important. But if those needs were not met well when you were a child, you may have an especially strong reaction to not getting them met as an adult. As a child you really did *need* certain things to thrive and grow. As an adult, you still want those things, and benefit from them, but if you react especially strongly to not getting them, you are needing them in a counterproductive way.

Trust vs. Mistrust

At the heart of this issue is your reaction to disappointment. Regardless of your level or preparation and skill, there will be times during your internship when you need help and support. Sometimes you will get it, and sometimes you won't. You may find yourself needing various things from your co-workers or supervisor. If they disappoint you, it is natural to be slightly upset, but a person with a shaky sense of trust may find these experiences unduly distressing. You may also find that, fearing your needs will not be met, you hide them, trying to appear confident and refusing to ask for help.

Autonomy vs. Shame and Doubt

Impulses and impulse control may seem like childhood issues and indeed they are, although you will meet clients who struggle significantly in this area. But an internship is often powerful, emotional work in which you are not sure what will happen next and yet have to react quickly. Sometimes you will handle these situations with ease and clarity. Other times you will find that your first impulse is to do something that in your calmer moments you know is not appropriate. At the heart of this issue is your ability to handle impulses. The issue is how you feel when you have an impulse, regardless of whether you act on it. Consider the example of an intern who is insulted in a particularly hurtful way by a client. Her first impulse is to insult him right back and put him in his place. She stops herself, though, and sets a limit with the client in a firm and calm manner. But her drive home, her journal, her conversation with peers, and her supervision time are dominated by how tempted she was. Rather than being proud that she controlled the impulse, she feels guilty for having it. The impulse, not the control, has confirmed her fundamental view of herself. Everyone wants to have impulse control; the intern in this example *needs* it to feel good about herself, and that can be a trap.

Initiative vs. Guilt

Picture this: You are in a meeting at your internship and you offer an opinion on an approach to use with a group of clients. Your supervisor thinks it's a great idea and asks you, in front of the rest of the staff, if you would like to develop your idea further, design some activities, and co-facilitate them with another staff member. You say yes, but inside you are terrified and can think of hundreds of reasons why this is a very bad idea.

At the heart of this issue is your ability to run with a task and develop it through to completion. If the idea of this happening has evoked feelings of anxiety, then know you are not alone. Here's another example for you. Imagine you have an idea for a project in the community and mention it to your supervisor. Your supervisor thinks the idea has some merit but does not think that the timing is right, or that you are quite ready to handle it. Later you find yourself angry and unable to let go of the anger. You feel like your supervisor is never going to "turn you loose" (even though this is really the first time you have asked).

Both of these examples may be touching the Eriksonian issue of initiative. Ask yourself how you feel now when you have an idea, make a plan, or need to take charge of something. Does it feel like an adventure or more like a minefield? Think about your current reaction to being told "no" or having your desires frustrated in some way. Inevitably, some of these issues will come up in your internship. What's important is that you recognize them for what they are so you understand what you are dealing with and how to move beyond them.

Industry vs. Inferiority

Your internship is your chance to shine, to see if you are really suited for work in this field or with this population. You may decide after the experience is over that it is not for you, but you want to feel like you *can* do it. A sense of professional competence is important to most helping professionals we know, and to most interns. Odd as it may seem, the heart of this issue is separating out whether or not you generally feel competent as a person—regardless of whether you are having successes or failures in your work. The internship offers countless opportunities for you to experience both. Whether it's an interaction with a client, an attempt to defuse an argument, a group activity, or a phone conversation: it could go awry. No one enjoys such occurrences, but interns who have a sense of "inferiority" (in Eriksonian terms) may find that instances that don't "go right" are especially troublesome and difficult to overcome. So, too, when it comes to reactions to criticism or evaluation. They may be very difficult to endure. How you respond to these events, which occur frequently in internships, tells you something about yourself. And, that something is whether you have a general sense of competence about who you are as a person, whether it's in the face of criticism, failure, or even in the face of success.

Considering Unresolved Issues

Each of us has struggled with different personal issues in our lives, and this self-examination may have uncovered or reminded you of some. These struggles can leave us all with issues that are unresolved or partially resolved. For example, you may have overcome your adolescent struggle, but the memory of those days may still be very painful. These issues could be thought of as your "unfinished business."

This unfinished business does not have to come from some traumatic event or a particular struggle. Family patterns can be another source. For example, if you were always the "mediator" in your family—the one who stepped in and calmed people down and helped them resolve their differences—you may have some very strong feelings about conflicts and find yourself wanting to step in and calm the situation in your office between co-workers or with a client. It may be hard for you to see someone in a conflict and not step in to help, even though it is sometimes best to let people work it out for themselves.

Your vulnerable areas are not just the result of your personality or of your childhood and family experiences. Your unfinished business can also come from your membership in certain societal subgroups. They are also the result of your experience in society as a member of racial, ethnic, gender, and other subgroups. If you are a member of a group that has been discriminated against, you may have a difficult time with clients who express prejudice, especially toward that group. If you are a member of a more privileged group, such as males, Whites, or heterosexuals, and have thought about issues of discrimination, you may feel guilty or hesitant around members of corresponding less privileged groups. Your responsibility is to be aware of your areas of unfinished business, understand how they affect the ways you respond, and take engaged approaches to moving beyond them in the workplace.

If you are interning in the helping professions, you have additional responsibilities to understand your unfinished business. Even if your unfinished business is not part of the reason you selected human services as a career, human service work can often stimulate those issues. Suppose you are working with a substance abuse population. As you work with a particular client, he begins to discuss his struggle with an overcritical father. Even though you have not had the experience of being a substance abuser, you have had a similar struggle with your father or some other parent figure. You are likely to be touched by this client in ways that seem mysterious at first. You may find yourself thinking about him when you are at home or when you wake up in the morning. This kind of preoccupation can be very draining.

In addition, some clients have an uncanny sense for your sore spots and will use them to manipulate you. If you have a strong need to be liked, for example, the client may withhold approval as a way to get you to relax a rule or go along with a rationalization. Your unfinished business can show up with colleagues and supervisors as well. If your father or mother was hypercritical of you, you may decide that in the internship you are not going to accept criticism without standing up for yourself.

The point here is not that you should be free of unfinished business; all of us have some. However, you need to be as aware as you can of what that business is and how those areas of vulnerability may be touched in your work.

Conclusion

This chapter has encouraged you to begin, continue, or develop the habit of self-exploration, and we have suggested specific areas that will be important to your internship. Making an investment in self-understanding will help you see your particular issues as you move through the stages of an internship. It will also help you identify and overcome obstacles and deal more effectively with clients, supervisors, and co-workers. Perhaps you will also see the benefits in your life outside the placement. There are other issues about yourself that you need to explore as well. These issues, which pertain primarily to your functioning as an intern, are the subject of the next chapter. For now, we leave you with a quote from John Schmidt (2002):

> *By seeking assistance for oneself and exploring our own development, we illustrate our belief in the helping process and demonstrate the same courage we expect of those who ask for our assistance.* (p. 12)

For Review & Reflection

Checking In

One More Time. You've spent time in this chapter exploring aspects of who you are in relation to being an intern. We'd like you to re-visit THINK About It box—Diagram Your Cultural Profile (p. 103). As you look it over, consider other categories that affect worldview and how they affect your daily life where you live/work and where you intern. We'll give you two leads: your educational status and your political stance.

Experience Matters (For the EXperienced Intern)

It's taken you a while (that could be an understatement!) to reach this point in your studies. And life. This may be "the big one" when it comes to internships (either because of the site/work or because of the amount of time in the field, for example, 30 hours a week for a full semester, 20 hours a week for the full academic year, and so on). In any case:

- What are the major reasons you chose the academic major you did, and in what way is that connected to this internship?
- What professional field do you plan to enter eventually, and how is that connected to the internship?
- Why this internship at this site at this time?

Only you can determine just how much is riding on this internship: Graduation? Security in an entry-level position even you've had past employment/careers? Insurance and other benefits? Here's where you take off the pink-tinted glasses, if you still have any tint on them, and think about the choices you made along the path to this internship. Can you see any way that those motivations could be troublesome for you in your internship? That past employment/career(s) could complicate the internship? If you can answer *yes* to either of these questions, what are your next steps to ensure that you keep moving in the direction want?

Personal Ponderings

Now that you've had a chance to read and reflect through this chapter, we'd like you to consider two things: What aspects of self-understanding do you believe are strengths and particularly useful to your role as an intern—and, why? And, are there aspects of self-understanding that give you reason for concern when it comes to the internship? If so, do you understand why that is and have you given thought a plan to move through it?

Seminar Springboards

Reality Checks on Worldviews. Although we may believe that we know are ourselves fairly well, that is not necessarily the case. This exercise asks you to think about the worldviews you hold, their affect and effects on your daily lives and possibly on your internships, and whether how you see yourself differs from how others see you. Within your community of interns, create conversational spaces (on campus or online) to think through some tough questions that will help you construct more knowledge about yourself. You are the source of those questions, but we provide you a couple of guidelines to begin your exploration.

- Review the cultural profile that you created earlier in the Checking In exercise.
- Choose the aspects of your worldview that you wonder about—whether in positive or in compromising ways—in terms of their influence on your ways of being.
- Develop a list of how those aspects of your worldview affect you—in your daily life and how you think they are or might be affecting

you in your internship. What questions do these worldviews raise for you?

- Now develop a list of the effects of those aspects of your worldview on your ways of thinking and acting. What questions do the effects raise for you when it comes to your internship? What changes do you want to see happen?

- Bring your questions to the conversational spaces. As you discuss them and learn about yourself, think about how the perspective you hold of yourself is being affected. Ask the tough question: Do others see you in the same ways you do? If not, what do you do with that knowledge?

For Further Exploration

On the Importance of Self-Understanding

Corey, M. S., & Corey, G. (2011). *Becoming a helper* (7th ed.). Belmont, CA: Brooks/Cole.

A very readable and thought-provoking book, devoted entirely to self-understanding and effectiveness in interpersonal relationships.

Levine, J. (2012). *Working with people: The helping process* (9th ed.). New York: Longman.

Excellent chapter on self-understanding.

On Values

Corey, G., Corey, M. S., & Callanan, P. (2011). *Issues and ethics in the helping professions* (8th ed.). Belmont, CA: Brooks/Cole.

The issue of values is woven throughout this book, but Chapter 3 in particular discusses the issue of imposing-vs.-exposing values.

On Family Patterns

Goldenberg, I., & Goldenberg, G. (2013). *Family therapy: An overview* (8th ed.). Pacific Grove, CA: Brooks/Cole.

Extremely readable text on families and family therapy.

Goldenberg, I., & Goldenberg, H. (2003). *Family exploration: Personal viewpoints from multiple perspectives* (6th ed.). Belmont, CA: Brooks/Cole.

The accompanying workbook serves as a guide for you to explore your own family dynamics.

On Learning and Thinking Styles

Belenky, M. F., Clinchy, M., Goldberger, N. R., & Tarule, J. M. (1986). *Women's ways of knowing: The development of self, voice and mind.* New York: Basic Books.

A pioneering work examining ways of learning, thinking, and viewing the worlds that appear to be more commonly found in women. At the time this book was written, a good deal of the research and writing on learning style was derived from all male populations.

Gregorc, A. F. (2006). *The mind styles model: Theory, principles and practice.* Columbia, CT: AFG.

Easy to understand and use, this theory looks at two dimensions of learning style: concrete vs. abstract and random vs. sequential.

Magolda, M. B. (1992). *Knowing and reasoning in college: Gender-related patterns in undergraduates' intellectual development.* San Francisco: Jossey-Bass.

Classic book that describes how ways of knowing change over the course of college years and gender influences ways of reasoning.

Perry, W. G. (1970). *Forms of intellectual and ethical development.* New York: Holt, Rinehart & Winston, Inc.

Another classic in the field. This work was done with a male population, but the basic formulations have held up well over time.

On Cultural Competence

Adams, M., Bell, L. A., & Griffin, P. (Eds.). (1997). *Teaching for diversity and social justice.* New York: Routledge.

A collection of readings that describes workshops on many forms of oppression. Useful as a resource for trainers with exercises to work through.

Adams, M., Blumenfeld, W. J., Castenada, R., Hackman, H. W., Peters, M. L., & Zuniga, X. (Eds.). (2000). *Readings for diversity and social justice: An anthology on racism, anti-semitism, heterosexism, ableism, and classism.* New York: Routledge.

Diller, J. V. (2011). *Cultural diversity: A primer for human services* (4th ed.). Pacific Grove, CA: Brooks/Cole.

Doane, A. (Ed.). (2003). *White out: The continuing significance of racism.* New York: Routledge.

Green, J. W. (1998). *Cultural awareness in the human services: A multi-ethnic approach* (3rd ed.). Upper Saddle River, NJ: Prentice Hall.

Covers both theoretical and practical aspects of working with diverse populations.

Lum, D. (2011). *Culturally competent practice: A framework for understanding diverse groups & justice issues* (4th ed.). Pacific Grove, CA: Brooks/Cole.

Payne, R. K. (2005). *A framework for understanding poverty* (4th ed.). Highlands, TX: Process, Inc.

Rothenberg, P. S. (2000). *Invisible privilege: A memoir about race, class and gender.* Lawrence: University Press of Kansas.

Rothenberg, P. S. (Ed.). (2004). *White privilege: Essential readings on the other side of racism* (2nd ed.): New York: Worth Publishers.

On Communication

Godwin, A. (2011). *How to solve your people problems: Dealing with your difficult relationships.* Brentwood, TN: Rosenbaum & Associates Literary Agency.

Kegan, R., & Lahey, L. (2001). *How the way we talk can change the way we work.* San Francisco: Jossey-Bass.

References

Axelson, J. A. (1999). *Counseling and development in a multicultural society* (3rd ed.). Pacific Grove, CA: Brooks/Cole.

Baxter Magolda, M. B. (1992). *Knowing and reasoning in college: Gender-related patterns in undergraduates' intellectual development.* San Francisco: Jossey-Bass.

Belenky, M. F., Clinchy, M., Goldberger, N. R., & Tarule, J. M. (1986). *Womens ways of knowing: The development of self, voice and mind.* New York: Basic Books.

Brammer, L. M., & MacDonald, G. (2002). *The helping relationship: Process and skills* (8th ed.). Boston: Allyn & Bacon.

Corey, G., Corey, M. S., & Callanan, P. (2011). *Issues and ethics in the helping professions* (8th ed.). Belmont, CA: Brooks/Cole.

Corey, M. S., & Corey, G. (2011). *Becoming a helper* (7th ed.). Belmont, CA: Brooks/Cole.

Doane, A. (Ed.). (2003). *White out: The continuing significance of racism.* New York: Routledge.

Erikson, E. H. (1980). *Identity and the life cycle.* New York: W. W. Norton.

Gollnick, D. M., & Chinn, P. C. (2005). *Multicultural education in a pluralistic society* (7th ed.). Upper Saddle River, NJ: Prentice Hall.

Gregorc, A. F. (2004). *The Gregorc style delineator.* Columbia, CT: Gregorc Associates, Inc.

Johnson, D. W. (2008). *Reaching out: Interpersonal effectiveness and self actualization* (10th ed.). Boston: Allyn & Bacon.

Kegan, R. (1982). *The evolving self: Problem and process in human development.* Cambridge, MA: Harvard University Press.

Kegan, R. (1994). *In over our heads: The mental demands of modern life.* Cambridge, MA: Harvard University Press.

Kolb, A. Y., & Kolb, D. A. (2005). Learning styles and learning spaces: Enhancing experiential learning in higher education. *Academy of Management Journal of Learning and Education, 4*(2), 193–212.

Kolb, D. A. (1984). *Experiential learning: Experience as the source of learning and development.* Englewood Cliffs, NJ: Prentice Hall.

Levine, J. (2012). *Working with people: The helping process* (9th ed.). New York: Longman.

McIntosh, P. (1989). White privilege: Unpacking the invisible knapsack. *Peace and Freedom, July/August,* 10–12.

Moon, J. (Oct. 20, 2011). *What you learn from experience relates to the quality of your thinking.* Keynote Address: 40th Anniversary Annual Conference, National Society for Experiential Education, Dallas Texas.

Payne, R. K. (2005). *A framework for understanding poverty* (4th ed.). Highlands, TX: Process, Inc.

Perry, W. G. (1970). *Forms of intellectual and ethical development.* New York: Holt, Rinehart & Winston, Inc.

Schmidt, J. J. (2002). *Intentional helping: A philosophy for proficient caring relationships.* Upper Saddle River, NJ: Merrill Prentice Hall.

Schram, B., & Mandell, B. R. (2006). *An introduction to human services: Policy and practice* (6th ed.). Boston: Allyn & Bacon.

Sweitzer, H. F., & Jones, J. S. (1990). Self-understanding in human service education: Goals and methods. *Human Service Education,* 10(1), 39–52.

Weinstein, G. (1981). Self science education. In J. Fried (Ed.), *New directions for student services: Education for student development.* San Francisco: Jossey-Bass.

BEGINNINGS

The previous section was the most theoretical, and for some students the most challenging part of the book. We have given you theory to understand and think about (experiential learning, developmental stages of an internship), content to grasp (engaged learning; development in the personal, professional, and civic dimensions of learning and growth; and civic professionalism), tools to draw upon, and ways to understand yourself in relation to the work and workplace of the internship. Now we get to the business of applying the theories, concepts, and ways of thinking to your internship; and you get to use the tools to move through the stages and meet the challenges of each one.

The beginning of an internship can be very exciting with many intellectual and emotional challenges. It can also be overwhelming emotionally. Experiencing the internship is much different from reading about it, imagining it, or even doing role-plays or simulations. There is a lot to think about and a lot to prepare for; there is a lot to respond to and a lot to be challenged by.

This section of the book will help you have as smooth a beginning as possible and set the foundation for further progress by focusing on the tasks of the Anticipation Stage. In Chapter 5, we focus on the major concerns that beginning interns often have about themselves, including acceptance, role clarification, capability, and relationships with co-workers. Chapter 6 is one of the Essential chapters in the book, focusing on the Learning Contract and the supervision plan. Because it is very important that the contract and plan be in place by the time you are reading and experiencing this chapter, we guide you in developing clear learning goals and understanding aspects of the supervisory relationship so that both of these tools are ready for the work ahead. Chapter 7 is the final chapter on the tasks of the Anticipation stage. It is intended for those of you who are interning in the helping professions and working directly with clients. In it, you will find important information and suggestions for beginning those helping relationships.

The Anticipation stage is a brief stage in most internships; typically, it lasts no more than two to three weeks. Consequently, it's important to read this section of the book (Chapters 5, 6, and 7) in a way that you are keeping pace with the experience you are having in the field.

CHAPTER *5*

Experiencing the *What Ifs*: The Anticipation Stage

Many of us seem to have a lot of self-doubt about our capacities and what is manageable for us. It would seem that after all the required academics we would not have these feelings, but apparently not.
STUDENT REFLECTION

Your internship is probably something you have been looking forward to for a while. Some of you have been waiting just a semester or two, others have waited their entire college careers, and some have been waiting a lot longer than that. You have been hearing about internships for some time; many of you have worked hard and sacrificed to get here, and now here it is. But as your friends and family ask whether you are excited, and you answer "yes," you may also feel some anxiety creeping in. It may be a little or a lot; it may be visible to your friends, or you may keep it hidden. You may even hide your anxieties from yourself; some interns become aware of them only after they start to dissipate (Wilson, 1981). In any case, you are entering an unknown experience, and that's always at least a little unsettling, even when it's exciting at the same time.

Stop a moment and think about other new experiences in your life. Do you remember your first day of elementary school? High school? College? How about your first day at a summer camp or a trip to see relatives you didn't know? How about an excursion into an unfamiliar neighborhood? You were probably excited about these things, too—and nervous. You have probably heard lots of stories from students who have already completed their internships; that is both a blessing and a curse. It builds your excitement, but you may also hear a little voice inside saying, *"Am I going to be as good at this as they were?"* You may also have heard some "horror stories" about internships that went sour in one way or another; even though these instances are usually few and far between, the stories seem to have great staying power.

In this chapter, we help you understand the concerns and tasks of the Anticipation stage. The concerns typically focus on your role as an intern and acceptance by key people in the internship. As we discussed earlier in the book, the word *concern* has two meanings— concerns of *interest* or *care*, and concerns of *worry* or *fear*—and you can experience both in each stage of an internship, sometimes even around the same issue. For example, from the interest side you might be thinking, *I am most concerned with having a good experience* while at the same time you are thinking *I am concerned that I don't know enough*, obviously from the worry side. There might be times when most of your concerns fall on the worry side, and there might be times when most fall on the care or interest side. Your responsibility is to move your worries and fears to interests and commitments, and we will guide you in doing so through each stage.

We refer to this first stage of an internship as the *"What if...?"* stage because many interns sum up this stage in just that way. Very early on in the internship, it is common for interns—even those who seem quite confident—to experience some anxiety. Not all anxiety is "bad" anxiety; think for example about the anxiety you experience when you are preparing for a very special or exciting event in your life. That kind of anxiety happens throughout the internship as well. There also are times when you will have concerns and hence anxiety in the form of *fear or worry*, as the examples in the following box attest. As you peruse the lists, think about those concerns that are familiar to you. How many more can you add? Some will be shared with your peers, but some will be unique to you.

The tasks of this stage are paths to addressing concerns that are normal for this stage—both interests and worries. Consistent with one of the themes of the book, we will encourage you to take an engaged approach to those tasks. If you do, the concerns you worry about can and do get resolved, and when that happens your general feeling may become one of care or interest. Some of you experience that stance at this point, and if you do, that tends to come from your *vision* of an internship. Once you address the tasks of this stage, it will come from a more experienced and informed perspective.

THINK About It

The *What Ifs*

Listed below are some of the "what ifs" we have heard. Although some apply to those in the helping professions, most are common across a variety of settings.

What if…

- I can't pick things up fast enough?
- They all think I know more than I really do?
- I make a mistake—and something really bad happens as a result?
- I don't grow up mentally and emotionally in time for graduation?
- I'm asked a question and don't know the answer?
- My supervisor doesn't like me?
- I don't like my supervisor?
- They do things differently from the way I was taught?
- The other employees resent or ignore me?
- I get sued?
- I hate it?
- I fall apart?
- I don't feel my service is valid?

- I can't handle all my responsibilities at home and at the internship?
- I don't have enough time for my loved ones?
- I don't have time for those who need me?
- A client insults me?
- A client confronts me?
- A client gets physical?
- A client attempts to sexually assault me?
- A crisis happens and I don't know what to do?
- A client lies to me?
- A client falls apart?
- My clients don't get any better?
- Clients ask me personal questions?
- I say something that offends a client?
- I have a client I can't stand?
- I have to discipline a client who won't listen?

Some of these concerns may seem silly, and no one (so far) has worried about all of them. Your particular anxieties will be shaped by your personality, your knowledge of your placement site, and your past experiences. Every one of the entries on the list, however, is something we have actually heard many times. You probably can add to them. When we encourage our students to share their anxieties with one another, they are often surprised and relieved to learn that they are not alone in their fears and concerns. Every one of these concerns is perfectly normal, and every one can be addressed, or at least the attendant anxiety can be lessened.

Becoming a More Engaged Learner

As you are aware from Chapter 1, engaged learning is one of the underlying concepts of this book. But what does that mean in practice? An internship is probably *the* biggest discovery zone that you will ever enter. Discovery zones are contexts that allow you to seek and learn new knowledge and skills, develop new insights and perspectives, and rethink what's meaningful to and valued by you. They are learning experiences in which *you* are the center of energy and in a position to take charge of what you learn, how your thinking develops, and what you value. By definition, the internship is a high impact learning experience (Kuh, 2008). It is also an example of engaged learning as described in Chapter 3, in that it encourages students to seek and discover new knowledge by exploring authentic questions and problems (Hodge, Baxter Magolda, & Haynes, 2009). By design, you become an *active* participant in *what* you learn and the *way* you go about learning.

At this point in your internship, you are *learning* to become an engaged learner. Of all the knowledge and skills you've been developing in preparation for this internship, *learning to be a learner* probably wasn't on your list. But you have already begun the process. By the time your internship began, you might have felt more than a bit exhausted getting through the hoops required by the academic or student affairs' programs (e.g., internship offices, career services). That is because you were being asked to *participate in* the process of your placement and be *actively involved* in decisions about it. At least, that is how it is on many campuses. Learning to empower yourself can be exhausting at times, and that's what you were doing preparing for your internship *if* you were taking an engaged approach. If you weren't, there's no reason to worry. By the time you reach the Competence stage, you will be demonstrating engaged approaches to learning and to your internship. So, how do you get from here to there—which is two stages from now? The boxes that follow inform you of ways to do that and how to gauge yourself in doing so.

THINK About It

What Does the REAL World Mean?

Some like to say that the internship is the "real world" and that's true. But so is your life as a student, and what you have learned in life will have added value to the context of your internship. This is especially so for those of you who have returned to campus after having another career, caring for family members, serving in our military, or spending time figuring out your direction in life.

The Engaged Intern: A Profile

The engaged intern is invested in a high-quality internship. Learning is the priority. The engaged intern strives to take charge of the learning by:

- **Having Foundational Understandings**
  - Understands the importance of reflection in experiential learning
  - Appreciates the central role of *discovery* as the context of an internship

- **Seeking Engagement Benchmarks**
  - Takes the lead integrating into the organization and developing key relationships in useful and effective ways
  - Determines the pace, direction, and shape of the learning by pushing and challenging oneself to reach beyond your comfort zone of current knowledge, skills, and competencies and seek out new understandings and perspectives
  - Invests intentionally in maximal learning using Kolb's existential learning cycle
  - Recognizes strengths and where to focus energy for improvement
  - Knows when supervision is needed, how to use it effectively, and how to seek it out appropriately and in timely ways
  - Embraces site's ways of being (culture) and working

- **Demonstrating Personal Qualities and Competencies**
  - Is aware of and understands self in relation to the role of intern
  - Is familiar with, embraces the tasks of, and experiences the developmental stages of an internship
  - Asks provocative questions, including those that challenge the "givens"
  - Is eager to and enjoys learning—seeks out learning opportunities that are meaningful and accepts those offered to them
  - Wants to be successful in all aspects of the internship
  - Recognizes the need for supervision, seeks it out appropriately and in timely ways, and uses it effectively

In each stage, you'll have plenty of opportunities to use engaged approaches to situations: problems to solve, crises to deal with, and dilemmas to resolve. Your supervisors and co-workers, as well as your peers, are there to guide you. Showing you is one form of guidance, but more likely you'll learn by watching others deal with situations at the site, listening as co-workers process what they are dealing with and the emotions they are experiencing, listening and supporting peers as they describe how they dealt with situations in their placements, and responding to provocative questions intended to make you think through problems, crises, or dilemmas. Such questions are powerful ways of learning

---

THINK About It

Visualizing the Engaged Intern

When we think about examples of engaged approaches to the internship and its many trials and tribulations along the way, we tend to frame our thinking in terms of an engagement-disengagement continuum, as shown below.

$$\textbf{Engaged} \longleftrightarrow \textbf{Disengaged}$$

Using this continuum, think for a moment where your *personal point* is, i.e., how you tend to approach situations in general; then think about how that point compares to what you know of yourself already as an intern (*internship point*). Then consider:

- Is there a difference between these two points and if so, why might that be?
- Are you where you want to be in both situations? If not, think about what it will take for you to move on the continuum in the direction of being more engaged.
- Take a moment to visualize yourself as an engaged intern. Literally, stop and see yourself as that engaged intern. What do you look like, i.e., how do you appear and how do you act? Do you like what you see? If not, why not?
- Now think about the engagement-disengagement continuum and note on it where you are as an engaged learner.
- Then, think about where you want to be as an engaged intern. It's important to always see yourself as having more potential than you think you have!

---

because they are intended to make you think deeply and perhaps differently about the situations at hand.

The Tasks at Hand

I have come to realize that we are all in the same boat although we are not all physically in the same sites. And, we are all going through the same feelings, worries, and anxieties. Knowing this and seeing it in black and white has helped me tremendously.

STUDENT REFLECTION

Even though you may be a little nervous, you are probably eager to get going. After all, you didn't sign up for the internship to read this book or write in your journal. You signed up because you want to start *doing* the work, acquiring new knowledge and skills, and even begin making a difference by being at the site. However, at the same time you start your work, we are going to encourage you to attend to other tasks of the Anticipation stage, which focus on you as an intern and on your relationships with colleagues:

- Examine your expectations and critique your assumptions.
- Acknowledge and explore your concerns.
- Clarify your role, purpose, and goals as an intern.
- Develop acceptance in key relationships at the internship site.
- Make an informed commitment.

These tasks, if done effectively, will help address both your interest and worry concerns; in turn, they will form a solid foundation from which to learn and grow. Without that foundation, you may falter sooner and harder. Because it is your worry concerns that are apt to create obstacles to your success in this internship, we will be guiding you in focused ways to move those concerns into interests and commitments.

The chart that follows describes a "close up" view of the tasks of the Anticipation stage, as well as examples of ways of approaching those tasks as engaged and disengaged interns. Remember that engagement is both a style of learning *and* an approach to tasks, and it is not a static or unitary characteristic. You may find yourself more easily taking an engaged approach to some of these tasks than to others. That doesn't mean that you are any less or more of an engaged intern; it is simply your approach to a given situation.

In each of the developmental stages of the internship, a "close up" view like the one on the next page is provided to guide you through the tasks of the stage.

Examining and Critiquing Assumptions

Just as you begin your internship with a plethora of interest and worry concerns, you begin with certain expectations as well. These expectations are products of assumptions made, correctly or incorrectly, about many aspects of the upcoming or beginning internship (Nesbitt, 1993). As we mentioned in Chapter 4, these assumptions may come from stereotypical portrayals in the media of certain groups or agencies or from your own experience. Your previous experience, however, does not always predict what the future holds, although it is natural to generalize from it. You probably have at least some assumptions about your co-workers, the populations and clientele with whom you'll have contact, the site, and your supervisors. The engaged intern acknowledges and then critically examines the assumptions by making them explicit, whereas the disengaged intern assumes that the assumptions are realities and generalizes from them, perpetuating them and their related biases without questioning them. Making these assumptions explicit and subjecting them to critical examination will help you develop the most realistic picture possible of the internship you are beginning.

Acknowledging Concerns

In Chapter 2, we discussed some of the kinds of concerns interns typically have at this stage. We will discuss just a few of the more common ones here.

One of the greatest concerns for beginning interns is competence (Brust, 1986), which we refer to as *capabilities*. This is a concern that many interns think about; some interns worry about it. Naturally, you want to do a good job and be recognized for it. Furthermore, if you think about all the examples of professionals at work that you have observed, read about, or seen in videos, you have probably seen very few examples of mediocrity; the idea was to show you how effectiveness looks in action (Yager & Beck, 1985). However, a distortion effect can occur. When we go to the movies, we see the scenes that were kept, not the ones that were scrapped

The Anticipation Stage			
		Response to Tasks	
Associated Concerns	**Critical Tasks**	**Engaged Response**	**Disengaged Response**

• Getting off to a good start
• Positive expectations
• Acceptance
• Anxieties
  Capability
  Relationship with supervisor
  Relationship with co-workers
  Relationships with clientele
• Life context
	Examining and Critiquing	Willing to challenge own assumptions by making them explicit
		
		Seek out accurate information to dispel differences in expectations
			Assumes expectations about internship are grounded in truth and reality
			
			Lives with untested expectations of relationships and inherent biases

	Acknowledging Concerns	Recognizes interest and anxiousness as normal and works toward resolution where needed	Accepts interest; accommodates or denies anxiety: The *Not Me!* fantasy
	Clarifying Role and Purpose	Seeks clarification of role and responsibilities to ensure that learning expectations can be met and that their interests are respected with opportunities for them to be realized	Accepts role and responsibilities without critically thinking through how it meets learning needs and interests

(continued)

The Anticipation Stage *(continued)*

Associated Concerns	Critical Tasks	Response to Tasks	
		Engaged Response	**Disengaged Response**
• Getting off to a good start • Positive expectations • Acceptance • Anxieties *Capability* *Relationship with supervisor* *Relationship with co-workers* *Relationships with clientele* • Life context	Developing Key Relationships	Explores differences between being liked and being accepted; considers others' perspectives rather than personalize reactions of co-workers	Confuses being liked with being accepted and personalizes reactions of co-workers
		Invests in mutuality in relationships and recognizes the role mutuality plays in the success of the internship	Sees no purpose to mutuality because relationships not seen as partners in their success
		Works toward acceptance in key relationships; sees value in sources of support and learning	Depends on others to initiate and/or is disconnected from colleagues; sees little value in relationships and no connection to their success in the internship
	Making an Informed Commitment	Pursues understanding of key variables in the internship and commits to going forward	Clings to naïve or uninformed notions of the internship

Sweitzer & King, 2012

or had to be re-shot (again and again). And, even the most brilliant baseball players fail to hit safely three out of five times!

Yet we tend not to think about the failed attempts; we concentrate on the successes. Similarly, watching those videos and reading those cases make it easy to forget that everyone undergoes a learning process and that everyone makes mistakes. We imagine that we could never do quite as well and recall our own fumbling efforts. We often see only one of the paths to success and imagine we may never find it, while fearing the many paths to failure. So the question is: *Can I really do this?* And, that is an important question to consider.

You may have been quite successful in your classes, even in those that focused on skills building. If you have had previous field experiences, they may have gone quite well. But interns often tell us that the semester long internship is the "big one," and that some of the feelings they have are new and different this time around in the field. Some interns, for example, struggle with what has been called the *imposter syndrome* (Clance, 1985). This phenomenon is a cognitive distortion preventing a person from internalizing any sense of accomplishment (Gravois, 2007, p. A1). They are vulnerable to believing that whatever success they have achieved is due to incredible good fortune rather than their capabilities. At each step, they fear being exposed for the pretenders they imagine themselves to be. In the case of the internship, they imagine that they did well enough to qualify only through luck and that they will surely be found out at their placement site (Clance, 1985).

Voices of *EXperience*	Being an Expert...and *Then* an Intern!

Some interns have expertise in a related field to the one in which they are interning. Their site supervisors and co-workers may even know them from that career or know of their reputations in the field. These interns may have returned to campus to complete their degree, earn another or advanced degree, or complete a specialty certificate. They faced unique challenges, and they offer these words of guidance to those of who are in the same situation today.

- Go out of your way to ask questions to learn additional information as an intern.
- Get as much exposure to different parts of the site as possible in your role as intern.
- Attend meetings as an intern that you might otherwise not be privy to.
- Learn as much as you can in the time that you have—this might be your only chance.
- Make yourself available to colleagues/supervisors when appropriate in your expert role.
- Go in and develop an agenda for learning—know where it can bring you educationally/professionally and make sure you have opportunities to learn all you want and need to learn.
- Be clear on the set of lenses you are wearing at a given time—say things like, "As an intern, I think...." or "Based on my expertise and not my role as intern, I would..."
- Maintain your existing skills but develop new skills—that is why you are there, not just to learn things but to learn how to do them from this other perspective.
- Supervision—it's the key to new and challenging learning goals. Make the most of it.

Sweitzer & King, 2012

The good news is that there is a cure. The bad news is that, in large measure, the cure is time and experience. You will almost certainly feel more confident and capable as you become involved in the work as time goes on.

In the meantime, if you check with your peers you will realize that you are not alone with these feelings and simply knowing this can have remarkable healing powers.

Disengaged interns tend to deny their feelings of anxiousness about the internship or accommodate them in ways that mask the real reasons for them and the depth with which they are felt. They often fall prey to the "Not me!" fantasy that the feelings others are describing cannot happen to them. Of course, that might well be the case if they are emotionally distanced enough from the internship experience that they are going through the motions and not investing at all in the experience itself. It can also be the case that if they are so under-challenged that there is no learning expected in the internship, just time off campus for resume fodder. Even though we said the cure is time and experience, that does not mean that you can just sit back and wait; that would be a disengaged and ineffective approach. In an engaged approach, you accept your concern about capabilities as normal and do what you can to engage the other tasks, knowing that they will lead you, eventually, to feeling capable.

Clarifying Your Role, Purpose, and Goals

Many interns think about the *nature of their role* in the field experience with much excitement; others think about it with some trepidation. Interns are concerned about how others will see them at the site and how they will see themselves: As a student? Volunteer? Observer? Researcher? Staff member? In fact, you could probably perform all of these functions, but none of these roles in and of themselves defines *who* you are at the internship. The role of an intern is unique. Initially, you may spend a good deal of time observing the work being done. Later, you will be a more full-fledged participant in the work, doing what you went there to do. In between, you have an intense period of "learning the work"—the knowledge, skills, attitudes, and values needed to do it well.

To complicate your understanding of the role of the intern, there may be volunteers at the site as well. This can cause confusion for you as well as for the staff. Volunteers and interns do some of the same tasks, but their roles and responsibilities are very different. Volunteers are essentially there to help out; you are there to learn. You will have specific learning goals the volunteers do not have. As a learner, you should be involved in the work, but you also have the responsibility to reflect, analyze, and critique what you are doing, which a volunteer does not have. Furthermore, although you may perform some of the same tasks as a volunteer, as an intern you have a responsibility to learn why a particular task is done, why it is done in a particular way, and how it relates to the bigger picture in the organization (Royse, Dhooper, & Rompf, 2011).

Additionally, interns wonder *what* it is they will be asked to do. Will they have to dive right in? Will they be given boring, mechanical tasks like copying, filing, answering the phone, or driving? Your Learning Contract should help allay some of these fears, as should conversations with your supervisors at the site and on campus. You should expect, however, to be given some of what is often referred to as *grunt work* (Caine, 1994). In many agencies, almost everyone pitches in and does some of the drudgery; you will probably be asked to do your share. You may even be asked to do a little more than your share so that more skilled staff can focus on other kinds of work. Bruce "Woody" Caine (Caine, 1994) points out that there is actually quite a bit you can learn from performing these tasks. Often, the learning is not in the task itself, but in the organizational meaning of the task. You get a chance to learn what it really means to be an entry-level worker in that organization. You become familiar with these tasks, which are critical if not always interesting, and how long it takes to complete them (which is easy for people who don't have to do them to forget). You can also learn how much it costs the organization in material and human resources to get this work done. For example, answering the phones is one task we encourage most of our interns to do for short periods of time at their sites. There is perhaps no better way for an intern to develop a sense of the public's perception of the purpose of the site and the work that goes on there. We have found that most interns who manage phone calls during the Acceptance stage are astonished by the differences between the stated mission of the site and the public's expectations of what they can get from the site.

The value of grunt work aside, it still needs to be managed. By that we mean that it needs to be clear to everyone concerned that the grunt work is not the focus of the internship and the learning outcomes. It should also be clear exactly who can assign such tasks to you. If you think you are being asked to do too much grunt work or are being asked by too many co-workers to do their tasks or grunt work, you should speak with your site supervisor and/or campus instructors as soon as possible.

A Word to the Wise.... There are three points to keep in mind when thinking about your role, goals and purpose.

- First, you are there to learn as well as to work, which can be quite different from the role of employees or volunteers.

- Second, because of the way we learn experientially, time needs to be built into your work so that you can fully engage in and complete the experiential learning cycle (see Chapter 1). That means taking the time to observe and reflect as well as develop your ideas abstractly and take action to create new experiences.

- Third, you must have a clear sense of the purpose of your role as intern. In other words, when you think of the role, purpose, and goals that comprise your internship, whether you were involved in determining them or not and they make you feel important (although more than a bit intimidated), then you know you're on the right track. If on the other hand the role, purpose, and goals leave you feeling exploited or otherwise not respected as a student who has serious learning ahead, then you are *not* on the right track and need to discuss it with your campus supervisor immediately. In the next chapter, we discuss in detail how the learning contract and the supervision plan ensure that these three points are embedded into your internship so that you will have the best possible of learning experiences.

When it comes to concerns about their *role* as an intern, engaged interns are visibly involved in the work and intentionally use the experiential learning cycle; they seek to understand the *why* of their tasks (its purpose, the reason for the way it is done, and how it relates to the bigger picture of the work of the site); grunt work is seen as an opportunity for additional and perhaps unexpected learning. Disengaged interns accept the role without thinking about or questioning aspects of it. If questioning goes on, it tends to focus on issues of fairness in the assignments.

Developing Key Relationships

Great discoveries and achievements invariably involve the cooperation of many minds.
ALEXANDER GRAHAM BELL

As an intern, the primary work of your organization, regardless of whether it is improving the environment, developing marketing plans, providing direct service, or working with communities, is probably your biggest concern. However, it is those you work with who can make or break your internship. We noted in Chapter 1 that relationships are the context or the medium for learning during the internship and a feature of all high-impact practices. You will be involved in many relationships during the course of your internship, both at the site, in the communities of the site, and on campus. We use the terms *co-workers* and *colleagues* interchangeably to describe the wide variety of people who will become part of the world of your internship. We focus the rest of this chapter on the concerns you may have as you begin to meet and work with your colleagues. Concerns about relationships with supervisors, though, are considered in depth in Chapter 6, where we also consider the Learning Contract, both of which are instrumental in guiding your internship to be all it can be.

Working with Co-Workers

We are bound together by the task that lies before us.
MARTIN LUTHER KING JR.

As with any relationship, your relationships with colleagues can initially feel like a good match or a poor one. And, as with any relationship, they can be worked at, nurtured, cultivated, and improved upon provided you take an engaged approach. Although your supervisor may be your most important colleague at the placement site, you may actually spend more time with other staff members. Sometimes interns have "informal" supervisors at the site, someone they go to for "on-the-job" help in the moment. This often is someone they learn a great deal from and may feel more supported by than the person assigned as their supervisor, whom they may only see once a week. Whatever the importance of co-workers to you, they are a great opportunity for learning and a potential source of support. They may be mentors, sponsors, and/or role models offering you informed knowledge and skills, an insider's experienced perspective on the agency and its work, and valuable feedback on your performance.

Most likely, you are in the throes of beginning relationships with many of these people. The relationships you are developing are interactive in nature; half of that interaction is you and the other half is another person as unique as you are. Many of the facets of yourself that you explored in Chapter 4 and probably more may come into play as these relationships develop. Although each relationship is unique, there are some things we can help you be aware of as you consider them. The three most common concerns are: expectations, acceptance issues, and feelings of exclusion.

Expectations

You may not have thought much about it, but you probably have an image of what your co-workers will be like. What level of education do you think is needed for working at your site? What level of education do you think most of the staff have?

You can, and should, spend some time getting *accurate* information about your co-workers and finding out information about them. You may be quite surprised, by the answers to questions as obvious as these:

- What is known about the hiring process at the site?
- What level of education is in fact required for the work you are doing? Your supervisor is doing?
- What other requirements and preferences does the organization have?

THINK About It

What Guides Your Co-Workers' Work?

Most people are guided by some standards or principles in their work. Your organization has a set of rules and policies and may even have a code of professional behavior or ethical guidelines. You may have studied ethics and other aspects of professional behavior in your classes as well. To what extent do you think your co-workers are aware of these issues and standards? To what extent do you think they adhere to them? What do you imagine happens to them if they violate those standards?

- How many staff people leave the organization in an average year?
- What do people have to do to get promoted?

Just as with clientele, if you have them, you probably have an image of your co-workers, and that image needs to be tested by learning what you can to close the gap between your expectations and the realities of your internship. You can start by creating questions that guide your thinking about the expectations and begin to test the assumptions that underlie them.

Acceptance

Most interns are immediately concerned about whether, when, how, and on what terms they will be accepted by the staff at their site. As you think about these questions, there are a number of issues to consider. Obviously, different individuals will react to you differently and you, in turn, will react differently to them. When the reactions are the same and positive, it feels relieving and affirming. When the reactions are the same and negative, well, that's a very different story.

Based on what you know now, who are the people at your site whose acceptance is most important to you? List them in your mind and try to do so in their order of importance. Some of you will need paper and pencil at this point, depending on the size of your site. Once the list is done, ask yourself why the relationships are important enough to be on the list. Next, ask yourself what happens if you and those key people don't share the same enthusiasm for the relationships? In thinking about the effects of not being accepted by certain people on the list, keep in mind that *acceptance is* not the same as *being liked*. Take some time right now to think through the difference and how that affects the way you look at the issue of *"What if I am not accepted by these key people?"* Is it possible that they accept you but don't necessarily like you?

Most of us want to be liked but, again, staff members do not have to like you in order to accept you. Think about this: If you *need* every staff person to like you, you may find yourself making decisions to make that happen, and those decisions may be at odds with what is best for the clientele, the work, or the site. You'll have to look deeply and spend time considering past experiences to determine if being liked is what you want, not necessarily being accepted. On the other hand, here is a reasonable, although not always achievable, desire expressed by an intern: *"What I want from them is to remember when they first started out and appreciate what I am going through."*

Mutuality   Keep in mind that acceptance is a two-way street. One challenge for you is to accept your co-workers for whom and what they are. Again, you do not have to like them nor even appreciate everything they do, but you do need to show them collegial respect. In your classes, you may have studied and practiced how to be accepting of clientele whom you may not like or approve of. Your co-workers are not clientele, and they are accountable to standards that your clientele are not, but they deserve the same accepting, nonjudgmental stance that you extend to your clientele.

Co-Worker Reactions Depending on the size of the organization hosting your placement, you will probably elicit a range of reactions from the staff, from warmth to indifference and even hostility. Some interns report feeling like guests, some like employees, and still others like intruders. These two quotes from student journals are a study in contrast:

- *I felt like an outsider at first. I didn't know anyone at lunch and they would all sit together.*
- *They are very helpful in that they take time to explain what they are doing and why. They don't seem to be annoyed with my questions and inexperience.*

Before you are too harsh on the staff members mentioned in the first quote, try putting yourself in their place. It is another busy day at work, and they are meeting an intern whom they may or may not have known is coming. They may be wondering what your presence means for them, what you want from them, exactly why you are there, and whether they can count on you. Some of them may even resent you. Energetic new workers can be threatening to those who are disenchanted with or exhausted by their work. A new intern's ideas—or even willingness to follow established procedures to the letter—can be threatening and some staff members may fear for their jobs.

> *After the visit, I began making notes on my clipboard. The worker I was partnered with asked me what I was doing and when I told her she said, "Oh. We don't have to do that." I said that my training manual says that we do, and she just shrugged. Later on I asked my supervisor, without mentioning what she said. I was right…. I found out that she was complaining and making remarks about me behind my back.*

Your co-workers' reactions to you depend on many factors, including their personalities, past experiences with interns, their understanding of (or lack thereof) what an intern is, and their relationship with your site supervisor (Gordon & McBride, 2011). For example, agencies sometimes use interns and volunteers as a way to stretch their staffing size and meet their workload needs without hiring additional people (Suelzle & Borzak, 1981). This practice may violate labor laws and/or ethical guidelines and often results in a lack of continuity within the agency, as well as resentment from frontline staff members who have to deal with the effects. Another example is the reactions of staff who may or may not have any experience with interns but who have heard and in turn perpetuated stories about interns. You might have been privy to one of those stories at your site. Those stories tend to live on long after the interns leave the site and often take on skeleton qualities; subsequent interns at the site—even years later—may pay the price of the interns who preceded them.

As you are learning first hand, when an internship works out well, it is a productive and exciting experience for everyone involved. However, if that is not the case, having interns at the site can be very taxing on staff time and energy and can be disruptive to the rhythm and flow of the work. As you look around the site, you may also find yourself surprised by the variety of approaches to the job taken by your colleagues. Some may be very different from what you expected. Some you might find very interesting; others may disappoint you and you may be justified in feeling that way.

Feelings of Exclusion

Gordon and McBride (2011) have noted that interns often struggle with feelings of intrusion and marginality. If you feel like you really don't belong at the placement, that you don't fit in and can't contribute much, you are struggling with feeling like an intruder. On the other hand, if you are feeling like the staff doesn't want you there and doesn't see that you can be of much help, then you are struggling with marginality. Of course, as the following student journal entries show, both can occur at once:

- *I feel like my co-workers treat me like an adolescent if I don't present myself properly.*
- *They tell inside jokes and don't clue me in. One of them in particular just looks at me, as if I don't belong there and she is waiting for me to prove it.*

In fact, you may feel like an intruder when you first begin your internship, and you may be treated like one as well. If so, it will take some time to prove yourself to the staff. But, in time, you will and these feelings and reactions will tend to dissipate. But they sure are troubling while they are present.

In Their Own Words	The Unexpected…Early On!

- *It seems that my main value to some of them is as a gofer. They give me jobs they don't want, so they can take a break.*
- *I was amazed to find out that several of the staff here are brand new in the field.*
- *With my previous field experiences and practicum, I have more experience than they do.*
- *During recreation period, I am out there trying to make contact with the kids, and the other staff members just sit and talk with each other—and they're the ones getting paid!*
- *In private, they make jokes about the residents. They have derogatory labels for almost every one of them.*
- *Some of them don't have degrees in (this field). Some of them don't have degrees at all!*

Focus on PRACTICE

Engagement in Relationships: What Does It Look Like?

Engaged interns work at relationships and see their value in the internship as sources of learning and support. They seek out accurate information to dispel differences between the expectations they have about relationships and what is real and possible. They also try to understand situations from the co-workers' perspectives when differences occur.

Disengaged interns live with expectations about the key relationships that are neither explored nor tested as based in reality. They tend to be disconnected from their co-workers and/or depend on and expect them to initiate and develop the relationships. They appear not to care about the relationships, although that may not at all be the case. Issues of marginality (staff doesn't want them there or see how they can be of much help) and intrusion (not feeling like they fit in, belong, or can contribute in meaningful ways) are tolerated or worse, accepted as the way things are. They often confuse being *liked* with being *accepted* and personalize the reactions of co-workers. They do not work at mutuality because they do not see relationships as partners in their success as an intern, because they are not interested in what others' perspectives are or what value they may have in the internship—they do not believe they matter.

Making the Commitment

Regardless of how well you are doing with the tasks of this stage—becoming aware of yourself as an engaged learner; examining your assumptions; realizing your concerns; clarifying your role, purpose, and goals; and developing key relationships (which we discuss in the next section)—without an *informed* commitment to the internship during this stage, the concerns that drive your learning will leave you with a sense of distance from the *experience* of the internship. A tepid, uninformed, passive commitment is one that is much easier to abandon if things get too difficult.

By an informed commitment, we mean that you see this internship as *yours* regardless of who or what influenced your decision to do it and that you are *willing and able* to commit to its work. It takes a conscious, active choice on your part to examine the key components of the internship and to say to yourself, *"Yes, this is what I am willing to do"* and/or *"This is what I want to do."* It means taking on responsibilities and being involved in shaping your internship experience. Now, if you are thinking that we've just given an example of an *engaged* way of approaching the internship, you are correct. As you move through and beyond the Anticipation stage, it is important that what you are doing is something you are *willing* to do and are *invested* in doing. As you think about the depth of the commitment you are being asked to make to this internship, our interns over the years offer you both words of advice and words of wisdom in the box that follows. Hopefully, any hesitations you may have about this commitment will be attenuated by these experienced voices.

Voices of *EXperience*   **Wisdom, Advice, & Lessons Learned**

- *It's overwhelming, scary, exhausting, frustrating, exhilarating, challenging, growing, and mind-boggling, illuminating. I couldn't do it again, but I'm SO glad I did it!*
- *Challenge your supervisor to challenge you.*
- *You can't possibly learn everything about your placement during the year, so just open up and drink in as much as you can.*
- *Get lost—finding your way is the funnest part and how you'll learn the most.*
- *Remain optimally uncomfortable for positive change/growth.*
- *This too shall pass.*
- *It is what you make it.*
- *It is what it is.*
- *Sometimes things will not be perfect or how you would like them to be.*
- *Every mistake is an opportunity to learn.*
- *Don't try to be perfect; just make sure you do enough to be proud of.*
- *No matter your experience, good or bad, you'll have changed for the better.*
- *Learn from your experience, positive, negative, or somewhere in between.*
- *Approach with humility. There is no way you can know everything. Always be open to learn. You will grow even if you don't believe it in the beginning.*
- *Just when you think you can't do it anymore or have nothing left, reach deep down inside and think of how far you've come.*

Sweitzer & King, 2012

Slipping & Sliding… Through the Trials and Tribulations of Anticipation

The "what ifs" of the Anticipation stage can be overwhelming at times for many students, if not for most. The vast majority of students, though, are able to move beyond the feelings of disengagement once they realize that their peers feel the same way and what they feel is normal. However, there are situations, although rare, in which these feelings are exacerbated by additional negative experiences that leave the intern feeling helpless and disempowered. The intern usually senses that something is happening that should not be happening in their internship. This feeling can occur due to personal circumstances that have pervasive effects, such as health or financial issues affecting a loved one or the intern. Or, it can occur due to concerns that are internship based, such as the initial reactions of co-workers, one's sense of capability, the pace or intensity of the work, work adjusted to the appropriate level, or discovering that there are expected skills that you neither have nor were informed about that can threaten the placement.

Interns who find themselves at this point are experiencing a crisis that is often accompanied by a feeling of incapacitation. This is the shape of the *disillusionment* experience in the Anticipation stage; it is a *crisis of commitment*. They are "shocked" to find themselves in the situation, especially so early in the internship. Their mantra becomes *"I can't believe this is happening,"* and they can't seem to stop thinking or saying it. They may feel embarrassed by or ashamed of their reactions, which may include low morale and low motivation, feeling indifferent about acceptance. They may even feel consumed by anxiousness and feelings of disappointment in themselves and/or in their internships. Their sense of personal agency and self-efficacy have been reduced to a point where they don't know what to do or how to move out of the situation. They feel in over their heads and as if they are descending a vortex with little understanding of how they got there so quickly or how to get out of it. This is not what the intern expected, it may never have been considered, and it may be nothing the intern has dealt with in the past. He can only ask, *"What is happening to my internship? How is this possible?"* and, *"What is happening to me?* With these feelings, a genuine commitment may not be possible, although you may stumble forward all the same.

Remember, though, that our concept of crisis, as explained in Chapter 1, combines both elements of danger and opportunity. Certainly, there are dangers here, as described earlier. If the experience of disillusionment does happen, it is imperative to seek support and guidance to change the circumstances through *engaged* approaches rather than waiting for the situations to magically change on their own—because they probably won't. Your internship is far too important to let negative experiences set the tone for your journey or to try and manage this kind of situation without the informed guidance of your campus and site supervisors, provided the latter are not part of the problem. When you engage and emerge from experiencing disillusionment, you not only move forward, you empower yourself as a learner. And that is the opportunity.

We offer suggestions for stopping the slide and managing the recovery of your internship in Chapter 9, *Internship Essentials: Advanced Tools for Staying Engaged*. If you think you might be heading toward the experience of disillusionment, now is the time to take a look at that chapter's resources before moving forward into the next stage.

Conclusion

While reading this chapter, you have been focusing your energy in the field on understanding yourself as an intern; staying on top of your profile as an engaged learner; working to dispel your expectations and assumptions; realizing the concerns that drive your learning; getting clarification on the purpose, role, and goals of your internship; and thinking about your capacity to make an informed commitment to the internship. That's a lot of work!

You have come to understand that anxiety is a normal feeling during the beginning of the internship (and before it starts for that matter). Sometimes the anxiety occurs in anticipation of something happening; sometimes it occurs while you are involved in a situation; and sometimes it occurs while you are reflecting on incidents. In some instances, it seems to be ever present. Regardless, unless it is getting in the way of your work, it is a feeling to understand and manage. If it is interfering with your work, you will want to talk with your campus instructor so that you can learn to manage it more effectively and get back to the excitement of the internship.

Besides the focus on you as an intern resolving your concerns, you are also paying a lot of attention in the field to the relationships you are developing. Beginning them well is an important part of meeting the challenges of the Anticipation stage. Becoming comfortable with a new group of people takes time; how much time depends on the situation you are in, the people that you work with, and the person that you are. Ideally, you will enter your relationships with co-workers with a clear set of expectations and be prepared for some of the challenges of being accepted by them. If you use an engaged approach, such as questioning your assumptions and gathering as much factual information as possible about them and their work, you will have a better chance of success. Additionally, examining your stereotypes about your co-workers will improve your chances of a smooth entry and should reduce your anxiety as well. Acceptance is an issue in all these relationships, as it is in so many areas of life. Thinking about the issues raised in this chapter and in Chapter 4 will let you think about and approach the acceptance process with a broader and deeper perspective—from both sides of the relationship.

Although people are of critical importance, they do not tell the whole story of an internship. To complete the picture of your internship and your journey through the Anticipation stage, you must consider two essential aspects of the internship that need grounding early on: the Learning Contract and the supervision plan. In the next chapter, we turn our focus to them and discuss ways in which you can get the most from both. We have found that by addressing the process of setting learning goals, activities, and outcome measures, the anxiety that comes from the general uncertainty of the Anticipation stage—the ever present low level of anxiety—continues to diminish. By the time you leave the Anticipation stage, the anxiety that has been a companion these first few weeks makes way for an exciting period of learning through exploration.

For Review & Reflection

Checking-In

The Critical Learning Log. Kolb (1984) instructs us that there is no guarantee that an experience necessarily will be a learning experience. In order for that to happen, there needs to be a framework within which one can process the experience; Kolb provides that with the Experiential Learning Cycle. The Reflective Observation phase of the cycle also needs a way to be processed, and the Critical Learning Log (King, 2012) is one way to do that. This exercise asks you to reflect on your week and think about your moments of learning, specifically the moment of learning that made a difference. Although most of the time the critical moment of learning is directly related to the internship, it doesn't have to be. Winning the lottery or losing your pet can have a profound effect on how you go about your work and your realizations about what you value.

1. *What* was the focus of your internship work this past week?

    What was the focus of supervision, and what was your role in it?

    What *insight did you develop about yourself in your role of intern?*

2. *What* was the most critical moment of learning during the past week?

    Why is it important to you?

    How *does it affect you? Your internship?*

    How *might it affect your work in the future?*

3. *What* will be the focus of your work this coming week?

Learning to Go Deeper. We are using the following questions as a way of guiding your thinking deeper into reflection. This is not an easy exercise, but with practice, it will become easier. It will also prove to be an invaluable addition to the toolkit you are creating for this internship and your career. This is a well-known exercise to which we have added an additional layer of thinking at the very end.

What? have you learned about your need for acceptance versus your need to be liked? It's important to spend time thinking about this "what" question...but why?

So What? This is an even tougher question to answer, but do your best. Give it some time.

Now What? This question might seem easier as this question asks you about what you are going to do. But that plan depends on the answer to the question *So What?*

Then What? This question forces you to think down the road into the future and the implications of your plan. Not an easy task, but give it a try.

Experience Matters (For the EXperienced Intern)

You are no stranger to having hopes and dreams and knowing what happens when reality is added to the mix. What hopes do you have about this internship? Importantly, what kinds of information have you gleaned so far that informs you about the likelihood of those happening? What are your next steps knowing the likelihood of your hopes being realized?

Personal Ponderings (Select the Most Meaningful)

- Create your own list of anxieties about the internship experience. What are you finding most effective in managing them? Which ones do you think will diminish over time—without you intentionally doing anything about them? Why do you think that is the case?

- Think about the concerns that you care about or deeply touch your curiosity and give you reason to pause and wonder. What is it about those concerns that resonates so strongly with you?

- You came into the internship with some expectations about your co-workers. What do those expectations tell you about *yourself*? How are you similar to and different from your co-workers? Is either important to you as the internship goes on? If so, in what ways?

- Think about the changes—personally and professionally—that you are experiencing even this early on in the internship. What did you *do* to bring about these changes? What did you *not* do that in turn allowed the changes to occur?

- In what ways is your profile as an engaged learner developing through this internship? Think about situations in which you have used engaged approaches and consider why were you able to use them in those situations. Now think about incidents in which you did not use engaged approaches and consider why you didn't.

Seminar Springboards

- *Hopes & Anxieties.* A Heads Up! This exercise involves your campus instructor, so be sure she or he is onboard before you move ahead with it. Make a list of both your anticipations—that which give you feelings of hope and excitement—and your anxieties—that which give you feelings of apprehension. Then e-mail your lists to your campus instructor so that yours and those of your peers can be compiled and distributed in class for a discussion of similarities and differences. With anonymity of course. Tuck away this exercise for the final seminar class and review it then. When was it that your anxieties diminished and became cares or interests *or* when was it that they resolved? Which if any returned at a later point in the internship and, if they did, why do you think they did?

- *Write a Letter of Expectations to Yourself.* In it describe all you want to get from this internship—what you hope for, what you expect, what you need, and what you want. Be specific when it comes to the goals of what you expect to do and learn. Again, tuck it away for the final seminar class. When that time comes, spend time thinking about the following:

  - How close were you in expecting what you could accomplish?
  - How realistic were you in what you hoped for?
  - How realistic were you in getting what you wanted? Needed?

- *Write a Letter of Encouragement to Yourself.* In it say to yourself what only you can say to keep you pushing on when the going gets tough. Seal it, then address it to a place where you will receive it and no one else will open it. Either give it to your seminar instructor to mail to you when you most need it or tuck it away and do the same to yourself. After you read it, put it away for another time. We always need those kinds of letters.

For Further Exploration

Colby, A., Erlich, T., Beaumont, E., & Stephens, J. (2003). *Educating citizens: Preparing America's undergraduates for lives of moral and civic responsibility*. San Francisco: Jossey-Bass.

Particularly useful for stimulating thinking about civic development.

Gordon, G. R., & McBride, R. B. (2011). *Criminal justice internships: Theory into practice* (6th ed.). Cincinnati, OH: Anderson Publishing.

Excellent discussions of issues relating to supervisors and co-workers with many useful applications beyond the focus on criminal justice internships.

Kuh, G. D. (2008). *High-impact educational practices: What they are, who has access, and why they matter*. Washington, DC: Association of American Colleges and Universities.

Describes educational practices that have a significant impact on student success which benefit all students and especially seem to benefit underserved students even more than their more advantaged peers.

Schutz, W. (1967). *Joy*. New York: Grove Press.

In his first stage of group development—Inclusion—Schutz talks a great deal about acceptance concerns and the various ways of handling them.

References

Brust, P. L. (1986). Student burnout: The clinical instructor can spot it and manage it. *Clinical Management in Physical Therapy,* 6(3), 18–21.

Caine, B. (1994). What can I learn from doing gruntwork? *N.S.E.E. Quarterly* (Winter), 6–7, 22–23.

Clance, P. R. (1985). *The imposter syndrome.* Atlanta: Peachtree Publishers.

Giles Jr., D. E., & King, M. A. (2012). *Reflection: Making experience educative.* National Society for Experiential Education. Milwaukee, WI.

Gordon, G. R., & McBride, B. R. (2011). *Criminal justice internships: Theory into practice* (7th ed.). Cincinnati, OH: Anderson Publishing.

Gravois, J. (Nov. 9, 2007). You're not fooling anyone. *Chronicle of higher education,* 54(11), 1 & A14.

Hodge, D. C., Baxter Magolda, M. B., & Haynes, C. A. (2009). Engaged learning: enabling self authorship and effective practice. *Liberal Education,* 95(4), 16–22.

Kolb, D. A. (1984). *Experiential learning: Experience as the source of learning and development.* New York: Prentice Hall.

Nesbitt, S. (1993). The field experience: Identifying false assumptions. *The LINK: Newsletter of the National Organization for Human Service Education,* 14(3).

Royse, D., Dhooper, S. S., & Rompf, E. L. (2011). *Field instruction: A guide for social work students* (5th ed.). Upper Saddle River, NJ: Prentice Hall.

Suelzle, M., & Borzak, L. (1981). Stages of field work. In L. Borzak (Ed.), *Field study: A sourcebook for experiential learning.* Beverly Hills: Sage Publications.

Wilson, S. J. (1981). *Field instruction: Techniques for supervisors.* New York: The Free Press.

Yager, G. G., & Beck, T. D. (1985). Beginning practicum: It only hurt until I laughed. *Counselor Education and Supervision,* 25(2), 149–157.

CHAPTER *6*

Internship Essentials
The Learning Contract & Supervision

There has been a loss of focus…in regards to my not doing what I went there to do. I should have discussed the issue sooner with my supervisor…it is written right there in the contract that I will be doing it…I need to face the issue with my supervisor…so I can feel like a member of the team again.
STUDENT REFLECTION

During the Anticipation stage, there are essential components of the internship that need to be in place so you can move forward in your learning. One of those components is the learning contract. This is a document that involves you, your campus, and the field site and ensures that the internship constitutes a learning experience as well as one that adheres to the expectations of your program and campus policies. Some students begin their field placements without the contract in place; others begin with a well-developed contract. Regardless, this chapter is an opportunity for you and your supervisors to make sure that your contract reflects high-quality learning outcomes. When it does, it is ready to go to work for you!

The second essential component that needs to be in place at this point is supervision. Supervision ensures that you stay focused on the learning goals and activities that are in the contract and that the contract keeps pace with what and how fast you are learning. Having a supervision plan in place with clear expectations of and by your supervisors is very important to being grounded in the internship.

In this chapter, you will explore the learning contract and the supervisory relationships—both at the site and on campus—and develop an understanding of how the supervision plan keeps the learning contract viable. You also will explore the issues interns typically deal with in these relationships and learn how to engage in and use supervision for personal and professional growth.

The Learning Contract[1]

I never thought that I would feel this way, but I am very happy I took the time to work on the contract…. It helped enormously to define my role.
STUDENT REFLECTION

At some point in the internship or the placement process, you, your site supervisor, and your campus supervisor or instructor negotiate the goals and activities of learning for the field experience. This written agreement is a document that articulates in detail what you will be learning, how you will go about learning it, how you will determine your progress, and how you will be evaluated on achieving your goals. This document may be referred to as a learning plan or agreement, depending on your academic discipline or campus policies. We refer to it as the *learning contract* and for us, the term is made up of two important concepts: The *contract* concept brings clarity and focus to the internship, while the *learning* concept brings creativity, reflection, and engagement to the experience. We'll explore both of these concepts in this chapter and how they are tied together.

---

[1]Many campuses use two documents for their internships: a *placement* or *internship agreement* and a *learning contract*. The *placement agreement* details the responsibilities of the involved parties (campus, internship site, and intern) and describes in general terms what those parties will do to carry out the terms of the agreement. The *learning contract* articulates what an intern is expected to learn during the internship and how that will happen (learning goals and objectives). This contract typically involves the campus supervisor, the intern, and the site supervisor. On some campuses, one document is used that reflects the intent of both the placement agreement and the learning contract. The question of whether these documents are legally binding is not simple to answer. Each of the documents may have different intent depending on the campus, with different parties involved in agreeing to the specified conditions, on campus as well as at the internship site. To make matters more complicated, the documents may or may not be written in ways that make them legally binding. The best resource for answers to each of these questions is the legal counsel for your campus.

The Importance of a Learning Contract

The learning contract is perhaps best described as a master plan for success in your internship. Developing a thoughtfully considered contract is essential to maximizing your learning and ensuring that your learning needs are met. For one thing, the learning contract minimizes the chance of misunderstandings. It will also keep you focused on your internship commitments, provide you with a sense of achievement, and serve as an opportunity to exercise imagination and creativity and to consider new possibilities. Most importantly, the contract will ensure the integrity of your academic work (Royse et al., 2012).

Some of you may be feeling a bit lost in your internship right now, not really knowing what you are going to be doing over the course of the field placement or perhaps concerned that the work you will be doing will be insignificant to you or to the site. If either is the case, it may help to know that a well developed learning contract will clarify your role and responsibilities and will give you the opportunity to make sure that your work is meaningful to both you and the site. That translates into you having a clear sense of purpose in role of intern. Without a sense of purpose, you will begin to question your value to the site and of the work to your learning. We cannot stress enough how important it is for you to be involved in developing your learning contract and to work conscientiously on it, as it ensures that you have a most satisfying experience.

Get Involved with the Contract!

As the student's reflection that introduced this chapter makes clear, the learning contract grounds the internship in many ways. The intern who experienced it took an engaged approach—recognized how the internship was off track and what needed to happen to put it back on track. He realized not only the importance of the contract, he also appreciated the importance of working with the supervisor to bring about change.

There are three key individuals who are involved in creating this master plan: you, your site supervisor, and your campus supervisor/instructor. An internship is most effective when all three parties work together to develop a field experience that meets everyone's needs and provides an opportunity for all to learn (Hodgson, 1999). You are one of the three, and your active participation is what engagement looks like when it comes to the learning contract. This degree of involvement may seem a little odd at first; however, our experience tells us that once you become involved in the process, you will get beyond these feelings, share in the process, and be more invested in the outcome.

There are three factors you should consider to maximize the possibility of a fulfilling field experience (King, 1988). First, your work in the field must be *worthwhile* for you to do it. That translates into you being given opportunities to accomplish tasks, be acknowledged and respected for your work, create your own work, push your limits to touch your potential, and enjoy what you are doing. You will need to take an engaged approach to ensure that these opportunities are part of your internship. Second, you must develop *responsible relationships* in working with your supervisors and co-workers during your internship. Those relationships translate into competent supervision and mutual collegiality. You already understand what this entails when it comes to co-workers (Chapter 5); you will learn later in this chapter what this entails when it comes to supervisory relationships. Third, you must be willing to cultivate an *active* and *conscious* involvement in the internship. That translates into taking on responsibilities that may be quite challenging initially and being actively involved in designing your field experience.

The Timing of the Learning Contract

Exactly when you, your campus instructor, and your site supervisor negotiate the contract tends to be a matter of policy—either that of the academic or student affairs program, your campus, or the field site. There are campuses or programs we know of that develop their agreements before the field placement begins—sometime between when the site/campus agreement is negotiated and when the internship itself starts. In some programs with year-long internships, students have up to a month after the internship begins in which to develop their learning contract, while interns in programs of a semester or less may have from one to two weeks to develop the contract.

It is important that you have some understanding of your field site, the populations you are working with if you are in the helping or serving professions and your learning tasks before negotiating the contract (Rothman, 2002). We generally develop our contracts within the first three weeks of the placement, regardless of the length of the field experience or hours in placement, as we believe you can take an active and informed role in the contract process by the third week.

Fundamentals of the Learning Contract

There are four general components to the learning contract, and they are pretty straightforward: *learning goals, activities to reach the goals, assessment measures,* and the *supervision plan.* Some learning contracts also include the term *objectives,* which refers to the activities to reach the goals. How carefully these components are developed and how well matched they are to your expectations make all the difference in whether you have an empowering, transforming field experience or a mediocre one.

Remember that it is best if the contract is written in such a way that it can be changed as needed and adapted to shifts and modifications in the internship. Just imagine if an opportunity for an exciting project develops after you negotiate the contract and your contract is nonnegotiable. You could miss out on what might be *the* defining aspect of your internship! By the same token, if a project you agreed to take on does not materialize, it is equally important that you are able to replace it with a more viable learning opportunity.

We now turn to three of the interlocking pieces of the contract: goals, activities, and assessment. Although we will discuss them in that order, and we suggest you approach them in that order initially, it is not at all uncommon for goals to be rethought once activities are considered or for both goals and activities to be rethought once you are focused on assessment. The fourth interlocking piece, the supervision plan, will be discussed along with the supervisory relationships later in the chapter.

The Learning Goals

There are many kinds of learning contracts, ranging from simple goal statements that apply to all internships to highly specific and individualized plans (Wilson, 1981). Your academic program may already have a format for developing learning contracts. If so, it probably uses of one of two general approaches: a *goals-focused contract* or an *objectives-focused contract*. Both of these approaches begin the contract process by identifying the *goals of learning* and end the process by setting forth *assessment criteria* to determine how well the goals are met. These two general approaches also identify *activities* as ways of achieving the goals. Differences between the approaches arise in whether or not *learning objectives* are used to guide the learning contract (Garthwait, 2011; Rothman, 2002). It is generally accepted that the more succinct approach—the goals-focused contract—is easier to implement and provides sufficient direction to be a worthy document for students and site personnel to use (Garthwait, 2011). Regardless of the approach, the learning goals should reflect in part the answer to the question: *What is your purpose as an intern in this organization?* In the sections that follow, we discuss some considerations in writing learning goals to help you clarify your role as an intern and provide examples to guide your thinking toward that end. For those of you who want or need to write a contract from a learning objectives perspective, see the Focus on SKILLS box, Using an Objective-Based Approach (p. 151), for additional guidance.

The contract itself should be demanding and ambitious, stretching you while enhancing your knowledge. It should also be grounded in realistic time frames as well as realistic expectations of the work and your capabilities. Keep in mind when negotiating your contract that the goals you set for the end of the placement should seem lofty when you set them, and feeling intimidated by them is not uncommon. Be reassured, though, that the feelings of intimidation will pass and you will attain your goals, but it will take time and work to make that happen.

Domains and Dimensions of Learning

Before you start drafting your goals, we encourage you to think broadly about the learning opportunities in front of you. In Chapter 1, we discussed at length the sorts of things you can learn from an internship; that abstract discussion can now be concrete because you have experience to refer to. We suggest two ways of categorizing goals for your contract. First, goals can focus on the learning domains of *knowledge, skills,* and *attitudes and values.* Second, goals can focus on the three dimensions of learning and development we have emphasized throughout the book: personal, professional, and civic. Each domain is part of each dimension, so you actually can construct a chart to identify your learning goals.

	Dimensions		
Domains	**Personal**	**Professional**	**Civic**
Knowledge			
Skills			
Attitudes/Values			

We focus briefly on each of these domains and dimensions, offering examples of specific goals you might include in your learning contract.

Knowledge Goals focus on learning and understanding factual information, terminology, principles, concepts, theories, and ideas of the profession. Typically, they involve learning new information that builds on previous learning, such as:

- Becoming familiar with how the criminal justice system in this state operates
- Learning the network of services available in this city to people who are physically challenged

Skill Goals focus on things you want to learn to *do,* such as:

- Learning to wait longer before speaking
- Learning to write a case report
- Learning how to approach a neighborhood group

Attitude and Value Goals focus on attitudes and values you believe to be important and that you want to improve in yourself, such as:

- Being more patient
- Being less defensive about criticism
- Seeing and appreciating the assets in all communities

Focus on THEORY

Levels of Learning: A Guide to *Learning Goals*

A framework that may help you think about learning goals is that of different levels of learning that need to take place along the way: (a) orientation, (b) apprenticeship, and (c) mastery (King, Spencer, & Tower, 1996). It is possible—even likely—that you will be at different levels of learning with regard to particular skills, be they personal, professional, or civic, and the same holds true for knowledge and even attitudes and values.

The *orientation* level is a time when you become acquainted with the work of the internship, or some aspect of that work. You begin to explore an area of knowledge, perhaps a theory, or a close understanding of a cultural group, or the political dynamics of the community. You are introduced to a skill, perhaps a particular way of doing client intakes, or facilitating a group. And you may be encountering an attitude you want to work on, such as being non-judgmental of certain people or groups.

The *apprenticeship* level is an intensive teaching and learning period, with much learning-by-doing under close supervision. Here you become more and more comfortable with a skill, explore knowledge in more depth, and deepen your work on attitudes and values.

The *mastery* level is a period during which you no longer have to focus so explicitly on the skill set(s) you are developing, the body of knowledge you are trying to acquire, or the attitude and values you are cultivating. Instead, you focus on assessing each situation and applying the skills, knowledge, and values that now come more naturally to you, relying on yourself more than others to reflect on your progress. You are not yet a "master" of the professional skills and knowledge in your field—that takes years, but on the spectrum of learning in the internship, you are mastering the skill sets expected at the end of the internship, along with a command of the knowledge and a demonstration of the attitudes and values set forth in your learning contract.

Personal Development Goals focus on areas in your life that apply to internships as well as other aspects of your life, and that you want to improve or change. For example:

- Discovering more about your personal strengths and weaknesses
- Becoming more able to operate in a constantly changing, uncertain environment
- Being better able to say no and set limits when you want to

Professional Development Goals focus on how you want to grow as a professional or on what you want to learn about a profession or discipline. They can also include goals about higher order thinking skills. For example:

- Learning specific skills of the profession, such as process recording, case management, or project management
- Becoming familiar with the major ethical issues in the profession
- Strengthening your understanding of various theories
- Becoming a more effective writer
- Conducting an effective interview
- Developing the skill of critical thinking

Civic Development Goals focus on the knowledge, skills, and values of citizenship in a democracy, and on the understanding of the public relevance of the profession in which you are interning. Jeffrey Howard, in his very useful *Service-Learning Course Design Workbook*, has developed a sophisticated chart showing examples of civic knowledge, skills, and attitudes (2001):

Purposeful Civic Learning

PURPOSEFUL CIVIC LEARNING OBJECTIVES

Categories of Purposeful Civic Learning	Knowledge	Skills	Values
Academic Learning	Understanding root causes of social problems	Developing active learning skills	There is important knowledge only found in the community
Democratic Citizenship Learning	Becoming familiar with different conceptualizations of citizenship	Developing competency in identifying community assets	Communities depend on an active citizenry
Diversity Learning	Understanding individual and institutional "isms"	Developing cross-cultural communication skills	Voices of minorities are needed to make sound community decisions
Political Learning	Learning about how citizen groups have effected change in their communities	Developing advocacy skills	Citizenship is about more than voting and paying taxes
Leadership Learning	Understanding the social change model of leadership	Developing skills that facilitate sharing leadership roles	Understanding that leadership is a process and not a characteristic associated with an individual or a role

(continued)

Purposeful Civic Learning *(continued)*			
PURPOSEFUL CIVIC LEARNING OBJECTIVES			
Categories of Purposeful Civic Learning	**Knowledge**	**Skills**	**Values**
Inter and Intrapersonal Learning	Understanding one's own multiple social identities	Developing problem-solving skills	Learning an ethic of care
Social Responsibility Learning	How individuals in a particular profession act in socially responsible ways	Determining how to apply one's professional skills to the betterment of society	Responsibility to others applies to those pursuing all kinds of careers

Source: From *Service-Learning Course Design Workbook* (www.umich.edu/~mjcsl). Copyright © 2001 by University of Michigan.

Tools for Writing Goals

Developing goals can be tricky business. One technique that we find useful is sentence-completion exercises like this.

For each of the categories (personal, professional, and civic development), complete the following question:

- If I could make it happen, what _____ goal would I like to achieve by the time I complete my internship? (deShazer, 1980, cited in Curtis, 2000).

- If I were given a magic wand, what would I like to see happen in my internship in the area of _____ goals?

Also essential to formulating a goals statement is knowing the language to use to frame the learning goals. The following verbs are frequently used in goals statements and may help you in writing your contract (Garthwait, 2005, p. 22).

acquire	*comprehend*	*learn*
analyze	*develop*	*perceive*
appreciate	*discover*	*synthesize*
become	*explore*	*understand*
become familiar with	*know*	*value*

Resources for Learning Goals

Before settling on goals, there are several resources that you, your supervisor, and your instructor may find useful to consult in considering the goals. Some of these resources are familiar to your supervisors, and some may be new to them.

Program-Specific Goals Many academic programs have learning outcome statements for their students and sometimes even for the internship in particular. This is a resource that you and your campus instructor are probably familiar with, but your site supervisor may not be. What many do not know is that many professional organizations have published statements about what professionals should know and be able to do. We've identified a few of those associations on p. 153.

Focus on SKILLS

Using an Objective-Based Approach

Learning-objective statements are behavioral in nature and tend to be more detailed, concrete, and precisely worded than goal statements. Learning objectives describe specific actions and activities; e.g., to demonstrate the ability to match needs of clients to available resources. It has been our experience that developing learning objectives is both an art and a skill. The art is reflected in the vision of the objective; the skill is reflected in the ability to incorporate the goals, activities, and assessment measures in one statement and in realistic and heuristic ways. The following verbs are typically used in writing learning objectives (Garthwait, 2005, p. 23). If you are using the learning objectives contract, this list may be helpful.

answer	*decide*	*obtain*
arrange	*define*	*participate in*
circulate	*demonstrate*	*revise*
classify	*direct*	*schedule*
collect	*discuss*	*select*
compare	*explain*	*summarize*
compile	*give examples*	*supervise*
conduct	*list*	*verify*
count	*locate*	*write*

Rothman (2002) offers excellent examples of how to write and not to write learning objectives—advice that will serve you in good stead. For example, she frames the contract in such a way that you move smoothly from the general (overall goals), to the specific (objectives), to the most specific (approaches, tasks, and methods) and then she suggests that you reverse the process when it is time to review your progress. Rothman also stresses the importance of using measurement criteria that are both relevant and measurable and she recommends that your objectives be grouped by time frame so that you can easily review what you have done in the designated time periods. She offers an excellent example of a statement using both relevant and measurable criteria: the "ability to use empathy appropriately as demonstrated in five instances recorded in process recordings of three client interviews in one month." This example will guide you well in developing assessment objectives.

Focus on THEORY

A Guide to Learning in the Civic Dimension

Earlier we introduced you to the levels of learning as a way to think about setting internship goals. If you adopt an objective-based approach to your learning contract, the levels of learning can be of additional help, as in the examples below:

Orientation Level

Learning Goal: *Will become acquainted with the concept of civic development in an internship*

Learning Objectives:
- Become aware of civic dimension of development in the internship
- Be introduced to concepts, vocabulary, and definitions
- Understand concepts of civic learning and civic professionalism
- Become aware of one's responsibility to civic development in the internship

Apprenticeship Level

Learning Goal: *Will convey a working knowledge of the principles of civic learning and civic professionalism*

Civic Learning Objectives:
- Develop understanding of the historical and current forces that contribute to the issues being faced by the population served at the internship site
- Develop awareness of the profession and organization's social obligations to the community
- Develop understanding of the ways in which the profession and organization meet social obligations to community

Civic Professionalism Learning Objectives:
- Begin to acquire the core civic knowledge, skills, values, and attitudes expected of the professional staff
- Begin to practice the practical reasoning skills required in the work
- Cite examples of the human context and the public relevance of the work
- Examine instances of the organization's ethical and moral obligations to the community and how they are connected to a larger social purpose.

Mastery Level

Learning Goal: *Will begin to apply the principles of civic learning and civic professionalism in an internship*

Learning Objectives:
- Demonstrate civic knowledge, skills, attitudes, and values in the internship setting
- Take actions that reflect an understanding of the public relevance of the profession
- Demonstrate knowledge of the organization's position on responsibility to issues in the community; advocate for change as necessary

Professional Associations   Many professional associations have recommended learning goals, or good resources for thinking about them.

- *Council for Standards in Human Service Education* has a detailed list of knowledge, skills, and attitudes (www.cshse.org) expected of human service education programs and/or their students.

- *National Society for Experiential Education (www.nsee.org)* has published the document *NSEE Principles of Best Practices in Experiential Education*, which identifies the principles that underlie the pedagogy of experiential education and the responsibilities of the student and facilitators in the learning process (NSEE, 1998, 2009). In addition, the NSEE has published a series written by Inkster and Ross titled *The Internship as Partnership*, the first in the series being *A Handbook for Campus-based Coordinators* and the second being *A Handbook for Businesses, Nonprofits, and Government Agencies*, both of which have value for faculty and other campus supervisors in developing effectiveness in working with placement sites (1995, 1998).

- *Association of American Colleges and Universities* has published a comprehensive list of learning goals for the 21st century, including intellectual goals such as creative thinking and problem solving; practical skill such as teamwork, and personal/social responsibility goals including ethical reasoning, civic engagement, and intercultural learning (Crutcher et al., 2007).

- *Council for the Advancement of Standards in Higher Education* (CAS) has published the standards and self-assessment guides for internship programs and for service-learning. This resource is intended for the campus instructor and offers a book of standards (2009a), *Frameworks for Assessing Learning and Development Outcomes* (2009c), and *Self-Assessment Guides* (2009b) to ensure institutional effectiveness, student learning, outcomes assessment, and quality assurance (www.cas.edu).

- *Service-Learning Community* The literature in service-learning focuses on, among other things, how to use service activities to develop specific academic and civic learning outcomes. Of particular note are three nationally recognized resources: a series edited by Zlotkowski (2006), which contains twenty-one individually edited volumes each focusing on a separate academic discipline; the *Michigan Journal of Community Service Learning* (www.umich.edu/~mjcsl); and Campus Compact and its *Introduction to Service-Learning Toolkit: Readings and Resources for Faculty* (www.compact.org).

Choosing Activities

Strategies, activities, methods, and *approaches* are all terms to describe ways of accomplishing your goals or objectives. Sometimes they are used interchangeably. Technically, objectives are *strategies* for achieving goals, but we like the word *activities*. They should be described as specifically as possible, specifying what you will do, where, when, and how. As you probably suspect, there is considerable room for personal preferences when it comes to selecting ways to meet your goals; just be sure the resources are available to carry out the activities. As with goal setting, choosing activities should ideally be a three-way process, involving active participation from you, your supervisor, and your instructor. In some settings, the site is actually quite prescriptive about what activities interns will engage in; there may even be a detailed position description. In those cases, the choosing of activities will be less collaborative; however, it is no less important that they be matched to explicit learning goals.

Each of your activities should connect clearly and specifically to a goal and you should make sure you have at least one activity listed for every goal. Sometimes an activity may be appropriate for the pursuit of more than one goal. For example, if you are going to be writing project or case summaries, that may connect to a number of knowledge and skill goals. We also encourage you to think hard about opportunities for civic development. If you are assigned case management duties in a social service agency, for example, you can use that opportunity to find out more about the community context of your clients. You might have to suggest that as a goal, though, or an extension of the activity.

It is possible that you will have some activities assigned to you that are not connected with a goal—when that happens, keep in mind our discussion about "grunt work." Also remember that there is much to be learned from such work, and you may want to develop a learning goal or two toward that end.

Focus on THEORY

Levels of Learning: A Guide to *Learning Activities*

Learning takes time, and your activities must be organized so as to build on your knowledge, skills, and attitudes and values as they develop and change over time. We find the three sequential levels of learning useful to guide the activities: (a) orientation, (b) apprenticeship, and (c) mastery.

The *orientation* level is a time when you become acquainted with the work of the internship. Days at the site are marked by lots of observing, reading, shadowing, and inquiring. These are far from aimless activities. You need to have goals during this time because you are building a foundation for a successful internship (Feldman & Weitz, 1990). In the *apprenticeship* level, your activities will focus heavily on learning under close supervision. The third level is a period of *mastery* during which you will be engaged in your own work and actually structuring and organizing your schedule to meet timelines and agency goals. You are more self-directed at this level of learning, and your need to understand subtle aspects of the work and yourself in relation to the work is heightened. Typically, you will need less direction and may or may not need much support from your supervisor.

Assessing Your Progress

The third step in developing your learning contract is to identify ways to measure how well you are doing in meeting the goals. Terms like *measurement, evaluation, outcomes,* and *assessment* are often used to describe the essence of this step. What is important is that you are able to objectively measure your growth using tangible activities or methods. These assessments, though, must connect clearly to the activities you have chosen. For example, you could include in your contract such measurable activities as giving presentations that will be evaluated, creating resource lists that can be reviewed, keeping journals that can be read, developing written documents for review (Kiser, 2012; Rothman, 2002), videotaping yourself, being observed and given feedback, and writing a case study or writing up a project for review by supervisors. If you are stuck, good questions to ask yourself are "What could someone read or observe that would help them tell me how well I am doing in meeting my goal?" or "If I had achieved this goal, how would someone besides me know?" In some cases, particularly if you are using an objective-based approach, you will be encouraged or required to state your assessment approaches in quantifiable terms. For example, instead of saying "present cases or projects to the supervisor," you would say "present at least five cases or four projects," and instead of saying "receive feedback," you would say "receive ratings of satisfactory or above on all of my case presentations."

Table 6.1 shows two learning goals with corresponding activities and assessments:

TABLE 6.1
Sample Learning Contract Sequence

Goal	Activities	Assessment
Function effectively as a team member	Observe a variety of teams and make notes on what seems to work (or not) and why	Present findings to instructor in journal entries
	Read about effective group behaviors	Review understanding with supervisor
	Practice specified behaviors in group	Supervisor observes group, looking for those specific behaviors
Improve my ability to identify personal and community strengths in the clients I work with	Review client files and summarize strengths	Present findings to supervisor in supervision session. Supervisor offers critique and keeps notes on improvement
	Accompany experienced workers on home visits and make notes on strengths visible in the home	Discuss impressions with the experienced workers and get feedback
	Conduct a community assessment	Present written assessment to instructor and site supervisor

An important factor in how well the learning contract guides you to the expected learning outcomes is the quality of your supervision, which often times is closely associated with the quality of the supervisory relationships. In the section that follows, we explore that relationship and its many aspects, including the evaluation process and the triadic relationship that exists in the typical internship.

Getting the Most from Supervision

Starting in a new place is a blank slate. While this could be positive for some, I felt apprehensive at the thought of starting that relationship of supervisor/supervisee all over again. I am sure it will become less difficult as time passes, though a sense of apprehension is probably positive because it would keep one attuned.
STUDENT REFLECTION

Certainly, one of the most important relationships at your site and in your internship experience is the one you have with your site supervisor.[2] You will spend a great deal of time with this person, who is responsible for creating opportunities that allow you to embrace many aspects of the work and strive toward your potential. The relationship with your site supervisor is a tremendous source of learning about the work, about yourself, and about the supervisory relationship itself (Bogo, 1993; Borders & Leddick, 1987), as well as about the organization and its community. This person—your supervisor-of-record—has the responsibility to ensure that you reach the learning goals of the internship in substantive ways. For many interns, the site supervisor is the person whose opinion comes to matter most and who has the most impact, in either a positive or negative way, on learning.

Your relationships with your campus supervisor(s) are also key to the quality and success of your internship. Many of the issues you already considered about your site supervisor also apply to your campus supervisor or instructor. We won't be repeating them here, but we will be reminding you to keep them foremost in your mind, as well as the triadic relationship among the campus, your site supervisors, and you. Supervisors indisputably have power over you, and that can have a real effect on these relationships. For example, they are responsible in

---

[2]In some cases, interns have more than one person functioning as an on-site supervisor. When this is the case, one personal typically oversees all aspects of the internship (usually the supervisor-of-record), while the other supervisor, who meets with the overseeing supervisor, handles the on-the-job supervision (direction, guidance, and support). In other instances, your instructor or supervisor on campus may also be fulfilling some of the functions of a supervisor. In this section, unless otherwise noted, we are assuming that you have one on-site supervisor.

part for your evaluations, are critical in most instances to your professional networking, and have veto power when it comes to the ever-so-cherished mentoring that goes on and is a necessary relationship in some professions.

You know from the previous chapter that your co-workers may be part of your supervisory meetings and/or be asked to evaluate your work for the site or campus supervisors. Whatever the case, your relationships with them can have a substantial impact on your internship. We'll be discussing your supervisory relationships in greater detail shortly. First, though, we want to focus your attention on that which keeps the supervisory relationships on track: the supervision plan.

The Supervision Plan

In addition to goals, activities, and assessment, the fourth and interlocking component of the learning contract is the supervision plan. You not only need a competent and qualified supervisor to oversee your internship, you need a plan in place to keep the supervision on track and you moving ahead. Like the learning contract itself, the supervision plan needs to be in place early during the Anticipation stage. There are many approaches to supervision and what you need most of all is clarity about what to expect, as well as some reflection on how well suited you are (in terms of your learning style) for what is to come. The supervision plan is the big picture of what supervision looks like for you—what you can expect of the supervisor and what the supervisor can expect of you. Although a commitment to supervision is agreed upon between the site and the campus when a placement is negotiated, the details of the plan rarely are; those are typically developed once the internship is in place and the intern is on-site. In all likelihood, your supervisor will discuss the supervision plan with you within the first few days you are at the site. If that does not happen, you will want to take a more proactive approach and ask for it in a supervision session. In some cases, the plan will be an actual document that is signed, either as part of the learning contract or separate from it. In others, it is discussed and agreed upon. What's important is that you are as clear as you can be on what to expect, that your campus supervisor is aware of the supervision plan, and that you have not missed opportunities to shape the supervision process.

A supervision plan covers a lot of ground, but most of it can easily be categorized into issues of *Who?*, *What?*, *Where?*, *When?*, *Why?*, and *How?*

Who?

You will want to know who is responsible for and involved in your supervision. That typically means knowing who is the main contact with your campus instructor as well as who does the on-the-job guidance, daily check-ins; "big-picture" discussions; weekly feedback and direction sessions; and periodic evaluations. Understanding the *who* also involves knowing the person(s) responsible for overseeing you and your work when your supervisor-of-record is away or otherwise unavailable for guidance at the site.

What?

The plan also involves knowing what is involved in supervision. For your part, you want to know what you can expect the focus of the supervision to be when you meet. Early on in the internship, the "what" of supervision often takes the form of direction about what to do, how and when to do it, and so on. In time, the focus of supervision changes as you move through the stages and toward greater independence in the work. Additionally, you want to know what is expected of you in the meetings and what you can expect of your supervisor.

Another important consideration is the structure of the meetings (Baird, 2011; Borders & Leddick, 1987; Garthwait, 2012; Leddick & Dye, 1987). Some supervisors use a highly structured format. They ask the questions, and they ask at least some of the same ones each time. They may also have a structured way that they want you to present your cases or your progress on projects. Others emphasize more didactic approaches, where they do most of the talking, teaching you about different theories, asking you questions about readings they have given you, and so on. There are supervisors who are very loose in structuring the session and respond only to what comes up in the conversation and those who come into conferences with a set agenda. Still others let you ask the questions and raise the issues. Some, of course, use combinations of these approaches.

Your supervisor's choice of structure and activities may be based on personal preference, deep beliefs, or assumptions about what will be most helpful to you. Some of them are more suited to particular learning styles than others. So, while you should stay open to all forms of supervision, spend some time thinking about how well suited each of these approaches is to your learning goals and your learning style.

Where?

It's important to know where supervision takes place. Sometimes it will be in the supervisor's office or the conference room, other times in the community room over coffee, sometimes on the floor while doing the work where you can step into an area that is relatively quiet. The specific location is not as important as clarity so you know where you need to be.

When?

You want to be sure that you have time set aside weekly to meet in a focused way with your supervisor. With advances in technology, these meetings can occur face to face or, if distance and schedules make that difficult, using software such as Skype or WebX. This is a time for you to report on your progress, ask and answer questions, and get feedback. Lots of supervisors will also meet with you spontaneously if there is something important to discuss, or just check in with you from time to time, and that is fine. However, we strongly recommend that some of your meetings be at regularly specified times. Find out from your supervisor how frequent and how long these meetings will be and schedule them ahead of time. It is easy, in the crush of crises and other responsibilities, for interns and supervisors to be tempted to cancel supervision conferences, especially if things are going well (Baird, 2011). Don't do it—few things are more important to your learning than regular and productive supervision. Supervision needs to be on your calendar and on your supervisor's. The schedule may change over time as you take on more responsibilities, need less direct on-the-job guidance, and grow toward independence in your work.

It is also crucial that you reserve some time to plan for these conferences (Birkenmaier & Berg-Weger, 2010; Garthwait, 2012; Kiser, 2012; Royse et al., 2012). At most placements, there is so much to do that there is a temptation just to work, work, and work some more. Conferences and planning for them force time for reflection, which is a critical part of completing Kolb's learning cycle described in Chapter 1. Planning for conferences can also empower you and give you more control and influence over the session. You should set aside time to think about questions, observations, and problems you want to discuss and to prepare supporting documents if necessary. You may need to put them in some kind of priority order, considering their importance to you and their immediacy in the lives of clients or the completion of projects.

Why?

Although the purpose of supervision may seem obvious, that is not necessarily the case. If you go into supervision expecting to be praised for your work and instead you find the focus on new information being disseminated to the staff, you are setting yourself up for disappointment and frustration. You will need to be forthcoming and ask the supervisor about the focus of the types of supervision you receive so you can plan accordingly for them.

How?

You will want to know about the different *ways* in which supervision will be conducted. For example, you will want to ask whether you will have supervision with your supervisor or with someone else; whether you have group supervision (with other interns), peer supervision (by upper level interns), supervision in meetings with or by co-workers, or other forms used at the site. All these forms of supervision generate information about how you are developing through your role and the work you are doing. Supervisors need other information in order to monitor your progress and development. This information guides supervisors in their conferences with you and in the evaluation process. There are several ways for supervisors to gather this information, including:

- *Live Supervision* Your supervisor either works alongside you or observes as you work. There is no substitute for live supervision; it is the only way the supervisor can see you work firsthand. However, live supervision is often not possible for any number of reasons.

- *Audio or Video Recordings* If you are making a presentation, leading a meeting, or working with a client, you can record yourself and go over it with your supervisor. While no one we know (including us) enjoys being recorded, it can be of enormous value.

- *Self-Report* Your supervisor may ask you to report, verbally or in writing, on your progress with clients or projects. Some agencies use a highly structured and prescribed format for these reports, and others are more flexible.

The approaches all have strengths and drawbacks, and your supervisor may use more than one of them. It should be fairly easy for you to find out which of these approaches will be used.

Focus on PRACTICE

Your Supervision Plan

The following checklist will guide you as you create your supervision plan. Use the checklist to ensure that the important aspects of the plan are in place. In that way, you can move forward knowing your supervision is on track.

The WHOs
- Know who is your supervisor-of-record.
- Know if other staff are involved in supervising your work and if so, *who*.
- Know who are the back-up supervisor(s) in your supervisor's absence.

The WHATs
- Clarify the focus of supervision.
- Clarify the expectations of you and your supervisor in supervision.
- Pay attention to the degree of structure and type of activities and their fit with your ways of learning.

The WHEREs
- Determine the locations of your various ways of being supervised.

The WHYs
- Clarify the intent of and your role in each of the modes of supervision.

The HOWs
- Identify the different ways in which you are being supervised.
- Inquire about the methods used to gather information about progress.

Your Relationship with Your Supervisor

Regardless of how much experience you have had being supervised, or even as a supervisor in another career, it is normal to be nervous as you approach the relationships with your supervisors. For one thing, internship supervision is different from other kinds of supervision. And even if you have had past internships, this is a new relationship. We find that concerns about supervision focus on both task and process issues (Floyd, 2002).

- *Task Concerns* center on the kind of work you will be given to do and how you will perform. You are not unusual if you are worried about what will happen when you make a mistake and your real and imagined short-comings become visible (Baird, 2011). You may also be concerned about being assessed and evaluated. Your supervisor's perceptions of you are important in and of themselves, but the added prospect of a formal evaluation adds extra weight to those perceptions. Your own issues and personality will determine how important or intense these concerns or others we've not mentioned are to you.

- *Process Concerns* center on the interpersonal relationship with your supervisor; perhaps primary among those concerns is acceptance. Interns often wonder whether supervisors will like them. More important, they wonder whether their supervisors will understand them, i.e., recognize their strengths, work with their weaknesses, and respect them for the students they still are and the evolving professionals they are learning to be. As in any new relationship, all parties wonder whether they are going to get along well. You may also wonder how much about yourself you want to share or are expected to disclose and whether you should divulge your feelings and reactions as you go through the internship. Since you do not yet know these people, you may wonder how they will react to your disclosures; some interns, especially in counseling settings, worry that their supervisors will take the opportunity to analyze them and expose personal and professional weaknesses (Wilson, 1981).

Again, the cure for some of this is time. As you get to know your supervisor, you may or may not like what you find, but at least it will be known. And a trusting, comfortable relationship takes time to develop. However, you can be as active and engaged in this relationship as you are in any other. Looking at yourself as an engaged participant, with rights and responsibilities, is a much more empowering stance than looking at yourself as a passive recipient or even a victim of the process or the supervisor's style. It can help you get the most from your supervision, even in those cases where the supervision you receive is less than adequate.

Being an engaged participant involves keeping yourself focused on the learning dimension of the internship. It also means developing realistic expectations of your supervisor and learning to see things from her or his point of view. Being an engaged participant means being proactive in assessing the match between you and your supervisor and making the match better if you can. It also means understanding how you will be supervised and evaluated. Finally, it means being mindful of the triadic relationship involved in many internships—the relationship among you, your site supervisor, and your campus instructor.

A Focus on Learning

As your relationship with your supervisor begins and develops, always keep in mind that the main focus of your internship and of the supervisor-supervisee relationship is on *learning, growth, and development.* In a work situation, the primary focus of supervision tends to be on performance of assigned duties, and in exchange, you get paid. Performance is important at an internship, but the primary focus of your supervision is—or should be—educational (Birkenmaier & Berg-Weger, 2010; Royse, Dhooper, & Rompf, 2012). Your supervisor helps you learn and grow, and in return, you provide important service to the site. This distinction can be a bit subtle—after all, it is through performing your duties that you learn—but it is also profound. For example, once you have reached a level of comfort and skill with a certain kind of task, should you continue to perform that task to be a contributor to the organization or should you move on to something else? There is no right or wrong answer, but the path to an answer is asking which choice contributes more to achieving the learning goals you worked so hard to formulate.

A focus on learning may also mean a somewhat different role for you as an intern. We know from experience that students approach their traditional coursework with varying levels of seriousness and focus. Their goals may be largely intrinsic or extrinsic, as when the goal is to achieve a certain grade. In the internship, the goal is to learn as much as you possibly can and to push yourself and your supervisor with that goal in mind.

A Focus on Realistic Expectations

In our experience, interns often bring certain preconceptions about supervisors to the internship (McClam & Puckett, 1991) as well as expectations, hopes, and desires. They expect their supervisors to function in a variety of roles and to do them all well (Alle-Corliss & Alle-Corliss, 2006; Baird, 2011; Borders & Leddick, 1987; McCarthy, DeBell, Kanuha, & McLeod, 1988; Royse et al., 2012). As *teachers,* interns hope that their supervisors will help them set goals and objectives, help them learn new skills, and be sensitive to their particular learning styles. They hope that their supervisors will be *role models* with great stores of experience and expertise. They also hope that supervisors will use their leadership skills to be a *support* to them and help them through difficult emotional times and decisions. In the role of *consultant,* interns want their supervisors to help assess a problem or situation and generate alternative courses of action. Supervisors also are in the role of *sponsor, advocate,* or *broker* when they take an active interest in the careers of interns, both at the placement and beyond. And some interns hope that their supervisor will become a *mentor*—someone who takes a deep and special interest in them as a person and a budding professional. The reality, of course, is that no one can do all these things equally well and that your supervisor is a person with concerns, foibles, strengths, and weaknesses just like you.

Try not to make assumptions about your supervisor's level of experience or expertise, which may not be a great deal more than yours. Remember, though, that experience is not always the best indicator of quality in a supervisor. Just as brilliant scholars may not make good teachers because students' struggles with the material are incomprehensible to them, so it is that experts in your field may not make the best supervisors. They may have forgotten what it is like to be new to the field. The gap between your skill levels may be so great that the supervisors cannot function as effective role models because the expertise seems unattainable. So, a less-experienced person is not necessarily a problem; and, regardless of her or his level of expertise, your supervisor has a different, more objective vantage point from which to view your experience and your struggles, which is what matters when it comes to supervision.

Your Supervisor as a Person and a Professional

In approaching the relationship with your supervisor, it may be helpful to think about it from her or his point of view. Supervisors have a range of reasons for agreeing to accept an intern. Many supervisors see it as a way to contribute to their profession or perhaps to return the favor that a supervisor did for them when they were students (Birkenmaier & Berg-Weger, 2010). There are some tangible benefits to working with interns as well. For example, some supervisors have sought out interns because of the positive effects the supervisors observed on the morale, commitment, and proficiency of their staff as a result of working with interns (King & Peterson, 1997). Others may be motivated by the desire to sharpen their supervisory skills (Birkenmaier & Berg-Weger, 2010). It is also possible that they had no choice in the matter and that the person who agreed to accept the intern is not the assigned supervisor (Milnes, 2001). Also, being an *internship* supervisor is a complex task, for which many supervisors receive little or no formal training (Baird, 2011). They

What Supervisory Roles Are Most Important?

Take a moment now and try to put the roles of a supervisor in priority order. By ranking them from 1 to 6, you indicate which ones are most and least important to you. Yes, they may all be important; that is why we ask you to do this. At times when you are perhaps frustrated with your supervisor for not functioning quite the way you hoped, one thing you can do is come back to this exercise and consider the relative importance of that role.

_____ Teacher

_____ Role Model

_____ Supporter

_____ Consultant

_____ Advocate

_____ Mentor

_____ Other

may have a lot of experience as a supervisor, though not necessarily supervising interns. Everyone has to have their first intern sometime, and it just may be you!

Your supervisor also has roles as a worker and perhaps an administrator or a supervisor of personnel at the site, in addition to the role of intern supervisor. Your supervisor probably has a supervisor as well, or maybe several of them. All of these roles place demands on your supervisor's time, and agreeing to supervise an intern is yet another commitment of time (Baird, 2011; Royse et al., 2012). Sometimes, your supervisor has been released from other responsibilities in order to supervise you, but not always. If the investment pays off and the intern does a good job, then everyone benefits, and the intern may even be able to take some of the workload off the supervisor. The internship sometimes requires a great deal more time and energy than anticipated on the supervisor's part and the return for the organization may not be as great as hoped or needed. These are concerns probably on the mind of your supervisor. Out of all these roles and responsibilities, your supervisor's top priorities are and should be the primary work of the organization. In service situations, the primary concern is always the clients, customers, and communities that the organization serves. All supervisors worry that interns may make a serious mistake, but in service situations, such mistakes can harm clients, co-workers, and even the intern.

The Match Between You and Your Supervisors

The goal in developing a supervisor-supervisee relationship is achieving what Birkenmaier and Berg-Weger (2010) refer to as an optimal teacher-learner fit. Of course, this fit is important in any learning situation, but in a one-to-one relationship, such as this one, it is both more crucial and more achievable; a classroom teacher, in contrast, is trying to manage multiple relationships. The fit, or match, is not something that is either present or absent; there is lots of room in between. It is also not static; both you and your supervisor can work to improve it. So, your task here is not just to assess the match but to think about how to work within that match for maximum learning.

Supervisory Style

In all the roles mentioned earlier, your supervisor has a unique style. Just as learning style is multifaceted, so is supervisory style; many different dimensions are involved. Some aspects of your supervisor's style may be the result of careful study and training and deliberate choice. Others may be the result of attempts to replicate the way she or he learned best or to emulate a favorite supervisor. Still others are matters of personal preference or personality. Some aspects of a supervisor's style may not even be deliberate; she or he may be quite unaware of it or assumes that everyone supervises in the same way. There are a great many theories about supervisory or leadership style; perhaps you are familiar with some of them. A systematic review is beyond the scope of this chapter, so instead we have selected orientations to the work, to management, and to supervision that we find especially useful to interns.

Theoretical Orientation to the Work The work of an individual, and sometimes of an entire organization, is usually guided by a particular theoretical orientation. There are, for example, many different theories about teaching, counseling, management, and community organizing. It is important that you know what theory or theories most inform your supervisor's approach to the work. It is also important to know how your supervisor reacts to other styles. Those reactions may range from an insistence that the intern do things according to the supervisor's orientation to allowing the intern to try and learn from a variety of theories and approaches (Baird, 2011). Costa refers to this continuum as one of collaborative versus hierarchical approaches (1994). A collaborative approach emphasizes mutual discussion between supervisor and supervisee and encourages divergent thinking. In a hierarchical approach, however, the supervisor serves as a communicator of "expert" knowledge. Expertise flows in one direction, as do the questions, except when the supervisee needs clarification on something.

So, for example, if you are in the helping professions and favor a humanistic or psychodynamic approach to counseling and your supervisor believes in a cognitive behavioral approach, you will need to know to what extent you are expected to use that style with your clients. Or, if you tend to focus on systems change and community advocacy and your supervisor focuses on individual work and adjustment, you will need to have a serious conversation about how best to work together.

Expressive Versus Instrumental Styles of Management The terms *expressive* and *instrumental* describe two general approaches to management or supervision (Russo, 1993). Supervisors who use expressive approaches are more people oriented; the primary concern is for people and relationships. Supervisors who use this style often cultivate friendly relationships with those they supervise. A supervisor with a more instrumental orientation, on the other hand, is primarily concerned with productivity and task accomplishment. Please note that this is quite possible even when the business of the organization is people. A supervisor at a social services agency can be concerned primarily with serving the maximum number of clients and achieving measurable results. For these supervisors, respect is important in the supervisory relationship but not necessarily affection and friendship. They are able to handle conflict and hostility from supervisees effectively.

Focus on THEORY

Situational Leadership and Supervision

Hersey, Blanchard, and Johnson (2008) have identified two dimensions of supervision, which can be combined in four different ways. *Direction* is the first dimension, which involves giving clear, specific directions, close supervision, and frequent feedback. *Support*, the other dimension, is a nondirective approach marked by listening, dialogue, and high levels of emotional support. So, if your supervisor tells you to write a report and says, "There are lots of good ways to do this, and I know you are nervous, but you can handle it. Please ask all the questions you need to and let me know when you are done," this is a high level of support and a low level of direction. A directive approach might tell you exactly what the report should look like and what it should contain, with frequent deadlines for the submission of sections and the final report. These two dimensions can be combined in four ways: high support/low direction, high support/high direction, low support/high direction, and low support/low direction.

(continued)

Focus on THEORY *(continued)*

You may think that high support and high direction would be the best all the time. Of course, it is frightening to be given a task and have no idea how it should be done. However, too much direction over a long period of time can discourage independence and deny you the opportunities to try something and learn from it. Sometimes, you need to take risks, and even make mistakes, and that is difficult when you are very closely supervised. Similarly, everyone needs some support and encouragement, but too much can keep you from seeing where you need to improve. It can also make you dependent on your supervisor in a different way; there may be times when you need to function without considerable external support.

Hersey, Blanchard, and Johnson (2008) suggest the most appropriate combination of support and direction depends on the maturity level of the supervisee, and they use the term *maturity* in a particular way. First of all, the term is meant to apply to a particular task or set of task that the supervisee is facing; it is not a global description or a personality trait. They identified four levels of maturity, which they describe as a combination of *willingness* and *ability*. Each of those levels is best suited to a particular combination of support and direction. For example, suppose an intern is going to be asked to lead a support group. It may be that the intern, at least at first, is neither willing nor able to do it. In that case, the supervisor provides a lot of direction about how to lead the group and monitors the progress carefully. At the second level of maturity, the intern may feel unable but willing to try. In this case, the intern needs high levels of both direction and support. On the other hand, if the intern has the skills but feels uncertain (able but not willing), then high support and low direction are needed. Finally, an intern who has the skills and the willingness does not need much support or direction with regard to the task of leading groups. In other areas and for other tasks, the maturity level and needs may be quite different for the same intern.

Other Dimensions of Style Garthwait (2011) has described some additional dimensions of supervisory style, including the *speed* with which decisions are made; how willing the supervisor is to share information; the degree of structure and routine used in supervision; the level of focus, from detail to big picture only; and the pace of supervisor, from fast to more "laid back." Keep in mind that each of the dimensions of supervisory style describes a continuum, and there is plenty of room in the middle. As you get to know your supervisor, you can locate her or him on each of these continua. Remember, though, that supervision is a relationship, and so far, we have only covered one side.

Making Sense of Mismatches

As you read through the last section, you probably identified your preferences and tendencies among the styles discussed. What were your reactions and preferences and what do you think influenced them? How do you make sense of the differences between your style and that of your supervisor or co-workers?

If you—or your supervisor—are expecting a certain kind of response and get a very different one, it can be very disconcerting. For example, a supervisor expecting a theoretical analysis who instead gets a long discussion of your emotional reactions may think that you either did not understand or are not capable of answering the question. These moments are never easy, but they are easier if they are understood as a *mismatch of styles*. Your supervisor is not necessarily insensitive to your needs any more than you are stubborn or inept. You may each be doing exactly what you think is appropriate and required. Misunderstandings can be avoided, though, and taking an engaged stance toward your supervision and accepting that some conversations will be challenging is an effective way of doing so.

As you think about some of the mismatches you have experienced with your supervisors or co-workers, think also about what you can do about them. Most supervisors do not mind being asked about their style of supervision. They may or may not be familiar with the specific theories discussed here, but you can find a way to ask about these aspects of style without using the particular jargon described in this chapter. Your supervisor may also be familiar with and use other theories about supervisory styles. If so, you can learn both about that theory and about your supervisor when discussing style. You may also want to consider discussing your style with your supervisor, including your preferences and needs. These needs do not have to be demands. You are merely considering your style, along with your supervisor's style, and looking for potential areas of match, mismatch, and compromise.

Your Reaction to Supervision

Even under the best of circumstances, the process of supervision often produces at least some anxiety. In your internship, you are being asked to try out new skills and approaches. You should be stretching yourself and taking some risks, and that means there will be times when you don't do as well as you had hoped. Your supervisor may also notice patterns of strengths and weaknesses that you are not aware of. In social service settings, a supervisor who feels you are running into some unresolved personal issues is probably going to let you know that. Under those circumstances, most of us experience two seemingly contradictory emotions. On the one hand, we are there to learn and grow, and supervision is a tremendous opportunity for that. On the other hand, we do not always want to see ourselves clearly or change our ways of doing things. We want to learn and change, yet we resist learning and changing (Borders & Leddick, 1987). Some of us are even resistant to compliments, as in this student's journal entry: *"Positive feedback is something I have a hard time accepting. In fact, I almost cry when I hear it. I am learning that my work in the world is important and well received."*

These conflicting emotions sometimes manifest themselves in behaviors that help you avoid really looking at yourself and your progress (Borders & Leddick, 1987; Costa, 1994). They include being overly enthusiastic, avoiding certain topics, and going off on tangents. You may also become forgetful, especially about projects, assignments, or issues that make you nervous. Some interns become argumentative, taking issue with every point the supervisor makes. They may or may not do so out loud, but privately and to their friends, they dispute every criticism. Others go to the opposite extreme and agree with the supervisor even when that is not how they feel. If you find yourself exhibiting any of these behaviors, think about the feelings and concerns that may be behind them. The point here is not that any single instance of one of these behaviors indicates that you are avoiding an issue; however, when they become patterns, you should look very carefully at them, discuss them with your instructor, and find a way to work through them.

The Evaluation Process

At some point in your internship, you will probably have a formal evaluation. At that time, the supervisor lets you know how you are doing and have done overall. In our experience, most internship programs request at least two evaluations: one at the midpoint and one at the end. It is natural to be apprehensive about the prospect of reading or hearing an evaluation. But try to remember that the evaluation is another opportunity for growth, empowerment, dialogue, and even assertiveness. Above all, it is an opportunity to learn. Again, the more you know about the process, the less mysterious and threatening it will be and the more you can direct your emotional energy toward making the most of these opportunities. As one intern said after reflecting on her evaluation, *"He made me curious about myself and want to work on the areas that need improvement."*

You can take the initiative to find out several things about how you will be evaluated. The structure of your evaluation is an obvious concern. Find out how often you will be evaluated and whether it will be oral, written, or both. Gather information on the specific format. Some written evaluations use scales, where you are given a rating from 1 to 5 or 1 to 10, to indicate how well you have met various criteria. Some evaluations let the supervisor write in a more narrative form. Still others use both formats.

It is also important to know what standards are being used in your evaluation (Wilson, 1981). Some supervisors will compare you to other interns, either current ones or those from past semesters. If you are among the best, you receive high ratings. Others evaluate interns individually, assigning ratings based on how much they have grown over the course of the semester. Still others have a standard set of expectations, criteria, or competencies that you must meet to get a high rating. If the supervisor is using numerical ratings, find out what they mean and what you have to accomplish to earn the ratings you want.

We know of more than one student whose supervisor had nothing but praise for them at their weekly conferences, yet, the mid-semester evaluations contained ratings no higher than 3 on a scale of 1 to 5. What typically occurs is that the supervisors explain that they never give anyone a 5 because no one is perfect, and if they gave 4s at mid-semester, there would be no room for improvement. To them, the 3s were an excellent mid-semester evaluation, but to the interns, they felt like an average rating, much like receiving a C on a paper. In instances like that, campus instructors often have a conversation about such grading practices and provide the on-site supervisors with materials and information so that they evaluate more objectively.

You will also want to know the function of the evaluations. Who can see them and for what purpose? Are the mid-semester evaluations recorded and counted as part of your grade? What portion of your internship grade is determined by the supervisor's evaluation? You should also clarify when and under what circumstances you will see your evaluation. Many supervisors will include the written evaluation as part of a supervision meeting. They may discuss the evaluation before filling out any forms or they may go over the written evaluation with you. Regardless of the specific arrangements, you should see the evaluation before it goes to anyone else and have an opportunity to review it and ask questions.

The Campus Instructor/Supervisor

Part of the success of your internship depends on the triad of you, your campus instructor, and your site supervisor. This particular triad can be a solid foundation for your learning. A triangle by definition can be a very stable formation; when set on its base, it is hard to knock over. On the other hand, when set on its point, it tips over all on its own. So, too, this triadic relationship can be prone to unproductive alliances, which we will discuss shortly. Once you have a clear understanding of the role of the campus instructor in this relationship, you'll have a deeper understanding of why the keys to success in this triadic relationship are mutual understanding and open communication. You can play an important role in making sure each of these keys is in place.

The role of the campus instructor/supervisor in an internship varies from campus to campus, but at the very least, this is the person who will be grading your overall performance. If a campus instructor is also leading a seminar, reading and reacting to your journal, or visiting you at the site, then in some ways that person is supervising you, too. And while the relationship is generally less intense and the contact less frequent than the relationship with your site supervisor, many of the same concerns and issues can and do arise. Acceptance is a concern in this relationship, as is assessment. It is important to have a sense of the styles of instructing and supervising used by your campus supervisor and see how well they match with your needs.

We have seen situations where the intern and the site supervisor form an alliance against the campus instructor, giving a glowing report when one is not deserved or agreeing not to divulge certain events that occurred. We have also seen the opposite situation, where the intern and the campus instructor form an alliance and talk together about their negative impressions of the site supervisor and how best to "get around" that person.

Mutual Understanding

It is vital for everyone's clarity, and your sanity, that all three of you have the same understanding of the mechanics of the internship. These mechanics include everything from relatively mundane matters, such as required hours, start and end dates, and dress codes, to more profound matters, such as your learning goals, the scope of your activities, the frequency and nature of visits to the site by the campus instructor, and evaluation methods. This last piece is especially important if there are actually two approaches to your evaluation—one used at the site and one used on campus. In some academic programs, mutual understanding about these issues is put in place between the campus and the agency long before any interns are interviewed or accepted. In other programs, due to staffing or other circumstances, these understandings are not in place, and it will be up to you to make sure they are established.

Open Communication

Open communication means that issues and concerns are addressed directly by the two or three parties involved. If your supervisor does not seem to like your instructor (or vice versa) or has a problem with something about the program, you can't always avoid talking about it, but you can encourage the supervisor to speak directly with the instructor. If your supervisor seems to want to give you higher ratings than you perhaps deserve or does not want to disclose to your instructor any of your areas for growth and development, it is tempting to let that occur. But to maximize your learning, you need to hear clearly and openly about your strengths as well as the areas you need to further develop. You can help your supervisor not play the "game of school" where the goal is the grade and instead help put your supervisor into the role of a collaborative partner in your learning. Everyone benefits when that occurs, especially you.

Conclusion

You've spent the chapter examining two important components of the internship: the learning contract and supervision. As you've come to realize, you can expect some anxiety as you work your way through this early phase of the internship because you have the responsibility to make sure that these components are in place and ready to work for you. For example, it may be concerning if your site supervisor tells you that you are the first intern she has ever had or your campus supervisor tells you that he's never supervised interns before. However, you are more aware now of the process you need to have in place to ensure the expected learning outcomes. And, you are better prepared to know when your role needs clarifying and your learning goals need adjusting. By taking an engaged approach, you know how to ensure that you have a successful internship—even if you are both of your supervisors' first intern!

The idea of a "supervision plan" may have taken on deeper and broader meaning now that you've spent time reading about the supervisory relationship, learning the differences between task and process concerns, realizing the supervisory style of your supervisors and the style that works for you, and having an understanding of approaches to supervision and evaluation. Learning to be a supervisee can be challenging, especially if the only supervision you've experienced is in a fast food restaurant or if you have a work or career history where you have been a supervisor.

With your learning contract and your supervision plan in place, you are prepared for the road ahead. You can continue on your journey knowing that you are grounded by the map that guides your learning and by the infrastructure that keeps the internship moving forward and staying on track. New experiences are just ahead in a time of intense learning and growth. Although you may be ready to move into that next stage, if you are interning in the helping professions, then the next chapter, which was written especially for you, should be read first. Otherwise, it is time for Chapter 8 and the Exploration stage.

For Review & Reflection

Checking In

Shoring Up the Contract. It's a good time to think about whether there's more work to do on your learning contract before moving forward. For example, does your contract reflect *substantial learning* on your part and enough *challenge* to leave you wondering if you are able to accomplish everything on the contract before the internship ends? It needs to do just that, so review each learning outcome in these terms. If any do not meet the criteria, then it's back to the drawing board with your supervisor. If you do meet the criteria, think about which of the learning goals intrigues you most and why.

Shoring Up the Plan. What does your supervision plan look like at this point in the internship? Use the plan checklist on p. 160 to determine what aspects of it need shoring up. What are your next steps to put them in place as you move forward?

Experience Matters (For the EXperienced Intern)

- Think about the supervision experiences you have had in past positions. In what ways has your past experiences prepared you for supervision in this internship? Are there lessons to be learned from the differences among them?

- Compare your *reactions to* supervision experiences in the past. It's not always the case that one is better or worse, but if they are different, they may affect you differently. Are you noticing that you are reacting differently this time around, and if so, how does it affect the way you are approaching your internship and supervision?

Personal Ponderings (Select the Most Meaningful)

- Take the time to list the strengths you bring to the supervisory relationship. Hold onto that list and make a point of reviewing it before—and after—your supervision meetings, until you no longer need to do so.

- If you were "shopping" for a supervisor, for whom and what would you be looking?

- In what ways is your supervisor's style well matched to yours? Are there any areas of mismatch? If so, are those areas important, and if they how do you plan to come to terms with those differences?

- How are you going to be evaluated by your supervisor? Are you aware of the criteria the supervisor will use for the evaluation?

Do you know what happens to the evaluation after it is written? Would you change anything about this process? Why? If you want to make a change in the process, what would you do and how would you do it?

- Supervision can evoke a wide range of reactions in both the supervisor and the supervisee. Do your reactions remind you of previous supervisory situations and how you reacted then? If so, and you were not satisfied with the quality of the supervisory relationship, what would you want to change this time around?

Seminar Springboard

Difference Strokes, Different Folks. Many interns are concerned about discussing their learning style and learning needs with their supervisors for a number of reasons. However, the internship is a *learning* experience and there is no guaranty that the site supervisor knows how an intern learns best. So, discussing what works for you when it comes to learning is an important conversation to have during supervision. Here's an exercise that might be helpful to bring about that conversation. Using triads, each intern takes a turn in the roles of *intern, site supervisor,* and *observer.* The observer's role is to give feedback to the intern on how effectively and clearly their learning needs were stated. The intern's role is to use engaged language and actions in the conversation. The supervisor's role is to describe what "worked" and what didn't when the intern used the engaged approach in the conversation. All three role-players then discuss how it *felt* using the engaged approach and how they would handle the next conversation about an uncomfortable topic.

For Further Exploration

Colby, A., Erlich, T., Beaumont, E., & Stephens, J. (2003). *Educating citizens: Preparing America's undergraduates for lives of moral and civic responsibility*. San Francisco: Jossey-Bass.

Particularly useful for stimulating thinking about civic development.

Corey, M. S., & Corey, G. (2011). *Becoming a helper* (6th ed.). Belmont, CA: Brooks/Cole.

Helpful section on supervision.

Crutcher, R. A., Corrigan, R., O'Brien, P., & Schneider, C. G. (2007). *College learning for the new global century: A report from the National Leadership Council for Liberal Education and America's Promise*. Washington, DC: AAC&U.

Excellent, comprehensive statement of student-learning goals in a wide variety of areas.

Gordon, G. R., & McBride, R. B. (2011). *Criminal justice internships: Theory into practice* (7th ed.). Cincinnati, OH: Anderson Publishing.

Excellent discussions of issues relating to supervisors and co-workers. Obviously, this book is aimed at interns in a particular kind of setting, but it has many applications outside criminal justice.

Garthwait, C. L. (2011). *The social work practicum: A guide and workbook for students* (5th ed.). Needham Heights, MA: Allyn & Bacon.

Lots of helpful information on learning contracts.

Hersey, P., Blanchard, K., & Johnson, D. E. (2008). *Management of organizational behavior: Utilizing human resources* (9th ed.). Englewood Cliffs, NJ: Prentice Hall.

Situational leadership theory is a very popular approach to management and supervision. Discusses combinations of support and direction and when each combination may be most helpful.

Howard, J. (Ed.). (2001). *Service-learning course design workbook*. Ann Arbor, MI: OCSL Press.

An invaluable tool that can be applied to developing goals for personal, professional, and civic development. Easy-to-read chapters and worksheets guide you through the process.

Inkster, R. P., & Ross, R. G. (1995). *The internship as partnership: A handbook for campus-based coordinators and advisors.* Retrieved from www.nsee.org

Inkster, R. P., & Ross, R. G. (1998). *The internship as partnership: A handbook for businesses, nonprofits, and government agencies.* Retrieved from www.nsee.org

A series of two comprehensive and invaluable handbooks for internship programs and field sites. The first publication focuses on the responsibilities and challenges of the work of the field coordinator; the second does the same for the work of the site supervisor. Both publications have significant value for faculty and other campus supervisors in developing effectiveness in working with placement sites.

King, M. A. (In press). Ensuring quality in experiential education. In G. Hesser, (Ed.). *Strengthening experiential education within your institution* (Revised). Mt. Royal, NJ: National Society for Experiential Education.

A chapter in a classic sourcebook that considers experiential educational practices that ensure high impact learning.

Kuh, G. D. (2008). *High-impact educational practices: What they are, who has access, and why they matter.* Washington, DC: Association of American Colleges and Universities.

Describes educational practices that have a significant impact on student success.

National Society for Experiential Education. (In Press). G. Hesser, (Ed.). *Strengthening experiential education within your institution* (Revised). Mt. Royal, NJ: Author.

A classic sourcebook recently revised; offers best practices intended for the entire campus community for institutionalizing experiential education across campus. Of special relevance for engaged learning opportunities such as internships, cooperative education, field studies, course practica, community service-learning, and other forms of experiential education.

Royse, D., Dhooper, S. S., & Rompf, E. L. (2012). *Field instruction: A guide for social work students* (6th ed.). Boston: Allyn & Bacon.

Excellent section on supervision.

Shumer, R. (In press). Evaluating and assessing experiential learning. In G. Hesser, (Ed.). *Strengthening experiential education within your institution* (Revised). Mt. Royal, NJ: National Society for Experiential Education.

A chapter in a classic sourcebook that considers trends in the approaches to and the focus of evaluation of experiential learning, as well as their impact on individuals, institutions, and communities. Tools and frameworks are described that include evaluation and assessment as an integral part of the process of experiential learning.

Shumer, R. (2009). New Directions in Research and Evaluation: Participation and Use Are Key. In J. R. Strait and M. Lima (Eds), *The future of service-learning*. New York, NY: Stylus Publishing.

This chapter is about the evolution of evaluation practice and the use of participatory approaches in the service-learning field.

Shumer, R. (2012). Engaging youth in the evaluation process. In Velure Roholt, R; Baizerman, M., and Hildreth, R. (Eds), *Civic youthwork: Co-creating democratic youth spaces*. Lyceum.

This chapter covers the use of youth participatory evaluation as part of the civic engagement process. It includes several examples of young people evaluating their own civic involvement programs.

Sommer, C. A., Ward, J. E., and Scofield, T. (2010, Fall). Metaphoric stories in supervision of internship: A qualitative study. *Journal of Counseling & Development* 88 (500–507).

Authors discuss their use of 3 stories during supervision of interns throughout one semester and the 3 themes it yielded which were related to the interns' experiences: recurring cycles of highs and lows; balancing external and internal influences; and struggles with self-awareness. Study suggests use of stories during supervision may contribute to self-reflection.

Wilson, S. J. (1981). *Field instruction: Techniques for supervisors.* New York: Free Press.

Useful discussions of supervision and evaluation.

References

Alle-Corliss, L., & Alle-Corliss, R. (2006). *Human service agencies: An orientation to fieldwork* (2nd ed.). Belmont, CA: Brooks/Cole.

Baird, B. N. (2011). *The internship, practicum and field placement handbook: A guide for the helping professions* (6th ed.). Upper Saddle River, NJ: Prentice Hall.

Birkenmaier, J., & Berg-Weger, M. (2010). *The practical companion for social work: Integrating class and field work* (3rd ed.). Boston: Allyn & Bacon.

Bogo, M. (1993). The student/field instructor relationship: The critical factor in field education. *Clinical Supervisor, 11*(2), 23–36.

Borders, L. D., & Leddick, G. R. (1987). *Handbook of counseling supervision.* Alexandria, VA: Association for Counselor Education and Supervision.

Council for the Advancement of Standards in Higher Education. (2009a). *CAS professional standards for higher education.* Washington, DC: Council for the Advancement of Standards for Higher Education.

Council for the Advancement of Standards in Higher Education. (2009b). *CAS self-assessment guides.* Washington, DC: Council for the Advancement of Standards for Higher Education.

Council for the Advancement of Standards in Higher Education. (2009c). *Frameworks for assessing learning development outcomes.* Washington, DC: Council for the Advancement of Standards for Higher Education.

Compact, C. (2003). *Introduction to service-learning toolkit: Readings and resources for Faculty* (2nd ed.). Providence, RI: Campus Compact.

Costa, L. (1994). Reducing anxiety in live supervision. *Counselor Education and Supervision, 34*(1), 30–40.

Crutcher, R. A., Corrigan, R., O'Brien, P., & Schneider, C. G. (2007). *College learning for the new global century: A report from the national leadership council for liberal education and america's promise.* Washington, DC: Association of American Colleges and Universities.

Curtis, R. C. (2000, June). Using goal-setting strategies to enrich the practicum and internship experiences of beginning counselors. *Journal of Humanistic Counseling, 38*(4), 94–206.

deShazer, S. (1980). *Clues: Investigating solutions in brief therapy.* New York: Norton.

Feldman, D. C., & Weitz, B. A. (1990). Summer interns: Factors contributing to positive developmental experiences. *Journal of Abnormal Behavior, 37,* 267–280.

Floyd, C. E. (2002). Preparing for supervision. In L. M. Grubman (Ed.), *The field placement survival guide: What you need to know to get the most out of your social work practicum.* Harrisburg, PA: White Hat Publications (pp. 127–133).

Garthwait, C. L. (2011). *The social work practicum: A guide and workbook for students* (5th ed.). Needham Heights, MA: Allyn & Bacon.

Garthwait, C. L. (2005). *The social work practicum: A guide and workbook for students.* (4th ed.). Upper Saddle River, NJ: Prentice Hall.

Hersey, P., Blanchard, K., & Johnson, D. E. (2008). *Management of organizational behavior: Utilizing human resources* (9th ed.). Englewood Cliffs, NJ: Prentice Hall.

Hodgson, P. (1999). Making internships well worth the work. *Techniques, September,* 38–39.

Howard, J. (Ed.). (2001). *Service-learning course design workbook.* Ann Arbor, MI: OCSL Press.

King, M. A. (1988). Toward an understanding of the phenomenology of fulfillment in success. Unpublished doctoral dissertation. University of Massachusetts Amherst

King, M. A., Spencer, R., & Tower, C. C. (1996). *Human services program manual.* Fitchburg, MA: Fitchburg State College.

King, M. A., & Peterson, P. (1997). *Working with interns: Management's hidden resource.* American Probation and Parole Association. Boston, MA.

Kiser, P. M. (2012). *Getting the most from your human service internship: Learning from experience* (3rd ed.). Belmont, CA: Brooks/Cole.

Leddick, G. R., & Dye, H. A. (1987). Effective supervision as portrayed by trainee expectations and preferences. *Counselor Education and Supervision,* 27(2), 139–154.

McCarthy, P., DeBell, C., Kanuha, V., & McLeod, J. (1988). Myths of supervision: Identifying the gaps between theory and practice. *Counselor Education and Supervision,* 28(1), 22–28.

McClam, T., & Puckett, K. S. (1991). Pre-field human service majors' ideas about supervisors. *Human Service Education,* 11(1), 23–30.

Milnes, J. (2001). Managing problematic supervision in internships. *NSEE Quarterly,* 26(4), 1, 4–6.

NSEE. (2009). *NSEE principles of best practices in experiential education.* Revised. Originally presented at the National Society for Experiential Education, Norfolk, VA, 1998.

Rothman, J. C. (2002). *Stepping out into the field: A field work manual for social work students.* Boston: Allyn & Bacon.

Royse, D., Dhooper, S. S., & Rompf, E. L. (2012). *Field instruction: A guide for social work students* (5th ed.). Boston: Allyn & Bacon.

Russo, J. R. (1993). *Serving and surviving as a human service worker* (2nd ed.). Prospect Heights, IL: Waveland Press.

Wilson, S. J. (1981). *Field instruction: Techniques for supervisors.* New York: The Free Press.

Zlotkowski, E. (Ed.). (2006). *Service-learning in the disciplines series.* Sterling, VA: Stylus Publications.

CHAPTER 7

Getting to Know the Clients: A Chapter of Special Relevance to Helping and Service Professionals

The interpersonal work is the critical aspect of any internship—graduate or undergraduate. It is also the most difficult for most people.
STUDENT REFLECTION

If you have chosen an internship providing direct service, it is natural that many of your thoughts and much of your excitement and anxiety are directed toward the population you are working with, whom we collectively refer to as *clientele* and who are your customers, consumers, patients, students, or clients, depending on the setting. Whether you are working in a government agency or a business, a human service agency, some other service or nonprofit organization, it is easy to become preoccupied with paperwork, policies and procedures, and office politics. As an intern, you will have all these concerns to deal with plus the additional concern of being evaluated by at least two supervisors. However, as J. Robert Russo reminds us in his book *Serving and Surviving as a Human Service Worker* (1993), the people your organization serves are the reason for your work. We have found that concerns about those who are served generally fall into two categories: the nature of the population served (which involves your assumptions) and your relationship with them. An area of emerging concern that is related to both of these categories is the personal safety of the helping professional.

Recognizing the Traps: Assumptions and Stereotypes

Since I started my internship, I can honestly say that my impressions have changed. When I first started, I did have stereotypes. I was wrong, and I am not ashamed to admit it. When I was exposed to the clients and what their situations were like, I started to change.

I'm embarrassed to say so, but the media taught me all my impressions of the clients.

STUDENT REFLECTIONS

As you begin your work, you no doubt have some expectations and assumptions about the people you will meet. You probably have some knowledge of the population as a group from the placement process and your first couple of days at the site. You know in general terms who the site does and does not serve and some of the needs the organization can and cannot meet. For example, populations can be described as homogeneous or heterogeneous with regard to various characteristics. The more *homogeneous* a population is, the more alike its members are; the more *heterogeneous* the population, the more diversity is found within it. So, for instance, a high school has a population that is relatively homogeneous in age but may be heterogeneous with regard to characteristics such as race, social class, and intellectual acuity. Beyond this factual information, though, it is important to be aware of the image you have of your prospective clients. You are bound to have one, and it is not all based on facts.

In spite, or perhaps because, of their inexperience, many interns make unconscious generalizations; they form stereotypes. The word *stereotype* has some pretty negative connotations, and you may be reluctant to consider that you have some. However, whenever you make a judgment about someone based on little or no information, you engage in stereotyping. And that is a major trap in being effective in the helping and service professions and in being a civic professional.

Uncovering the Roots

Your assumptions and stereotypes can come from several places. Perhaps you have met a person with the same needs and issues as those faced by clients at your site or heard one speak. Perhaps you have been such a client yourself or know someone who has. The media are another powerful source of stereotypical images and assumptions. There are scholarly books and documentaries on various client groups, but a more powerful source of images is the mainstream media, especially television. Think, for example, about how many mentally challenged adults you have seen depicted on prime-time television. Were any of them leading anything like normal lives? How many of them had committed a crime? Couple the images of this population portrayed in the media with your own lack of direct experience, and you can begin to see where your image and stereotypes may originate. As one student said, "*Given all the media attention to crime, in all the genres, it would be difficult not to form an impression of the criminal population.*"

THINK About It

Checking Out Your Assumptions

Imagine for a moment the people you will be helping or serving. Each one will be somewhat different, but you can describe them within certain parameters. Think about their backgrounds, their reasons for coming to your office, their personalities, life histories, and typical behaviors.

Think about their race, ethnicity, social class, religion, and sexual orientation. How heterogeneous do you imagine the population to be? In our experience, interns often imagine that the clients will be very similar to one another or they go to the other extreme and see each one as totally different, missing some of the important commonalities among them.

Think about how similar and/or different they are from you. What do you have in common with them? What aspects of their lives and experiences are totally unfamiliar? Here, we often find that interns imagine they have almost nothing in common with their clients. That is, of course, not entirely true, but it is easy to feel that way. On the other hand, interns who have struggled with the same problem that brings people to the agency (such as alcoholism or other addictions) may assume that they know just how it is for them. A little thought and some reflection on classes you have taken will tell you that is not true either, but it is another tempting assumption.

Think also about the source of that image. Probably very few of you have had extensive and varied experience with the people you will be working with. If you've not worked with this population before, what might be generating those images for you?

Engaging Your Stereotypes: Getting Beyond the Traps

The first step in getting beyond stereotypes and assumptions is to admit you have them. Check with other interns; they have them too. Theirs may not be the same as yours, but they have them. The next step is to gather as much factual information as you can and hold your assumptions up to the light of objectivity. At the beginning of this section, we asked you several questions about your clients or customers and what you were assuming about them. Now may be a good time to go back and try to find factual answers to those questions. Don't be discouraged if the answers don't come easily; you have held on to some of your assumptions for a long time.

Focus on SKILLS

Recognizing Your Stereotypes

Try imagining two people. One is a slim, slightly pale man in a tweed jacket wearing wire-rimmed glasses and carrying a beat-up briefcase. The other is a tall, heavy man with a large belly, long hair, and a big beard, dressed in a T-shirt, sunglasses, dirty jeans, black boots, and a black leather vest with a Harley-Davidson insignia on it. In spite of yourself, are you making assumptions about what each of them does for a living? About how educated they are? Their personalities? Which one is more likely to have read existential philosophy? To have been in a barroom brawl? It seems to be a human tendency to generalize, and the fewer people in any group we actually know, the more we are likely to generalize from the few that we do know or have read about.

This example is based on physical appearance, and this is one way people make assumptions. Another way of getting trapped by your assumptions is by becoming too focused on a client's behavior. Some clients served by human service agencies display fairly unusual or even bizarre behavior. It is easy to jump from the behavior you see to all kinds of conclusions, especially if this is your first time working with these behaviors. Until you are more informed about the nature of the behavior, it's best to reserve judgment about what you are actually seeing. However, you may find that in spite of yourself, the reality of what some clients have done, especially those who have committed violent acts, makes it hard to see past their behavior to the unique features of each person. What's important here is that you are aware that you are making assumptions and that you engage those assumptions by consciously making a commitment to objectivity in spite of the assumptions that live or linger in your mind.

Rethinking Client Success

When you think about being successful in your work with clients (whether they are called *clients*, *patients*, *students*, or *customers*), what does that *actually* mean? Be as specific as you can. Think about the kinds of goals that make sense for the population you serve. Think about the attitudes, skills, values, and knowledge that you need to help them empower themselves. This may be a good time to talk with your co-workers and site supervisor about this idea of being successful in your role as helper. Their experience will serve you well in understanding your assumptions.

In the sections that follow, we consider many aspects of the relationship interns develop with those they serve. For much of the discussion, we use the term *client* without specifying the various names that clients are called in different internship settings. If you are in the helping professions, then all of what follows applies to your work. However, if you are interning in other kinds of service professions, you'll need to read the following sections with your population in mind.

Acceptance—The First Step

I want clients to respect me. It would be nice if they liked me, but it's not necessary. Most important is that they trust me to know that I will do whatever is in my power to help make their lives easier.
STUDENT REFLECTION

As interns think about the kind of relationships they want to have with those they serve, several concerns often arise. One of the most common, especially for those in the helping professions, is how clients will react to them (Baird, 2011). In talking with undergraduate students who were going to perform community service with homeless individuals, Ostrow (1995) reported that one of their major anxieties was how they, as relatively affluent and fortunate young people, would be perceived by the homeless population. You may be wondering, *"What kind of reception am I going to get from these people? Will they respect me? Will they listen to me? Or will they just write me off?"* The theme in these concerns is acceptance, which is a crucial part of the foundation for any relationship. You have probably read about how important it is for you to accept those you serve, but they need to accept you, too. You need to find ways to help them accept you, and that's not always easy.

THINK About It

What Does Acceptance Mean to You?

Think for a moment about what the word *acceptance* means. Don't try to think of a definition for it; instead, think what the word *means* to you personally. Ask a peer to do the same, and you quickly realize that you aren't necessarily in agreement. Most people, though, know what it *feels* like when they are accepted and what it feels like when they are not. Think about how acceptance *feels* to you as well as what it means to you. Being aware of how you think about acceptance allows you to better understand your reactions to acceptance issues.

Being Accepted by Clients

Acceptance can mean different things to different client populations. Some clients show their acceptance simply by being willing to talk with you; until they accept you, they simply ignore you. Other client groups may show their acceptance by including you in their conversations, confiding in you, considering your suggestions, following your directions, or accepting the limits you set. How do you imagine you will know whether your clients have accepted you? Be as concrete and specific as you can.

Tuning in to the Clients' World

I couldn't survive a minute in their world, and they know it. I didn't grow up in the city or in the street culture. I haven't been in a fight since grade school. And I'm going to give these guys advice?
STUDENT REFLECTION

Getting those you serve to accept you can be a real challenge. You will have an easier time with it if you think about the situation from their perspectives. The main challenge is to enter, but not become lost in their worlds. Schmidt (2002) writes of the power of perception in shaping our day-to-day experiences and the importance of entering the perceptual worlds of your clients—another dimension of the human context of your work that is so important to a civic professional. Entering that world means going beyond facts and figures, reports, testing profiles, and school grades found in client files, even beyond catalogues of experiences that clients may tell you about. It means trying to understand how their experiences have shaped the way clients see the world. In the book *Crossing the Waters*, Daniel Robb writes about his experience working in a residential setting with troubled adolescent males (2001). Here, he recounts an insight gained while working with one of the boys to fix a car: *"I saw as we went that Louis had never had a man show him how to do anything. So, he expected to do it wrong, expected to be shamed for it, because unknown territory is not filled with angels"* (p. 173). Here is another example from a student journal: *"Most of these kids build a stone wall as a façade because it is the only way they have learned to survive. Behind that wall is an individual with feelings and emotions."*

Meeting with Resistance

Your clients have probably had many frustrating and stressful experiences. In addition, they may be for the first time in a position where they cannot meet their own needs, where they have to let strangers into their personal business, and where their welfare is not in their hands (Royse, Dhooper, & Rompf, 2011). They may be expecting a miracle or a quick solution to their problems, and they are not usually going to find one. Or maybe they have just figured that out. If they are clients with a long history of services that did not help them, they may be expecting more of the same (Royse et al., 2011).

In some cases, clients may be particularly wary of interns. Some respond better to volunteers, reasoning that they are there because they want to be, as opposed to interns, who are fulfilling a requirement for school (Royse et al., 2011). In addition, interns generally do not stay very long. Some may be there for one academic year, and some may even be hired, but many are gone after one semester. Many clients have had unfortunate experiences with people leaving them or letting them down, and they are reluctant to invest in a relationship that will end soon. One of our students who was working at a group home for adolescents reported that one of the residents was quite hostile and rebuffed all attempts at contact. On further inquiry, the intern learned that this resident had become very close emotionally to an intern the prior semester and, although she knew all along that the intern would leave, was heartbroken when it happened.

Cultural Competence: Clients' Cultural Profiles

Part of their perceptual worlds, of course, is shaped by your clients' cultural profiles. These profiles must, to a large extent, be understood one client at a time, just as your own profile (explored in Chapter 4) is different from almost anyone else's. Try to find out, then, not just what subgroups your clients may belong to but how they have shaped their behaviors and perceptions. And try to avoid the trap of only looking for the weaknesses or liabilities in a client's cultural background; see the strengths as well. For example, a client whose ethnic background encourages her or him to seek help from family and friends, but never from strangers, may be reluctant to let you in. But remember that this same client may have a wonderful resource in family and community—one that can be used to help. Many of you have probably studied this topic already, and this book is not the place for a lengthy discussion of multicultural issues. However, we *do* provide additional sources for further reading at the end of this chapter.

Putting Client Behavior in Context

Another advantage to understanding the perceptual world of your clients is that you can try to understand their behavior in that context (Baird, 2011). This will help you not to be overly flattered by early compliments or expressions of trust and not to be overly flattered by clients who turn away from you and your attempts to connect. Remember that clients often have an agenda they are not going to tell you about but one that you can understand once you come to know them. A client who approaches an intern and says *"You are the only decent person in this place—I can trust you"* is either trying to gain an advantage or rushing into closeness on very little information, or both. Similarly, a client who gives you a hard time may be doing it precisely because you are new; they want to see whether you will back away as so many others have done in the past.

Focus on THEORY

Transference and Overidentification

In counseling and psychotherapy, the client's behavior just described is referred to as *transference*. Clients transfer to you feelings and reactions that are rooted in their past experiences. Marianne and Gerald Corey (2011) emphasize that transference can take many forms, including wanting you to be the parent, husband, friend, or partner they never had; making you into some sort of super-helper who can fix anything; refusing to accept boundaries that you set; and easily displacing anger or love onto you.

While you are working to enter the world of your clients, take care not to get caught in it. If their experience is similar to yours, it is easy to have old feelings of yours evoked, i.e., come to the surface, and you react to those feelings. It is also easy to become so absorbed in the experience of the clients that you take on the feelings and reactions that they have. This is called *overidentification*. Once that happens, it is very easy to cross the line from *wanting* your clients to do well to *needing* them to do well. Maintaining distance, even while developing rapport with clients, will let you navigate the overidentification trap. It will also help you in confronting clients when necessary and helping them see how their perspectives may be creating problems for them.

Seeking Common Ground

Part of developing acceptance with clients is finding some common ground. During the Anticipation stage, doing so is especially important to creating the foundation for the helping relationship. A frequent concern for interns is their belief that they have nothing in common with their clients. A young woman interning in a shelter for single women and their children said:

> *I don't have any kids. I have no idea what it's like to be a single parent, or to grow up with one parent. I have no idea how to survive on the streets or what it's like to be abused by a husband. What do I have to offer these women? Why should they listen to me?*

These are real concerns, but remember that even having the exact same experience as the clients is not always a guarantee of success. Remember also that you don't give advice, so when clients are listening to you, it is not about telling them what to do. This may be of special relevance to the EXperienced intern in the helping professions who has a career or job history with clients with whom they share some common life experiences or worked on the financial side of social services where directing—not advising—clients is expected. Such experiences can be helpful in establishing acceptance, but they can also be a hindrance. You may have had some of the same things happen to you as have happened to your clients, but assuming or saying you know how your clients *feel* can be alienating for a client whose experience was quite different from yours. You may have experienced the feeling itself, such as loss, but not be able to relate it to the context or circumstances of their losses.

If you don't have much common experience, what can you do? You can't manufacture experience you haven't had, and pretending you know how it is will only make things worse. However, common ground does not always mean common experience. Showing respect for your clients' experiences is one way to gain their acceptance. Furthermore, even if you have not had the particular experience, you may be able to think of something in your background that gives you a hint of what your clients may be feeling. Again, we turn to the words of Daniel Robb (2001), who recalls being cut from a baseball team despite having superior skills:

> I turned and walked away from him, my good, dusty glove tucked under my arm, cursing the whole corrupt system.... It was clear to me at that moment that there was no trusting the men in charge. And I felt different.... I the kid who wasn't part of the team, the kid without a father. I had been cut.... Which all sounds maudlin and sappy until you remember the leverage of those days, the ferocity of your little peers, their cruelties and name callings.... The dangers of trust, the betrayals, the sweetness of being accepted and the tangles of rejection. These boys out here on the island, they were given no acceptance in the deeply psychological sense.... They're here because they were cut. (p. 56)

Learning to Accept Clients

Your ability to accept your clients is just as important to the relationship as their ability to accept you. In fact, it is more important. If you cannot accept your clients for the people they are, they will know it in some way and on some level, and they will be far less likely to accept or trust you. Really, we are talking about a process of mutual acceptance here, and much of the work you did to get your clients to accept you will help you to accept them as well, especially the work you did to understand their perceptual world.

Working with Troubling Behaviors

As we mentioned earlier, one important part of acceptance is to be able to see clients as more than their behaviors. Listed next are some concerns drawn from our experience and that of interns and beginning helpers (Cherniss, 1980; M. S. Corey & Corey, 2011; Russo, 1993). As you review these behaviors, think about how you might react to clients who:

- lie to you
- manipulate you to get something they want but cannot have
- are never satisfied with what you have to give and always seem to need more

- become verbally abusive and physically threatening
- blame everyone else for their problems
- are sullen and give, at most, one-word answers or responses
- ask again and again for suggestions and then reject every one
- refuse to see their behavior as a problem
- make it clear they don't like you
- refuse to work with you

Can you see these behaviors as understandable, although not helpful or praiseworthy, given what you know about your clients? Can you stay open to clients who do these things and see their behavior as only one part of who they are? If not, take some time to consider why.

Understanding Your Reactions

The most important thing you can do is to try and understand your reactions. Remember, just as clients bring their past experiences and emotional tendencies to the relationship, so do you. *Countertransference* is the term counselors and therapists in the helping professions use to describe situations in which your perceptions of and reactions to clients are distorted by your own past experiences and hurts (M. S. Corey & Corey, 2011). Look back to Chapter 4 and reconsider some of the things you discovered or affirmed about yourself. Do any of those things help explain your reactions? For example, think about the reaction patterns you have. If you have trouble confronting others, then clients who challenge you overtly may be especially difficult for you to work with. Knowing this will help you avoid assigning all the blame to the client. If your psychosocial identity includes a shaky sense of competence at this time, then clients who are not making progress may actually anger you more than they otherwise would, because of *your need* for them to make progress.

Cultural Competence: Knowing Your Identities

Some clients evoked feelings of mistrust and prejudice as well as feelings of sorrow. I am most shameful of the feelings of prejudice.
    STUDENT REFLECTION

Your cultural identities may help explain some of the reactions you are having as well. Consider the preceding quote from a student's journal. This quote shows an intern who is willing to look to herself, as well as to the client, to discover the source of her reaction. The willingness to admit prejudice is especially impressive. In time, this intern may well move past feelings of shame and be able to work on ways to overcome the prejudice and mistrust she lives with. Admitting these feelings, instead of pretending not to have them, is an engaged first step.

Dealing with Self-Disclosure

Another important issue in developing relationships, and a frequent concern for interns, is self-disclosure. How much about yourself should you reveal? There are actually two kinds of self-disclosure to consider: personal information and personal reactions.

Disclosure of Personal Information   It is natural for clients to want to get to know you, and they may ask you questions about your life and relationships. Some of these questions will seem quite comfortable and easy to answer; others may not. You may feel that the questions are too personal or that they concern matters you would rather keep private. You may feel that certain information is inappropriate for certain clients.

A Word to the Wise...   These are issues you should think about but not decide on your own. Your agency may have strict policies, and very good reasons for them, about how much you can disclose, and they may not always seem obvious to you. In some placements, interns are told not to have personal photos on display or divulge their last names. Matters of personal safety (which we will consider later) and client boundaries sometimes dictate limits such as these. Be sure to discuss these issues with your supervisor.

Personal Reactions & 3DQs   The second kind of self-disclosure is more immediate. You will no doubt have opinions about and emotional reactions to things your clients say or events at the site and wonder whether it is appropriate to share your thoughts with clients. The question *"What should I share with clients?"* is not possible to answer and, in fact, is not a helpful question. It fails to consider the wide variety of clients, situations, and goals that form the context of your work. It is better to ask a "three-dimensional question" (Hunt & Sullivan, 1974), such as *"What sort of self-disclosure is appropriate with which clients and for what purpose?"* If you are working with battered women, for example, it may be appropriate to share some of your relationship struggles as a way of establishing an empathic connection; clients sometimes think their counselors have no problems of their own. However, if you are working with a heterosexual teen of the opposite sex, such disclosure is probably inappropriate, as it can blur an important boundary and create confusion about your intentions. Asking three-dimensional questions (3DQs) will help you, in collaboration with your supervisor, to find answers that work for you and your clients.

Managing Value Differences

In Chapter 4, we stressed the importance of being aware of your values. In their book *Issues and Ethics in the Helping Professions*, G. Corey, Corey, and Callanan (2011) stress the importance of values awareness in relationships with clients. There will be times when you are dealing with a client whose values are very different from yours, and that can be perplexing and stressful for both of you. Let's suppose that honesty and straightforwardness are strong values for you. Of course, you are dishonest occasionally; you might even believe it is wrong when you are dishonest, but it makes sense to you under the circumstances. Let's also suppose that you are dealing with a client whose values differ from yours. In her experience, being honest, especially with human service workers and others in authority, means being taken advantage of by a service delivery system she perceives as unfair. For example, in some states a woman on welfare will have her benefits reduced if she is married, even if her husband is not employed. So, she may lie to you as a way of solving an anticipated problem.

THINK About It

How Will You Respond to Value Differences?

Let's take the example of sexuality and family values—hardly light topics, but certainly ones you will encounter in your work with clients. You certainly may encounter clients whose values are at odds with yours in these areas. Think about how you might respond to each of the following case situations and why you would respond in that way.

- You are working with an adolescent client who is sexually promiscuous and thinks it is just fine. Furthermore, he uses no birth control and says a pregnancy would not be his responsibility.
- You are working with another client who thinks that a marriage must stay together at all costs, and you are in favor of divorce in some cases.
- You are working with a client who thinks that unmarried couples should not have children.

One of the challenging aspects of situations in which it is obvious that there are differences between you and your client is deciding how to respond. And, important to deciding on how to respond is understanding the options that you have. G. Corey et al. (2011) make a distinction between *exposing* your values to clients and *imposing* your values on them. One choice you do have is exposing differences by telling clients about them. Another choice you have is imposing your values by attempting to influence clients to change theirs. And a third choice is to do neither, but to work with your differences.

Specific Client Issues

As you negotiate acceptance with your clients, there are a number of issues that seem to arise frequently as you launch your relationships with them. These issues tend to permeate the helping relationship and thus can be formidable companions in the helping relationship. Being aware of them will help you prepare and not be surprised when they appear.

Authority Issues

One particularly challenging issue concerns authority. You have certainly been in a position where you were subject to authority and so have those you serve. Authority is often an important issue in forming a relationship with them. Shulman (1983) has pointed out that clients tend to perceive you as an authority figure. This is especially true if you or your organization has some kind of legal authority over clients, such as the authority a benefits worker has to deny benefits or a probation officer has to surrender a client to the court. However, even when the authority is not explicit, you should be prepared for clients who see and react to you as an authority figure (Shulman, 1983).

In some settings, clients are in a "one-down" position. They need something they cannot get for themselves. It could be something tangible, such as food or shelter, or something less concrete, such as control over a substance abuse problem or help in understanding some bureaucratic procedure. You and your agency have what they need; you are holding the cards. Think about when you have been in a similar situation. How did it feel? Your clients will bring their fears about past experiences with authority to their relationship with you, and those factors will shape how they respond.

Finding Equalizers

One way clients try to reduce their one-down feelings is to assess your background and experience, often with a goal of finding a flaw or an equalizer. We refer to this phenomenon as *credentialing*, and it takes place in nearly every internship. Sometimes, clients will literally ask about your credentials, as when they ask about your education and training or experience with clients. Other times, the credentialing is more subtle, such as, "Do you have children?" or "Have you ever been arrested?" It may be as simple as asking about your age. Sometimes, clients are just trying to get to know you, but you may be surprised at the persistence with which these questions are asked. You may also be surprised by their reaction when they find the information they are looking for, and you may feel dismissed. Knowing that these assessments are a normal part of building a relationship will help you decide how to respond.

Testing the Limits!

Clients also will often test your limits. They want to see where your personal boundaries are and whether and how you enforce agency rules. Because you are an intern, they may be genuinely unsure of your role and what you can offer them (Shulman, 1983). Some clients have experienced many workers, not to mention other people in authority, and have been treated in many different ways. They need to know what to expect from you, and although they may not like it when you set a limit, when you do so with firmness, compassion, and consistency, it ultimately helps them trust you. Other reasons for testing you may include a need for recognition and attention or an attempt to gain status with their peers (Shulman, 1983).

You may think of this phenomenon as occurring more with children, as they try repeatedly to pick up something they have been told to leave alone or try to poke or push you. You may also associate testing with people who are in an involuntary situation, such as juveniles in a detention center or runaway shelter, who refuse to do chores or curse in front of you to see how you react. And, in fact, it is a very real issue in some criminal justice internships, where interns are often referred to as *officers in training* precisely so they are not seen as easy targets for limit testing. However, other kinds of clients can test you, too. An elderly client can press you to stay longer than you are able, a client at a soup kitchen may try to jump in line twice, or a parent may ask you to stay later, beyond the appointment, to watch the children until their ride arrives.

Interns are sometimes intentionally tested because they are initially seen as "not real staff" (Gordon & McBride, 2011). These behaviors can be exasperating, especially if you are not expecting them. Try not to imagine that really effective workers never have these challenges; of course they do. They may have learned over time to handle them a bit more smoothly and quickly—and you will, too, someday. Taking these challenges personally only makes it harder to meet them effectively. Remember that you once did this kind of testing, and you may still do it occasionally (Russo, 1993). Think back for a moment on your behavior with substitute teachers or babysitters. Perhaps you have even tested some of your college faculty in this way.

Personal Safety of the Professional

Safety is definitely a concern, since I work in a courthouse with metal detectors and my office is locked at all times. I know my level of risk is higher, but then again, I sometimes feel safer knowing that there are detectors, guards, locked doors, etc. I guess how safe I feel really depends on how I choose to look at things.

STUDENT REFLECTION

There are a number of people interested in your personal safety in the internship. Your safety is foremost in the minds of your campus and site supervisors. It is the concern of your co-workers and your campus administrators. Increasingly, we hear interns themselves express concerns about their personal safety, and we are not alone. Birkenmaier and Berg-Weger (2007, p. 57) cite several studies of real and perceived risk to social work interns and note that such concern is warranted because, among other reasons, (a) social workers are second to police officers in the risk of work-related violence directed at them, and (b) the "number and lethality of safety risk incidents on the job has increased for social workers."

Certainly, the potential for risk in the helping professions can be high. Human services is such a broad field of professions, though, that it is impossible to make general statements about risk. Here, however, are some examples, drawn from our experience, of internships with various levels of risk.

- Steven interns at a men's shelter in the city. It is the only shelter in the city that allows men who are actively abusing substances to be sheltered, and many men come in under the influence. Others congregate outside, seeking services only in a medical emergency. Furthermore, Steven is asked to ride along with an outreach worker who travels to abandoned buildings to try and help addicts be safer in their behavior and perhaps to come for treatment.

- Carlos interns for an agency that monitors adjudicated youth in the community. He travels with a staff member all over the city, to schools, playgrounds, and apartments, to check on these adolescents and talk with them.

- Su-Je interns at a Student Assistance Center at a huge, urban high school. The school has students of every race and from dozens of ethnic backgrounds, and intergroup tensions run high. Confrontations and fistfights are not uncommon. The center offers peer mediation, among other services, but sometimes these sessions erupt.

- Kavita interns at a women's shelter. She helps a staff member facilitate a support group for victims of domestic violence. One night, one of the clients tells the group that she is scared. Her abuser found out she was coming to the group and got very angry. He found her journal, and in it were names of clients and staff members. She snuck out to group that night but is frightened about what he will do when he gets home and finds her gone.

Hold these examples in your thoughts; we return to them as we continue our discussion of safety issues.

Assessing and Minimizing Levels of Risk

The personal safety chapter couldn't have come at a better time. A social worker in my office was attacked this week by one of her clients. She was hit on the head and has permanent damage to her eyesight. I hadn't thought much about my personal safety until this point.

STUDENT REFLECTION

Your level of risk depends on several variables, some of which are evident in the preceding examples. Although you should be made aware of any and all potential risks to your safety before you start working with clients, safety issues do arise that cannot be anticipated. What is important here is that you are as fully informed as possible about your safety.

A Matter of Law

Campus Liability for Internships

Student safety is the issue getting the most publicity when it comes to off-campus programs. Why? Because students are injured or killed while enrolled in such off-site programs as internships and study abroad. If an internship is required of a student for completion of a degree requirement, as it often is in the academic programs of the helping professions, then the college may face liability if the student is injured or killed while involved in the internship.

A Case in Point: A 2000 ruling by the Florida Supreme Court has implications for campuses requiring an internship in an academic program. The ruling resulted from the Court's review of a decision on this question:

"Whether a university may be found liable in tort where it assigns a student to an internship site which it knows to be unreasonably dangerous but gives no warning, or adequate warning, to the student, and the student is subsequently injured while participating in the internship?"

The campus in question is a private institution, Nova Southeastern University (NSU), and it offers a doctorate program in psychology with a required 11 month internship referred to as a "practicum." The approved internship sites for the program are provided to the doctoral students in list form, and the students must select 6 of the approved sites for the internship; the campus then places the student at one of the 6 sites. The doctoral student who brought the lawsuit was a 23 year old female who was accosted in the parking lot of the site after getting into her car one evening—"abducted from the parking lot, robbed, and sexually assaulted." Evidence indicated that the campus had been aware of a history of "criminal incidents which had occurred at or near the...parking lot." A lawsuit resulted and the court eventually ruled in the student's favor, noting that the campus had a duty "to use ordinary care in providing educational services and programs to one of its adults students" and that it "also assumed the Hohfeldian correlative duty of acting reasonably in making those assignments." Nova Southeastern University Inc. v. Gross, 758 SO. 2d 86 (Fla. 2000).

Assessing Client Risk Levels

Before beginning your internship, you should have a good sense of the likelihood of being exposed to violence. If you are interning in a residential setting, a locked facility, or in the field of criminal justice, the likelihood is higher than if you are interning in the maternity wing of a hospital. And you certainly should have been informed of that by the campus person who placed you with your agency. As we mentioned in Chapter 3, when it comes to clients, three factors should be considered: the client's developmental stage, motivations, and immediate situational factors and conditions (Baird, 2011, 157–159). Those factors include access to weapons, mental status, medication/controlled substances, stress level and precipitators, and history of violence (pp. 159–160). It's important to keep in mind that some clients are habitually violent, whereas other clients may become violent only under certain circumstances, such as the influence of drugs or the lack thereof. And unfortunately, some clients are in terribly frustrating and even desperate circumstances (Garthwait, 2010) and are not necessarily predictable when it comes to their behaviors. For example, if one of the adjudicated youths that Carlos is monitoring has gotten into more trouble and believes that he will be turned in, he may go to great lengths to stop that from happening. Non-clients are sometimes a risk as well. Parents whose children have been removed from the home, partners of victims of domestic violence, and friends of an adjudicated youth are all examples. If your placement site has not educated you about these issues, it is especially important that you be proactive and find out about them.

Minimizing Client Risk Levels

As we emphasized in Chapter 3, if you suspect the possibility of violence by clients or their family and friends, it is critical that you take an engaged approach and consult with others immediately, starting with the campus placement coordinator who arranged your placement, both of your supervisors, and your co-workers as to the history of violence at the field site. You may have ignored this suggestion the first time we made it; if so, we remind you that history remains the most dependable predictor of future behaviors. If there is a history of violence, it is time to meet with your supervisors and develop a safety plan. Because it is a safety factor, it needs to be noted on the Learning Contract with specific safeguards described.

Assessing Site and Community Risk Levels

There are organizational variables as well that can pose safety risks. As we said in Chapter 3, organizations vary considerably in the extent to which they have assessed risk to interns and workers, as well as have implemented safeguard procedures. For example, the high school where Su-Je interns gave her and all the staff careful instructions about what to do in case of a violent incident, when to respond, and how to get backup when needed. Organizations have the responsibility to be thorough in their assessment of risk factors and levels, and respond accordingly with policies and procedures for staying safe. Your responsibility is to determine how well your site meets these criteria for safety.

Location and hours of operation are another set of variables. Some human service agencies—and schools—are located in risky neighborhoods (i.e., at risk for violence or crime), especially for someone who does not live in that community. The neighborhoods that Steven travels to are more dangerous after dark. If your work takes you out into the community, as Carlos's and Steven's does, you may find yourself in such neighborhoods as well and will need to put necessary precautions in place.

Minimizing Site and Community Risk Levels

If you in any way feel unsafe or otherwise not comfortable traveling to and from your site due to location, hours of operation, or security measures in or near the building, again, take an engaged approach and consult with the site supervisor and co-workers immediately to determine how they ensure their personal safety working at the site. Be sure to inquire about the history of safety as it relates to your concern, requesting specific data about crime in the area at night and past crimes committed in the parking area or on the premises of the site. Of course, you can always contact the local police department and request such information if the staff is too new to the site to have it. If your concerns prove to be a safety factor, they need to be noted on the Learning Contract with specific safeguards described.

Facing the Fears

In order for you to embrace your field experience for the learning experience that it is, you will need to face your fears and assess the level of risk at your internship in a proactive way—even if your agency has not done that or not discussed it with you. It becomes your responsibility by default because it is *your* internship. And we know that may not be easy to do. Many interns are reluctant to discuss their concerns about safety for fear of being seen as overly timid, not ready, or not committed to the field (Birkenmaier & Berg-Weger, 2010).

Fear is a powerful emotion. In this instance, it can keep you from confronting your prejudices, if you have or are aware of them, and from making realistic assessments of your safety. It may well be that some of your fears are unfounded or exaggerated because of prejudices and stereotypes. For example, some persons with mental illness or developmental delays may exhibit some unusual behaviors, but that does not necessarily make them dangerous (Baird, 2011). A Caucasian intern in a predominantly Hispanic high school is not necessarily at greater risk for violence, but he may *feel* like he is. The only way to deal with these stereotypes is to acknowledge and confront them. If you are not comfortable discussing these issues with your supervisor, you may want to talk

Focus on THEORY

Russo's Patterns of Adjustment (1993)

Over the course of many years in the field, there are adjustments to be made. There are several ways to cope with the ongoing demands—both physical and emotional—of the work and the day-to-day strains and frustrations that are as much a part of the work as the joys and satisfactions. Russo has described several patterns of adjustment that are found in experienced workers: those who identify with the clients, with their co-workers, and with the organization. Russo emphasizes that these are not static categories; people often move from one to another and back over the course of their careers. It is probably far too early in your own career to determine which category fits you, but knowledge of them may help you make sense of what you are seeing at your placement.

Identifying with Clients   Workers who identify primarily with clients can be further divided into four subcategories. *Reformers* tend to be impatient with anything that they believe interferes with their ability to serve the clients. They often neglect paperwork and will try to change the organization's policies and procedures to better meet the needs of the client (i.e., people before paper). In our experience, this is a very common stance for interns, although they do not actively try to change the placement site. *Innovators* still promote change, but are more patient with and understanding of the change process. They listen, they ask questions, and they work with others to try to find the best way to achieve change. *Victims*, on the other hand, are frustrated by systems they see as inadequate and even harmful. They see themselves as the protectors of the clients and are likely to battle the administration openly, sometimes enlisting clients (or interns) in their struggle. Finally, *plodders* identify with the clients and may have some of the same concerns and frustrations as some of the other types described, but they seem to have given up on change. They work quietly with their clients, doing the best job they can. They often have their own way of doing things, and they work without making waves, and do not try to influence others or the organization itself to change.

Identifying with Co-Workers   Workers who identify primarily with their co-workers also care about clients, but in addition, they feel a strong allegiance to their profession as a whole. They may be active in unions and/or professional organizations and look to these groups, as opposed to their particular workplace, as their primary guides. Some relatively new workers are attracted to this stance because they are unsure of their own skills and knowledge. They follow the rules of the professional organization rigidly. Other workers in this category are more sure of themselves and regularly consult their union or professional organization for guidance, but consider these groups as one of several sources of wisdom.

(continued)

Focus on THEORY *(continued)*

Identifying with the Organization Workers who identify with the organization look to the organization and its policies and procedures as their primary source of guidance. Even though those rules may sometimes work against the needs of a particular client or be in violation of standards or ethical codes issued by professional organizations, these workers believe that following the rules will do the most good for the most people in the long run. Some may be hiding behind the rules so they do not have to think hard or take risks. Others have adopted this stance after careful thought and reflection on their experience. Russo also points out that some of them are conflict avoiders. They realize that the needs of clients, the organization, and the profession can sometimes conflict, but they want to resolve those conflicts quickly, and adherence to the rules is one way to accomplish that.

---

with your campus instructor instead or with your classmates. What's important is that you do discuss them with someone.

The likelihood is that once you have thought through your fears and gathered the factual information you need, you will feel ready to face what challenges there may be. If that is not true, if you feel like you are in over your head, you must discuss that with your campus instructor as soon as possible.

Conclusion

This section on clients may have raised more questions than answers for you. If so, we accomplished our goal, for that was the point of the chapter. It is very important that you are aware of and prepared for some of the challenges in the early stages of working with clients. The exact shape and pace of your experiences with clients we cannot know, but you can learn about them. You have materials for reflection and for discussion with peers, and you have unanswered questions to explore.

We have encouraged you to explore the basic dynamics of your clients, and the work you need to personally do to be prepared to work effectively with them. In so doing, you have addressed a number of the concerns and issues of the Anticipation stage that arise when working in the helping professions. With effort, and a little luck, you will feel more settled in your placement, more committed to the work, and more ready to move on to the Exploration stage of the internship.

For Review & Reflection

Checking In

Changes and Room for More. In what ways has your work affected you personally and professionally? What expectations are being upheld and what surprises are you having? Which of your responses suggests that more work needs to be done by you?

Experience Matters (For the EXperienced Intern)

If you have had other internships, a career or a job with longevity in the helping professions, contrast your reactions and experiences to working with the current group of clients with your reactions and experiences working with populations in the past. What's different? How are those differences affecting you and your work? What lessons from the past do you take onto your internship today?

Personal Ponderings (Select the Most Meaningful)

1. Now that you have gotten to know better those you serve, in what ways have your initial impressions of your clients as a group changed? Do any of your initial impressions bother you now as you look back on them, and if so, why?

2. As you think about yourself in relation to the work, which of your strengths will be useful to the work that you will be doing?

3. The following two inquiries may seem simple on the surface, but they are not. Give thought to how you and your clients differ. Then think about the ways in which you and your clients are the same. What surprised you about your responses?

4. Take this time to think about the word "empowerment." Now consider this statement: *You cannot empower anyone; only a person can empower him or her self. The most you can do as a helping professional is create a context for someone to empower himself or herself.* What are the implications of this statement for your approach to the work you are doing?

5. In what ways have your credentials been challenged already? When you think about the ways you reacted—noticeably and privately—what did you learn about yourself? How will you handle such a situation differently in the future given how you handled it up to this point?

6. What challenges you most about setting the necessary limits your clients need? How much of that difficulty is because of what you know

about the people you serve? How much is because of what you know about yourself?

7. How compatible is your own position on personal disclosures to clients and their families with the policy of your field site? Compare how you've responded to requests for disclosure up to this point with how you will respond in the future.

8. Think about the challenges you face beyond the ones identified in these questions. Why do you think they are challenges to you and may not be challenges to your peers? What engaged approaches can you use to move them from being challenges to being manageable issues?

Civically Speaking

Given that your choice of internship and perhaps life work is in the helping and service professions, you work in or with community issues very frequently if not all the time. What are you realizing about your civic awareness and knowledge now that you are actually in the field and living the work you want to do?

Seminar Springboards

- *Knowing Stereotypes.* As a class, make a list of the populations that you are working with. Brainstorm a list of the more common stereotypes that society has about each group. Remember that you are not being asked whether *you* subscribe to these stereotypes, but rather to identify that to which society at large subscribes. These you must know to do your work effectively. Once you have made a list of the stereotypes, which can be painful to look at in print, then think of the implications of those biases for your work and for your clients' lives.

- *Managing Safety.* As a group, put aside some time to talk about the safety issues that each of you faces. Frame your discussion by focusing on risk factors, risk levels, and safeguards that need to be in place. Remember, all risk levels that warrant safeguards need to be brought to the attention of your campus supervisor. An important piece of that discussion is acknowledging the feelings that the risk levels evoke and how you will manage them so you thrive instead of surviving each day in placement.

For Further Exploration

Birkenmaier, J., & Berg-Weger, M. (2010). *The practical companion for social work: Integrating class and field work* (3rd ed). Needham Heights, MA: Allyn & Bacon.

Thorough chapter on personal safety, with helpful exercises and inventories.

Corey, M. S., & Corey, G. (2011). *Becoming a helper* (7th ed.). Belmont, CA: Brooks/Cole.

Helpful sections on client issues.

Corey, G., Corey, M. S., & Callanan, P. (2011). *Issues and ethics in the helping professions* (8th ed.). Belmont, CA: Brooks/Cole.

Excellent section on the ethics of imposing values on clients.

Gordon, G. R., & McBride, R. B. (2011). *Criminal justice internships: Theory into practice* (7th ed.). Cincinnati, OH: Anderson Publishing.

Excellent discussions of issues relating to clients. Although this book is aimed at interns in a particular setting, it has many applications outside criminal justice.

Levine, J. (2112). *Working with people: The helping process* (9th ed.). New York: Longman.

A comprehensive text written for social work students that focuses on the theory of human service practice, the development of practice skills, advocacy work, and the importance of reflection in the work of the helping professions.

Northwestern Mutual Life Insurance and Financial Services. (2012). Community Service Award (www.northwesternmutual.com).

Northwestern Mutual financial representative interns don't just get to test-drive the full-time career; they can also get engaged in the company's commitment to volunteerism. Through its annual Community Service Award (http://www.northwesternmutual.com/about-northwestern-mutual/our-company/ northwestern-mutual-foundation.aspx#Employees-and-Field) Northwestern Mutual interns like Chad Monheim (http://www.youtube.com/watch?v=1ne4qft8gy4& feature=plcp) can get recognition—and charitable funding—for the philanthropic causes they're passionate about. Go to http://www.northwesternmutual. com/career-opportunities/financial-representative-internship/default.aspx to learn more about Northwestern Mutual financial representative internships.

Pedersen, P. (Ed.). (1985). *Handbook of cross cultural counseling and therapy.* Westport, CT: Greenwood Press.

Another good resource on multicultural issues. Triad model is helpful.

Pedersen, P. (1988). *A handbook for developing multicultural awareness.* Alexandria, VA: American Association for Counseling and Development.

A very "hands-on" short book with lots of exercises to help you.

Schutz, W. (1967). *Joy.* New York: Grove Press.

In his first stage of group development—inclusion—Schutz talks a great deal about acceptance concerns and the various ways of handling them.

Shulman, L. (1983). *Teaching the helping skills: A field instructor's guide.* Itasca, IL: F. E. Peacock.

Especially helpful regarding authority issues.

Sue, D. W., & Sue, D. (2007). *Counseling the culturally diverse* (5th ed). New York: John Wiley & Sons.

Classic text by one of the leaders in the field; informative and provocative.

The Washington Center for Internships and Academic Seminars. (2012) (www.twc.edu).

Students at The Washington Center (www.twc.edu) intern at government agencies, non-profits, think tanks and businesses such as the Department of Justice, Amnesty International, Merrill Lynch, No Labels and Voice of America. They also have the opportunity to become positive change agents by participating in civic engagement projects on important domestic and international issues, including domestic violence, homelessness, veterans, torture abolition and the environment. Students learn about these issues by interacting with national and local community leaders, then become involved in a direct service and/or advocacy project. They can choose to join a TWC-guided project or design their own individual one. For more information about civic engagement at The Washington Center, please see http://www.twc.edu/internships/washington-dc-programs/internship-experience/leadership-forum/civic-engagement-projects. Articles about recent civic engagement activities can be found at TWCNOW http://www.twc.edu/twcnow/news/term/civic%20engagement.

References

Baird, B. N. (2011). *The internship, practicum and field placement handbook: A guide for the helping professions* (6th ed.). Upper Saddle River, NJ: Prentice Hall.

Birkenmaier, J., & Berg-Weger, M. (2010). *The practical companion for social work: Integrating class and field work* (3rd ed.). Needham Heights, MA: Allyn & Bacon.

Cherniss, C. (1980). *Professional burnout in human service organizations.* New York: Praeger.

Connell, M. A., Franke, A. H., & Lee, B. A. (2001, February) Issues related to off campus programs. *Nobody said this was going to be easy: Legal and managerial challenges for department chairs and other academic administrators.* American Council on Education. Originally prepared for Stetson University Law School conference on law and higher education.

Corey, M. S., & Corey, G. (2011). *Becoming a helper* (6th ed.). Belmont, CA: Brooks/Cole.

Corey, G., Corey, M. S., & Callanan, P. (2010). *Issues and ethics in the helping professions* (8th ed.). Belmont, CA: Brooks/Cole.

Garthwait, C. L. (2010). *The social work practicum: A guide and workbook for students* (5th ed.). Upper Saddle River, NJ: Prentice Hall.

Gordon, G. R., & McBride, R. B. (2011). *Criminal justice internships: Theory into practice* (7th ed.). Cincinnati, OH: Anderson Publishing.

Hunt, D., & Sullivan, E. (1974). *Between psychology and education*. Hinsdale, IL: Dryden.

Ostrow, J. M. (1995). Self-consciousness and social position: On college students changing their minds about the homeless. *Qualitative Sociology*.

Robb, D. E. (2001). *Crossing the water: Eighteen months on an island working with troubled boys—A teacher's memoir*. New York: Simon and Schuster.

Royse, D., Dhooper, S. S., & Rompf, E. L. (2011). *Field instruction: A guide for social work students* (7th ed.). Boston: Allyn & Bacon.

Russo, J. R. (1993). *Serving and surviving as a human service worker* (2nd ed.). Prospect Heights, IL: Waveland Press.

Schmidt, J. J. (2002). *Intentional helping: A philosophy for proficient caring relationships*. Upper Saddle River, NJ: Prentice Hall.

Shulman, L. (1983). *Teaching the helping skills: A field instructor's guide*. Itasca, IL: F. E. Peacock.

RHYTHMS

In the preceding section, we helped you prepare for the beginning of an internship. You spent time developing your understanding of the Anticipation stage and the importance of the learning contract and supervision plan to the quality and vitality of your internship. For those of you interning in the service or helping professions, you also developed an understanding of the important issues when working with clients and were given suggestions for beginning your relationships with them. You emerged from that stage with a clearer vision of your work, a commitment to it, and an increased sense of confidence. In the process, you worked your way through a number of challenges that tend to arise early on in the internship, and now you know the sense of satisfaction and empowerment that comes from dealing with them effectively through engagement, effort, reflection, and reaching out for guidance and support.

This section of the book focuses on the Exploration stage of the internship and is designed to keep you and your internship moving forward. During the Exploration stage, you are involved heavily in learning to do what you went into the field to do. Many liken it to an apprenticeship, working closely with supervisors to learn the whats, hows, and whys, as well as the ins and outs of the work. To ensure that you are ready for this stage, you will examine the issues and challenges in key areas of your internship: the work, the people, the site, and the community. While doing so, you'll be developing your own rhythms...for the work, for how to get it done at your site, for key relationships, and for the connections you and your site have with the community. The process of finding your rhythm may be smooth and it may be bumpy; most interns experience some of both. But the bumps are just as much a learning opportunity as the smooth times; in fact, in many ways they are more powerful, although less enjoyable.

Chapter 8 focuses on the Exploration stage itself and the growth and development that are occurring as you embrace the learning of the work. You will revisit the Learning Contract to ensure that it remains viable through this crucial time in your internship. Informal and perhaps formal evaluations also begin occurring during this stage. In preparation for both fine-tuning your contract and your approach to the evaluation, we encourage thoughtful consideration of your personal, professional, and civic growth and those areas you plan to strengthen. We also explore the experience of disillusionment, because, in our experience, this is the stage where it is most likely to happen.

Chapter 9 is the Essentials chapter for this stage. It provides you with a number of advanced tools to revive, maintain, or enhance your engagement in the internship. The chapter includes a meta-model for creating change when disengagement occurs or if you have the experience of disillusionment. Chapters 10 and 11 help you widen your focus and your sense of context. Chapter 10 helps you become aware of and address your concerns about the organization as a whole, from both a systems and an organizational development perspective. Finally, in Chapter 11, you spend time thinking about the dynamics of the community in which your internship is being conducted and the relationship between that community, the internship site, the work you are doing, and yourself as an emerging civic professional.

Moving Ahead:
The Exploration Stage

I knew I would learn an incredible amount because of the actual hands-on experience that you cannot get from a book. But I was surprised at the impact the internship had on me and to learn how much I need to work on.
STUDENT REFLECTION

Y ou have dealt with the issues and concerns that accompany the first few weeks of the internship and moved successfully through the Anticipation stage. Now what? Do you relax and coast through to the end? Well, you might be able to do that (although it's not likely), but not only will you miss many learning opportunities if you do, you will miss a sense of satisfaction from completing a challenging internship. Do you wait, apprehensively, for problems to develop? Well, not if you want to shape your own learning experience. The problem with waiting is that it is a *disengaged* stance—passive and reactive, and one focus of this book is to help you become an *engaged* learner, shaping your own learning experience as much as possible.

In this chapter, we begin with some thoughts about growing and learning. We then encourage you to reconsider your Learning Contract. Even the best of contracts looks different after a few weeks at the site. You most likely have had your first experiences with being assessed and evaluated, formally or informally, and we offer some ways for you to think about that process. Then, we ask you to think about issues and challenges in the major arenas of your internship: the work, the people, the site, and the community. Finally, we look at the experience of *disillusionment* in this stage. In our experience, it is more likely to happen here than in other stages and if it happens to you, we want you to be prepared so you can respond efficiently and move through it effectively.

Focusing on Growth & Development

One of my anticipations was "growth." This I can feel every day.
I can feel myself growing as a person as well as a professional.
STUDENT REFLECTION

During this second stage of your internship, the focus is on growth and development, change, adjustment, and problem solving. Certainly these themes are present in every stage, but in this stage, they tend to be more in the foreground. The internship is no longer new, and at least some of the challenges of a good beginning have been met. As you move from being new at this to being experienced, you will need to take action to keep yourself moving forward; and you need to learn how to react to aspects of the internship that may not unfold in the way you expected.

A Time of Change

The setting may change but I continue to be who I am for the most part while I can sense some change within myself. Others are who they are. The field experience is rich because what may seem familiar experiences are reviewed with a different perspective....They are getting a twist and that is good.

Conventional wisdom says work smarter, not harder. Word of the day—resiliency....Feel the emotion, bounce and continue. Change is happening.
STUDENT REFLECTIONS

Moving on and changing are integral parts of the internship process. You expect them to happen, as do your site supervisor and campus instructor(s), so much so that if they do not, it can be a cause for consternation. *Moving on* suggests taking on additional responsibilities that further test and develop your knowledge and skills, as well as shape and test your attitudes and values. You seek out greater challenges and strive to meet them, often with considerable satisfaction. *Changing,* on the other hand, suggests that an internship affects you in more noticeable and pervasive ways. During the first seminar class, just after internship begins, we often ask our interns the question that we pose in the box below.

THINK About It

Becoming You!

What would you say if we told you that the person you are today, reading this passage, is probably not the person you will be when you complete your internship?

Our interns tend to look at us very strangely and quietly when we pose this question. How about you? What is your reaction? Would your reaction be different if the question were asked at the very beginning of your placement? Maybe the question strikes you as a little dramatic, and perhaps it is, but change involves something more substantial, more far-reaching, and more challenging than just taking on new responsibilities, learning new theories, or improving your skills. It involves looking at yourself, your work, the placement site, other organizations, civic responsibility, and even society in new ways. This integration and reintegration of attitudes, values, skills, and knowledge across the personal, professional, and civic dimensions are the hallmarks of a *civic professional*, and it is not solely an intellectual process. Change of this sort can be enormously exciting and frightening at the same time.

Focus on THEORY

The Essentials of the Change Process

There are natural tensions involved in change, and you are bound to feel them. According to Jon Wagner (1981), you deal with two sets of perspectives during your internship: the perspectives you *bring* to the field and the ones you are *developing* as a result of your work. On the one hand, you are drawn to the reality of the work, its vitality, its unpredictability, and its dynamic yet problematic character; on the other hand, you seek the comfort of "home," where you know the situations and the problems, and where you are known to others. Your challenge is to bring these two sets of perspectives together in such a way that learning goes on. If you keep both your feet safely planted at home—concentrating only on how you currently think and what you already know—you underextend yourself in the experience and risk not growing; however, if you bury yourself in unfamiliar soil, you overextend yourself and risk sinking in the experience.

Resolving these differences requires that you embrace both perspectives at the same time, "straddling them" in Wagner's (1981) words, so as to create a perspective or vision that incorporates the past as well as the emerging ways of looking at things. Sounds great, right? It can be, but William Perry reminds us of two opposing human urges: the urge to progress *toward* maturity and the countering urge to *conserve* (1970). These competing emotions accompany the two perspectives just discussed and can interfere with their integration. Similarly, Robert Kegan believes that all change involves letting go of old familiar ways, and that can be frightening (1982). So, expect some excitement as well as some trepidation, and remember that both these emotions are normal to have and can help you grow.

A Time of Adjustment

As your internship moves forward, there are bound to be adjustments for all kinds of reasons, many of which we will discuss later in this chapter. Sometimes an adjustment is simple; other times it is upsetting.

Focus on THEORY

Schlossberg's Transition Theory

The internship is a recognized high-impact learning activity (Kuh, 2008) that creates a context in which students are able to learn, grow, and develop in a number of ways, especially on the personal, professional, and civic dimensions. Internships often are the capstone experiences in academic programs and usher in transitions that students feel are occurring but may not have expected. Schlossberg, Waters, and Goodman (1998) offer a way of framing such transitions to understand them and ways to connect to helpful resources. We've summarized this theory into its "bare bones" and encourage you to explore it further on your own. As you peruse this summary, think about the transitions you've experienced or are experiencing and how this theory of transition can be useful to you in your internship (Evans, Forney, & Guido-Dibrito [1998]; Schlossberg, Waters, & Goodman [1995]).

- *Transitions are a process that take time*   They can be events or non-events that result in changes (to relationships, roles, routines, and assumptions), and they can be positive or negative and evoke associated feelings. There are *anticipated, unanticipated,* and *non-event* types of transitions, which are determined by the student's relationship to the setting and to the transition itself. The impact is measured by how the changes affect the student's daily life. Transitions usually occur as a series of phases referred to as *moving in, moving through,* and *moving out.*

- *The Four Factors Affecting the Ability to Cope with Transitions*
  - **Situational factors,** which involve a number of variables, including triggers, timing, control over situations, role changes, duration, previous experience with similar situations, other stresses at the time, and source of responsibility and have an affect on students.
  - **Self factors,** which involve *personal* as well as *demographic characteristics* that affect the student's perspective and *psychological resources* such as outlook, ego development, commitment, and values.
  - **Support factors,** which involve *types* of relationships (personal, institutional, communities, familial), the *functions* they serve in the student's life, and their *stability.*
  - **Strategy/Coping factors,** which involve *categories* (those that change, control the meaning of the problem, or aid in managing stress afterwards) and *coping models* (seeking information, taking direct action, inhibiting action, etc.).

One common impetus for adjustment—and source of distress—is the gap between expectations and reality. There is almost always a difference between what you anticipated about the internship and what you actually experience in the field. Fortunately, if you acknowledged and dealt with your concerns in the Anticipation stage, you will be less likely to have a wildly different reality now from what you expected. However, there will be discrepancies between your expectations and the realities of the placement, some of which may be troubling. It may be that the issues and personalities are not what you expected or that you are reacting differently than you thought you would. In addition, issues will arise in your placement that you simply never considered or never knew existed. These issues derive from that dimension of knowledge about which you don't even know you don't know...if you know what we mean! One student said, *"What I don't know I don't know, changes ALL the questions."*

After you have settled into the internship, people often expect more of you, and that can be hard at times. Take, for example, the case of the business intern who has stellar performance early on. Because her work was so outstanding in the first five weeks of her placement, she ended up being well over her head with the volume of responsibilities she was willing to take on. As her peers observed, she needed to put the brakes on and regain control of the internship: *"I feel as if they will look down upon me (if I slow down), and I will get a bad grade. Plus, I don't know what my priorities are.... It's making me stressed. There are too many things and I have no focus."*

Remaining Engaged

Remaining engaged during a time of change, adjustment, and deepening feelings is a challenge for many interns, even the most sophisticated. To help with this challenge, we take you back to the three dimensions of being an engaged learner developed by Swaner (2012) and include in our discussion of engagement in Chapter 3: *holistic, integrative,* and *contextual.*

Holistic Engagement: Thinking and Feeling

Holistic engagement, you may recall, involves balancing the cognitive and the affective (the thinking and the feeling) aspects of your internship. One way to make sure you are attending to the *human* side of the internship is to remain open to people by listening and considering carefully, resisting the temptation to categorize and label people. You also need to remain open to the relationships in your internship and the joys and challenges they bring. Finally, you need to remain open to your own emotional experiences and the opportunities they present. The more open you are to your feelings, even those you wish you didn't have or think you're not supposed to have, the more you will be open to the emotional experiences of others and the more you can use all those emotions as engines for growth.

Integrative Engagement: Reflection and Action

Integrative engagement means keeping yourself moving forward using Kolb's (1984) Experiential Learning Cycle described in Chapter 1. By all means, mobilize your own resources and those around you so you can seek and have new experiences. Remember, though, that action and experience are only part of the picture. The need to *do* is very high for most interns, as you have probably discovered, and the needs of placement sites can also be high due to the sheer volume of work. However, the activities of processing and reflecting are equally critical to your professional development and learning. They need to be consistent and structured parts of your internship experience. The balance between the *action of* the internship (the work tasks of the internship and related activities/relationships) and *reflection on* the internship (thinking about that "action") needs to be maintained throughout your internship and these two processes should affect and inform one another. When that balance is compromised, so is the overall quality of what you are learning in your field experience.

THINK About It

Just How Engaged Are You?

In Chapter 3, we outlined some characteristics of an engaged learner (p. 50). Now might be a good time to look back at that list and assess the extent to which you are displaying those characteristics as you move through your internship. For some interns, it is valuable to assign themselves a rating on each characteristic and congratulate themselves on what they have done well. It's important for all interns to make a plan for improving in areas where they may be struggling.

Contextual Engagement: Widening the Context

Depending on the nature of your placement, it may be time to think about *contextual engagement*—the larger contexts in which your work takes place. Some of you are working directly in the community; others are working with communities as advocates, organizers, contributors, or investors. For you, the widening of the context already has begun. If so, this may be the time to consider the larger political contexts—regional, national, and global—for the work you are doing. For those of you who are not working in the community directly, it may be time to think about the organization you work with and how organizational dynamics affect your work. Chapters 10 and 11 will explore organizational and community dynamics in some depth, and if you are curious there is no reason you can't read ahead.

Through the Lens of the Civic Professional

Remember, understanding the wider context of the work you and the staff do at the internship site is an integral part of becoming a civic professional. Each profession has an implicit contract with society. Some professions exist only to serve society, and they are funded largely by society because of the public value placed on that service. Even those professions, however, must grapple with the nature of their social mission or contract. For example, there is a history of debate within the field of criminal justice about whether its primary purpose is to protect society in the short term by incarcerating, monitoring, and/or punishing those who have committed crimes or to try and rehabilitate those who have committed crimes so that they may become productive, contributing citizens in the long term. Regardless of what you believe about that issue, an internship in criminal justice is an opportunity to explore it. Your internship, whether in literature, advertising, visuals arts, environmental science or public service provides you with an "authentic occasion" to discuss with members of the internship community issues concerning the public good (Newmann, 1989) and what your responsibility is to it as an intern at your site.

All professions also have ethical and moral obligations to the society in which they function, and the work of each professional is by definition connected to a larger social purpose. Journalism should be about more than entertainment; a free press should be an anchor of a healthy democracy. Even the intensely private domain of business can be seen as a public good as well as a private benefit (Colby, Erlich, Sullivan, & Dolle, 2011; Waddock & Post, 2000). Business educators have argued that those entering the business world need to understand that business is about more than maximizing profits and creating wealth; that corporations ought to contribute to community issues such as social justice and ecological stability (Godfrey, 2000). The internship, then, is a chance for you to learn about the public relevance and social obligations of a profession and about how those obligations are (or are not) carried out at your internship site. See the *For Further Exploration* section at the end of this chapter for examples of corporations and their contributions to their communities.

The Tasks at Hand

In the overall context of growth, change, and adjustment, there are a number of issues that you will face during the Exploration stage. As you move past the beginnings, your capabilities will be tested and stretched, and you need to be an active part of that process. You are also going to encounter formal and informal feedback on your progress, perhaps for the first time. Your relationship with your colleagues, especially your supervisor, will continue to evolve. In all these areas and more, there is plenty of territory for you to encounter challenges that you were not expecting; we will help you think about them and how you might respond.

The Exploration Stage

Associated Concerns	Critical Tasks	Response to Tasks	
		Engaged Response	**Disengaged Response**
Building on progress	Increasing Capability	Analyzes skills and knowledge needed to achieve goals	Does not try to overcome obstacles; lives with them
		Considers and adjusts learning goals	Settles for work unrelated to goals
Heightened learning curve		Seeks new learning opportunities	Ignores/declines new learning opportunities
Finding new opportunities		Seeks to expand understanding to embrace larger organizational and social dynamics (integrative engagement)	Is satisfied with focus on the individual dynamics of the work
Adjusting expectations	Approaching Assessment & Evaluating Progress	Seeks support and guidance; makes productive use of formal and informal feedback	Does not ask for or accept support; resists or rationalizes feedback
		Reflects on progress individually and with others	Pays minimal attention to reflection
Adequacy of skills and knowledge	Building Supervisory Relationships	Seeks to improve relationships	Distances from relationships and supervision
Real or anticipated problems	Encountering Challenges	Acknowledges and clarifies difficulties and fears	Conceals fears; denies or rationalizes difficulties
		Treats problems as an opportunity for learning and empowerment	Is upset that problems occur and expectations aren't met

Sweitzer & King, 2012

Increasing Your Capabilities

With thoughts about change, adjustment, learning, growth, and engagement in mind, we turn to a self-assessment of your current achievements and potential opportunities, something we encourage you to do on a regular basis, regardless of whether it is required (as it often is) by your site, your campus instructor, or both. This process involves considering general questions as well as specific issues.

Keeping the Contract Alive

If the Learning Contract is to serve as a valuable guide and point of mutual clarity for you, your site supervisor, and your campus supervisor/instructor(s), all of you need to work together to keep it from becoming obsolete. It may be that some new goals or activities are in order. Perhaps you have completed a goal faster than you thought you would. Perhaps an activity that you thought you would do is not available after all; that does not mean abandoning the goal. And perhaps some new and exciting opportunities have presented themselves. Even if none of these things have happened, this is a good time for systematic consideration of your progress in the learning domains (knowledge, skills, attitudes and values) and dimensions (personal, professional, and civic) discussed in Chapter 6.

Are you engaged in the activities called for in your contract? If not, why not? If these questions raise a flag for you, then it is your responsibility to bring it to the attention of your supervisors on campus and/or in the field. Whether it is a contract between the campus and the internship site or your individual Learning Contract, the activities identified on it were deemed academically worthy at one point; if you are not being guided in undertaking those activities, then it is time to reevaluate them and agree upon a viable contract for learning. *Why* you are engaged in the activities is the important question for you to think about before having such a discussion with your supervisors.

Are the activities you *are* engaged in the ones in your Learning Contract? If not, should they be? Exciting opportunities do arise, such as when you fill in for someone in an emergency and discover you like what you are doing, and it's not in your Learning Contract. There is nothing wrong with such a change as long as you keep your goals in mind. If those goals are changing as well, then a shift in tasks makes sense. If not, then adding new activities should be discussed with your site supervisor and campus instructor.

Balancing your needs as a learner with the site's needs for an intern can be a challenge. Even if you have tried to be as clear as you can at the outset, problems can arise from conflicting expectations among the site, the campus instructor, and the intern (Voegele & Lieberman, 2005). Responsibilities not previously discussed sometimes suddenly appear and are pronounced as expected by the site staff or even your supervisor.

But are these responsibilities a good match with what you most need or want to learn? If so, they need to be incorporated into the Learning Contract with the approval of your site supervisor and your campus supervisor. If not, they need to be discussed by all parties before you take them on. The dimensions of learning discussed in Chapters 1 and 6—personal, professional, and civic—are integral throughout an internship and the added responsibilities must contribute substantively to one of these dimensions. Similarly, the domains of learning—knowledge, skills, attitudes and values—are worth considering whenever you are re-considering your goals, and the added responsibilities must be integrated into at least one of these domains.

THINK About It

How Are You Doing Keeping the Learning Contract Viable?

When was the last time you looked at the contract or thought about it? In Chapter 6, we described the learning contract as a living, changing document, but we also know how easy it is to set it aside and forget it once it is handed in. Now is the time to take stock of this document. Give some thought to these questions and the engaged approaches suggested when change is needed.

- In what ways will the current learning contract let you realize your potential as an intern? In what ways will it hold you back? If the latter is possible, you should consult with your site supervisor about what needs to change in the contract.

- Are there substantial differences between what you planned and what you are doing? If so, they need to be discussed with your supervisors.

- Have some of the goals been reached and now seem uninteresting? If so, you should consult with your site and campus supervisors and develop new ones.

- Are there activities you know that you will not get to do? If so, they should not be in the contract and new ones need to be created.

- Have changes in personnel, needs, or priorities affected the contract? If they have and there are implications for the Learning Contract, then discuss this situation with your supervisor and campus instructor.

- Have new and interesting opportunities presented themselves? If so, and you want to see them in the contract, that needs to be discussed and approved by site and campus supervisors.

- Does the contract reflect productive, worthwhile work (for you *and* the site)? Responsible activities? Clear goals that *can* be accomplished? If not, then you want to meet with your site supervisor.

| Voices of *EXperience* | Interning Where Employed: What to Know and How to Cope |

Interns at both the graduate and undergraduate levels who were employees where interning have this to say about what an intern needs to know before beginning taking on this dual role.

Clarifying the Internship Role at Work

- *Understanding what this "dual" role's expectations are to both the site internship supervisor and the site employment supervisor, the administration, the staff, and the community*
- *Understanding implications of this dual role on your family and close relationships*
- *The reality of additional hours at place of employment*
- *Use the (Critical Learning) logs as your personal therapeutic moments*
- *Clear boundaries and limits for yourself in your dual roles*
- *Expectation: Be careful—watch for growth—fast, slow, linear, large—it is all okay as long as you grow.*
- *Be effective with time: you'll become better at what you do*
- *Approach with humility.*

Coping with the Dual Roles of Interning at Work

- *Stay organized*
- *Don't procrastinate*
- *Know which hat you're wearing*
- *As much as possible, leave work at work*
- *Learn to say NO!*
- *Take time to eat lunch!*
- *Make sure administration understands both your roles*
- *If possible, have two different locations: one for work and one for internship*
- *Exercise*
- *Make friends with custodian, lunch ladies, and secretaries*
- *Get a "vent friend" … someone who will listen*
- *No extra roles (like club advisor, etc.) or extra classes*

© Cengage Learning 2014

Balancing your needs as a learner with the site's need for an intern becomes even more complicated when you are employed at your field site or offered employment while interning. Potential problems can be managed if you clarify your needs in both roles (intern and employee) by discussing them with your supervisors. It is wise to keep supervision separate for these two roles. This usually means two different supervisors in the field, one for the internship and one for employment. Sometimes internship responsibilities shift because the site needs help and you are there and can help. Of course you will want to help, and it is understandable if you are nervous about saying no to your employment or internship supervisor, depending on the circumstances of your internship.

In the box on p. 219, those who have interned where employed pass along their words of wisdom to you, as well as their suggestions for coping with this potentially complicated situation.

Expanding Your Knowledge

Some of your goals are knowledge goals, and you want to be sure you are challenging yourself, and being challenged, to acquire more concepts and think in new ways and at deeper levels. The exact nature of these challenges is, of course, an individual matter. For example, when you consider the civic dimension, civic *knowledge* might mean learning not just about the challenges faced by the people your profession serves, but about some of the historical and current social forces that bring about those challenges. It also might mean learning that people who are hungry, poor, or underemployed are not necessarily lazy or unintelligent, but that their condition results at least in part from social conditions over which they have no control (Godfrey, 2000). In the professional dimension, there are two general issues to think about: integration of theory and practice and beginning to develop your own guiding principles.

Integrating Theory and Practice You probably have studied lots of theories in your classes and even given some thought to their applications. However, as you leave the role of observer and become a more active participant in the work, your theoretical knowledge changes as the result of trying to apply it (Benner, 1984; Garvey, 1992; Sullivan, 2005). This process can be troubling; you may find that your theories don't seem to work! Could all of your professors and textbooks have been wrong? Not at all. You have started down the road from *theoretical* reasoning to *practical* reasoning, as we discussed in Chapter 1. The integration of theory and practice is a normal challenge in a number of professions because not all theories work equally well in all situations. Furthermore, it may take you a while to recognize your theories in action because they might look different when applied to real problems. What will not help you here is to cling to a theory, no matter how elegant or impressive it was when you learned it, when the experiences you have seem to contradict it. Likewise you would be ill advised to abandon theory altogether. Remember that both theory and practice are transformed when the two meet.

Developing Guiding Principles for Your Work These principles are derived from your experiences in class, in the workplace, and in other field and life experiences (Benner, 1984). Often, interns can tell us about what they did in a certain situation and why, but struggle to answer questions such as:

- Can you formulate some general ideas about what works for you and what works in an organization like yours?
- Can you identify accurately and in specific ways your strengths and weaknesses in given situations?
- What are you thinking about as you approach a client, customer, or co-worker in crisis?
- How do you know when to press for a response and when to let the individual be silent?

You might be able to quote textbook answers but do not feel a personal connection to those answers because they are not coming from your experiences. If so, you may want to consider focusing on this aspect of professional development.

Expanding Your Skills

When we discussed learning contracts, we mentioned that there are three levels of learning. You are moving into the second phase or level of learning now, where the focus shifts from learning *about* the work to learning *how to do* the work. When you wrote your Learning Contract, you probably wrote about skills in fairly general terms. As you begin to acquire new skills, you can think about skill development in more specific terms and specify smaller steps for yourself.

Also, remember that the skills you are developing or can develop at your internship may be personal, professional, civic, or some combination of the three. In the area of *personal* development, you may have discovered a response pattern that you want to work on changing. The *Professional* skills you develop will depend on the internship placement, your career goals, and your academic discipline. *Civic* development skills might mean learning how to advocate successfully for change in a workplace, a neighborhood, or a community to make conditions there more equitable.

In the Focus on THEORY: Skills Acquisition box on p. 222, we offer some thoughts on how to sequence some of your skills development.

Living Your Attitudes and Values

The practice of every profession rests on a foundation of values and ethics; without them, skills can be used to all kinds of ends. And in most professions, attitudes are as important to success as skills and knowledge. It is no longer enough to know *about* these values and attitudes, or be able to recite them; now you need to begin to focus on how to live them. As you look back at your Learning Contract, did you set some goals for yourself in the area of attitudes and values? If so, how are you doing with living those values and acquiring those attitudes? If not, is it time to do so?

And don't forget civic values, such as the belief that understanding social issues is an obligation for everyone, not just those in politics, journalism, or social/public services.

Adjusting the Contract

We have been encouraging you to reflect on your experience in a variety of ways. Now we encourage you to take some intentional and direct action, a critical step for *integrative engagement* with your internship. It is time to make a decision about the viability of your learning contract. To do that, you must affirm what is still valid and consider the adjustments you need to make to keep this contract alive. Remember, though, that the Learning Contract was developed collaboratively by the site supervisors, your campus instructor or internship coordinator, and you; any changes should be approached in the same way.

Focus on THEORY

Skills Acquisition: What Is There to Know?

Dreyfus and Dreyfus (1980) studied how health care professionals acquired and developed the skills to carry out their work effectively. They identified five levels of skills acquisition: novice, advanced beginner, competent, proficient, and expert. These levels of proficiency reflect changes in three aspects of skills performance:

- A shift from reliance on abstract principles to past experiences
- A change in perceptions from seeing a series of parts to seeing things as wholes with relevant parts
- Movement from being a detached observer to being an involved learner

These findings and Benner's guiding principles are particularly useful when thinking about this new phase in your internship. As a novice, you had little understanding of the situations in which the work must be carried out. The basic skills that you brought to the field were generic and in need of *contextual meaning*, that is, knowledge of how to use your skills effectively in a particular situation. For example, in the helping professions, knowing the names of the resources in a community and being familiar with the indicators of potential suicide do not necessarily mean that you know how to interpret a client's behavior and prioritize your responses to deal with a life-threatening crisis. By the end of the internship, though, the student intern may be expected to develop the skills to respond effectively to just such a crisis. Certainly, it would have been helpful to have taken a course to understand the contextual meaning of the work, but that may be neither practical nor available.

Assessing Progress

We remind you that feedback, in all its forms, is an important component of a high-impact educational practice. Certainly this is the part of the internship that can be the most anxiety producing for an intern, yet can yield the most valuable information for success now and in the future. How you are progressing needs to be determined throughout the internship, and while we have encouraged you to engage in self-assessment, the process will also involve your supervisor(s) at the site, usually your supervisor on campus, and in many instances colleagues with whom you work. The assessment process can be done formally or informally, in writing or in a discussion, and it can be done frequently or periodically. Also, there will be some times when the process focuses primarily on helping you identify your strengths and accomplishments, as well as areas for improvement, there will be other times when you probably get ratings, or even a grade, showing how you measure up to expectations. Both experiences are normal and valuable, but the latter is a bit more anxiety provoking.

Feedback and Evaluation

You may or may not realize it, but your work is being assessed frequently in an informal way if you have a supervisor who is engaged in your internship. You may be getting feedback on a regular basis from both campus and site supervisors, and you will be having some type of evaluation of your performance at the site. Some interns accept both positive and negative feedback gracefully. Others are uncomfortable with it; they become defensive about negative feedback and think of it as criticism, and/or they give someone else the credit for their work when they are complimented. If you fall into the latter group, it makes sense that you talk about this in supervision or in your seminar class to learn how to deal with it in effective ways. Thinking of feedback as a tool that allows you to grow will serve you well; you will focus on the comments rather than only on the grade, and you will know the specific areas to focus your energy for improvements.

THINK About It

How Do You Do with Feedback?

- What opportunities have you had for feedback?
- How have you responded to the feedback you have received?
- What has been your experience with evaluations of your performance, and how do you respond to them?

Most interns have a formal, written evaluation at least once (and hopefully more often). Being nervous about this prospect is normal; we both feel nervous about it even after many years as professionals. But this is an opportunity, and to maximize that opportunity keep in mind that you have a right to a high-quality evaluation. Suanna Wilson suggests that a high-quality evaluation contains both positive comments and comments on areas for improvement (1981). Consistent with the principles of effective feedback that we mention at times in this book, Wilson further suggests that a high-quality evaluation is concise, specific, and describes behavior (as opposed to using labels). If you do not receive an evaluation that is guided by these qualities, you should check in with your campus supervisor. It may be consultation that you need, or it may be a discussion about the supervision you are receiving and the feedback you are getting about your performance on-site. Remember, if you don't understand why you received a comment or rating, it will be very difficult to learn from it. Working with your campus supervisor, you can learn the most by learning how to question feedback as well as question evaluations in productive ways.

Interns have a variety of reactions to and experiences of evaluations. Some feel satisfaction, joy, and relief; others feel anger, disappointment, and frustration. In both cases, there may be tears and disbelief. Evaluations can be a painful experience because of the introspection, self-doubt, and self-confrontation that can arise from them. Not everyone is comfortable scrutinizing their skills, abilities, and commitment to the field of study, or perhaps questioning the value of the work itself and their choice of profession. For some interns, the experience can be disorienting. Some find out that they excel in the classroom but not in the field, at least at this point in their internships; others find out that an excellent grade in one internship does not guarantee an excellent grade in a subsequent internship. Some interns experience competitiveness during evaluations and have trouble staying focused on learning from their own evaluations. This is especially so if other interns are at the site and there are differences in how the evaluations are conducted, what criteria are used, and the supervisor's style of "grading."

Encountering Challenges

I had arrived to class feeling okay about my internship. I did have some doubts, but I felt that with time, things could work out. As I listened to my classmates talk about their experiences, I felt my insecurities grow. It seemed that I was the only one in the group with any doubts about their internship and what was happening to it.
STUDENT REFLECTION

Just as there were issues that most interns face in the Anticipation stage of the internship, there are new issues and challenges that emerge as you move though the Exploration stage. And, while challenges can be both exciting and unnerving, some challenges are not necessarily enjoyable, especially the ones that become headaches. Internships must be challenging for learning to occur; and it is through problem solving that growth occurs—an observation made eons ago by the "father" of experiential learning that still holds true today (Dewey, 1938). In our experience, nearly all interns experience some problems, but how many, how severe, and when they will occur are difficult to predict. We are not talking about something minor that is quickly resolved. Rather, we are referring to situations that are more substantial, troubling, and that stay with you for a while.

Problems can be important or inconsequential, and as an intern you probably won't know how to make that distinction early on. The first problem, though, usually has an added impact, precisely because it is the first one. Once you have encountered—and resolved—the first problem or two, subsequent problems may be harder to resolve, but the shock of the first few will be behind you. You will also have the added confidence of knowing that you are capable of addressing problems and taking a proactive role in your internship.

The first thing to remember is that what you are experiencing is quite common; in that sense it is normal. Second, these experiences, while unpleasant, open the door to learning. It may sound strange for us to say that we hope you hit some bumps, but that is the truth. Without that experience, skills like critical thinking, problem solving, and conflict resolution remain academic exercises; when you hit a bump, you may have the opportunity to test and sharpen these skills in context, which we have stressed is key to development as a person and a professional.

Reactions and Responses

Just as there is a range of problems, there is a range of reactions to them—some more engaged than others. Some interns seem to take the problems in stride; others are really thrown by them. As you read this sentence, try not to fall into the trap of saying to yourself, "*Well, I am going to be the first type.*" Perhaps you'll be very accepting and supportive of those who do seem to be thrown, but you won't allow yourself that experience. Remember, allowing yourself to *feel* your problems, as well as to catalogue and analyze them, is one way of remaining engaged with the experience of the internship (you may remember this as *integrative engagement*). There is no right way to feel. Your responses and reactions—and those of your peers—will depend on your emotional styles, the particular intrapersonal issues touched by a problem or situation, your willingness or unwillingness to be open with yourself and others about your feelings, how you perceive moving beyond the problems, and, of course, the nature of the problems themselves.

In spite of our efforts to reassure them, interns often believe that if they talk about their difficulties or problems, or discuss their concerns or mistakes, then their grade will reflect these shortcomings. It is true that the grade you earn in your internship is influenced by your strengths as well as by the competencies that need further development. However, your campus instructor or supervisor recognizes that all interns have shortcomings, as do practitioners and faculty members. Your willingness to recognize and face the issues helps the faculty instructor gauge your growth in placement. In fact, not doing so may give the faculty reason to question how much you are gaining from your placement (Gordon & McBride, 2011).

Recognizing and acknowledging to yourself and others when you hit a bump is the first step in maintaining your course, but it is not the only step. We said there is no right way to feel, but there are engaged and disengaged ways to respond. Ignoring, minimizing, or shrugging off a problem, or bearing up nobly under the discomfort, may offer short-term relief or even validation but it will not make you stronger, it will not empower you, and it will not advance your learning. It may be, of course, that after careful reflection and consideration of alternatives, you choose not to try and change something that is bothering you. The important word there is "choose"—a choice born of deliberation and active engagement. Of course, you may be willing to engage when a problem occurs but uncertain how to do so. Chapter 9 offers you a variety of approaches and tools that may help.

We cannot say when or how often you will hit bumps and potholes. What we can do is tell you that they fall into several broad areas: the people, the internship site, and your life context. We invite you to think about each of these areas as you read further.

Focus on THEORY

Difficulties & Problems: Do You Know the Difference?

An interesting and useful perspective on problems and their resolution is offered by Watzlawick, Weaklund, and Fish (1974). The terms *problem* and *difficulty* are often used interchangeably, and indeed we have done so in this book, but these researchers draw a distinction between them. Difficulties, they say, are normal, if unpleasant, events that either have a fairly simple solution or no solution at all. However, in your attempts to solve these difficulties, you may make them worse or bring on additional troubles. It is then that you have created a problem. Many of the issues you face as an intern are normal and common. Knowing that they are part of the internship may make it a little easier for you, but probably not easy enough. However, if you choose to ignore or misdirect your energies in trying to address difficulties, they can snowball into full-blown problems.

Issues with Key People

I wanted the work, not the system. I realize now how significantly affected I am by the negativity of the staff. I want to separate myself from them so I don't have to deal with it. How do I do that? Can I do that?

Before beginning the internship, I had these ideas about what the people would be like, what the procedures would be like. Some of those expectations were met and some blew my mind. It was the reality of fair-does-not-equal-justice, and the judge-does-not-equal-unprejudiced.
    STUDENT REFLECTIONS

For most interns, the internship unfolds in the context of a web of relationships, with supervisors, co-workers, peers, and for some interns, with clients, or customers. Once the acceptance concerns have been addressed, the relationships continue to develop and new issues will undoubtedly present themselves.

Concerns with Supervisors   You may be having concerns with your site supervisor, whether it's the work you have been given by the supervisor, your impression of the supervisor as a professional, the supervisor's style of supervision, or aspects of the supervisory process. In many ways, the same issues that arise with site supervisors arise with the campus staff or faculty who oversee your placement, conduct the seminar class, grade your papers, and issue your grade. Although you do not spend anywhere near as much time with these people as you do with your site supervisor, they are important relationships nonetheless.

THINK About It

You Are Up to the Challenge BUT Is the Challenge Up to You?

Think about the volume, the depth, and the scope of what you are asked to do. Are your responsibilities over-challenging you, under-challenging you, or right on target? Too much responsibility can be overwhelming and leave you feeling ill equipped to do the work; too few demands result in being under challenged, which can be humiliating, frustrating, and disappointing (Burnham, 1981; Royse, Dhooper, & Rompf, 2011) and leave you feeling bored and not connected to the energy of the work or the field site. If the responsibilities are not right-on-target, and you choose not to address the problem, you can easily disengage from the internship.

Idealizing the Supervisor   Some interns initially are so impressed with their supervisors, both on campus and on site, that they idolize them. If you find yourself feeling this way, remember that the bubble eventually bursts. The supervisors will make mistakes, might even snap at you, miss an appointment, tune out in supervision, or exhibit one or more of any number of foibles that all of us can fall prey to at one time or another. Realizing that supervisors are human yet still professional with substantial responsibilities can be an eye-opener.

Supervision Styles   By now, you know a good deal about your supervisor's style and your own supervision needs. You have a sense of what is suited to your needs, what is not, and what needs have changed since the internship began. All these factors can affect the supervisory relationship for better or worse. This is a good time to ask yourself how your emerging needs match up with different styles of supervision and what you need to do to get your needs met, as did this intern: *I have learned a lot about her style. I have had to change my style in some ways to meet hers. I know that she runs things in a certain way and I have finally learned to make my needs match her style. I make sure all of my needs are met, but I make sure that they are on her grounds, so to speak.*

Even if you and your supervisor are well matched, you may have differences of opinion about strategies, projects, or other issues. Acknowledging and discussing differences are expected practices in many professions. However, you may feel anxious about doing so with your supervisors, especially at this stage of your internship (Wilson, 1981). Remember the section in Chapter 3 about the *power of discussion*? In most cases, open discussion, not confrontation, strengthens the supervisory relationships so long as your supervisors respond empathically and educationally to your concerns.

THINK About It

How Do You Match Up?

- How do your needs match up to your supervisor's style?
- How do you address differences between you and your supervisor?

Concerns with Co-Workers   Perhaps your relationships with co-workers are healthy and productive; we hope so. You should be on the lookout, though, for some common problems. For one thing, if you find yourself being given work by anyone other than your site supervisor, and you were not informed about this possibility up front, then it's time to discuss the situation with your supervisor so you are clear about what to expect. If you find co-workers seeking out your friendship, know that they may have hidden motives for befriending you, and it may take you some time to figure out what is going on. Sometimes, these employees are most marginal and most likely to reach out for alliances. In addition, if there are factions at the site, people may become friendly in order to recruit you to their "side" or clique. If you are not feeling welcomed as a team member at the internship site, or if you have had an interpersonal conflict with a co-worker, these can be major sources of stress (Yuen, 1990) and need to be discussed in supervision with your campus instructor or your site supervisor. Such a discussion is also needed if co-workers act in ways that give you pause or leave you feeling uncomfortable, or if they give you work or otherwise exert influence over you without your supervisor's awareness or approval. In the box that follows, interns pass along their experiences with ways interns are treated by co-workers.

You may have seen some behavior for which there is no good explanation. In any field, there are professionals who are lazy, jaded, harsh, or unethical. Experiencing such circumstances can leave you feeling quite alone in your reactions. In addition, the support you need may not be forthcoming from your co-workers. These issues can arise in an internship and should be addressed as early as possible. The Voices of EXperience below and The Interns' Problems on p. 236 provide inside perspectives on these issues.

Voices of *EXperience*   **Are You Being Treated as an Intern, or as JUST an Intern?**

Ask a group of interns if they are being treated as *just* intern and chances are they will tell you what our students tell us when we ask the question:

You Know You're AN INTERN When…

- *Staff are friendly.*
- *Staff keeps the intern informed.*
- *Intern feels grounded in the placement.*
- *Supervisor gives the intern "space to do" what the intern needs to do.*
- *The staff gets along with the intern.*

You Know You're JUST AN INTERN When…

- *The intern wants to learn and the co-workers don't want to include the intern.*
- *The intern is not involved; circumstances get in the way and the intern cannot contribute and there's nothing the intern can do (about it).*
- *The intern feels overwhelmed when she or he is not competent in doing the work given to him or her and are looked down upon for that.*
- *The intern gets "dirty looks" from co-workers who don't know that she or he is an intern.*
- *The intern is discriminated as an intern with bias and disrespect by staff.*
- *You just KNOW when they think you are JUST AN INTERN!*

It is also the case that what behaviors look like may not be what the behaviors are about. For example, perhaps you are interning at a state child welfare agency and you overhear some workers coolly discussing the murder of a child, which you recall from the morning paper. As the conversation goes on, you realize that the child was a former client of the agency. Other than this conversation, it is business as usual, and few outward emotions are evident. If you did not know that such an outward appearance of insensitivity and callousness is one way workers cope with heinous acts of violence that are part of their workday, then you might become very angry with the situation and very disillusioned about the people who work in the field.

Concerns with Peers   Your peers are your classmates and other students who are interning at your site. These are the people who know best what you are experiencing and they can be an enormous *source of support* throughout your internship. On the other hand, they can also let you down by talking too much (or too little), offering unwanted advice, focusing only on their own issues, or simply appearing indifferent to your needs. You'll have to decide what you expect from your peers in terms of support and assess whether or not it is realistic.

Another concern is that of *competition and rivalry* as it relates to same-site interns. The rivalry can emanate from either the staff or the interns themselves and usually results in interns being labeled winners or losers. Competition in and of itself is not unhealthy nor necessarily should be discouraged, as all may benefit from it. However, when competition becomes the focus of your energy and each of you is watching the other's assignments, level of support, progress, and involvement with staff, difficulties can develop.

A third issue that creates challenges for same-site interns as well as interns in the same seminar group is *differences in progress*. There are lots of reasons why placements turn out as they do. However, when there are two interns at the same agency and they are having very different experiences, the peer relationship can be affected. No longer are the interns sharing similar experiences, and it is difficult to keep both involved in supporting each other. Such differences in progress need to be addressed as potential issues by you, your supervisor, and your instructor.

THINK About It

Peer Connections: Are They Working for You?
- Can you depend on your peers for support?
- Do you feel competitive or contentious with your peers?
- Do you think that your progress is lagging behind that of other interns, either at the site or in your seminar class?

Challenges of Special Relevance to The Helping and Service Professions

Concerns with Clients

When you are engaged in direct service work, a great deal of your psychological, intellectual, and emotional energy is directed toward this aspect of your internship. As you move from an intellectual way of understanding the clientele to connecting with them on an emotional level, you are bound to have reactions to them and their situations. The challenge here is to understand your reactions and learn from them rather than letting them interfere with your work and your general well-being.

Reacting to Circumstances

As you get to know those you are helping or serving, learning about their individual and collective life situations can be emotionally demanding and possibly even damaging to you. For example, if you have already had your first encounter with issues of disease, death, or abuse, you already know that the work is never easy with those who are so socially vulnerable (Levine, 2012; Wilson, 1981). This also is true when dealing with and understanding the behaviors of those among us who are abusive, violent, or mentally ill. Destructive, irrational, or violent behavior directed at peers, family, or loved ones can be difficult to hear about or witness. If the behavior is directed at you, as it might well be if you are interning in the criminal justice system or in a psychiatric setting, the behavior can be most unsettling (see Chapters 3 and 7 for a discussion of personal safety issues). In all of these instances, supervision is critical to your understanding of how these situations are affecting you emotionally, and how to ensure your mental and physical safety.

As you come to know your clientele and their situations better, the issue of empathy can surface again. In Chapter 7, we discussed the problem that some interns in the helping professions have finding common ground and establishing empathy with clients. If you are successful in establishing empathy, you can face a new set of problems. Understanding others' viewpoints about and experiences in the world challenges your own perspectives, and such challenges can be threatening at times. For example, one intern from a middle-class background was working with juvenile offenders. She could not understand how anyone could commit crimes and think the behavior was "right," although she understood that people make mistakes and act impulsively. As she listened to clients, though, and came to understand how hopeless many of them felt and how angry they were at a system that never seemed to benefit them, she began to understand their contempt for the legal system and the reasons for their own code of "street ethics." She was quite rattled to discover that something she thought she knew for sure was not as certain anymore.

THINK About It

How Do You React to Life Situations?

- Have you thought about the emotional effects on your personal life of direct service work with those who are violent, mentally ill, abusive, victimized, or otherwise socially vulnerable?

- What happens when you try to understand others' viewpoints and experiences of the world? In what cases are you not able to do so? Not empathize with their situations?

- How have you responded in the past to emotionally stressful situations? Do you find yourself preoccupied with them and unable to think of anything else? What can you do to stop that response?

Your emotional reactions to this work can be problematic when they interfere with your learning or when you are unable to carry out your field responsibilities. If you find that you cannot get the clients out of your mind and cannot fully concentrate on other tasks, a problem exists. *"I'm overwhelmed by the sad stories of their lives; I can't get them out of my head. This is too much for me."* We typically hear comments like this from students who work with violence and other forms of extreme emotional content. These interns tend to develop needs early in their field experience that are different from those of their peers. For example, they may need more "air time" in seminar class. They also may be more cynical in their perspectives and preoccupied with loss, pain, and violence (King & Uzan, 1990).

Needing to See Progress

When I try to talk about the jam he got into the night before, a difficulty at home, or his drinking problem, he changes the subject. If I keep it up, he shuts down, or just gets up and leaves.
STUDENT REFLECTION

If you have developed a reasonably comfortable relationship with your clients, you may be wondering how and when you can move beyond what many interns call "chitchat" into some "real" work. This concern is not limited to those who work one-on-one with clients. Interns at sites such as shelters, soup kitchens, or drop-in centers often report that clients are more accepting of them at this point in the internship. They have stopped ignoring the interns and even sit and talk with them—about everything, that is, except their problems. Does that mean that they are not progressing? Not necessarily so.

As you spend more time with and around those you are helping or serving, you may find them engaging in all sorts of behaviors that seem unproductive or even counterproductive. For example, they may refuse to talk; or, they may talk on and on about their problems but show no interest in solving them; or, they may agree to try new behaviors and then not follow through. Initially, interns who are not making the progress with their clients that they think they should be making are often frustrated with such behaviors.

Addressing Boundary Issues

The clinical area of setting and maintaining boundaries can be very challenging for interns. Appropriate boundaries are very important in a helping relationship, although clients do not necessarily accept or appreciate them. When you set boundaries, you are making your clients aware of the limits of your relationship with them and what is and is not permissible. Clear boundaries will help your client understand your role and what to expect of you. They also let the client know where in your relationship there is flexibility. Ultimately, they help the client feel safe.

Clients often test boundaries, and that can be hard on the intern. Some boundaries are very clear and unquestionable, such as sexual relationships with clients. If a client is attracted to you, even if the attraction is mutual, there must be no sexual relationship. Other boundaries are more flexible and determined by the situation. For example, the issue of socializing with clients can be a complex one, especially if services are delivered in the community (as opposed to in an agency) or clients are from a culture that expects social contact with helpers.

By now you should be aware of the policies at your field site; if you are not, it is time to ask about them. You need to know what is expected when clients ask you for personal information that seems both irrelevant and inappropriate, ask to call you at home or stay beyond their time limit, or invite you to dinner, to their homes, or to their weddings. Without realizing it, you could find yourself involved in any of these situations. If so, they might prove to be very difficult to manage and extract yourself from. As one of our students pointed out, "*I realized this week that the staff and I are not at this facility to be friends with the clients...our main goal is to assist them in their recovery...we must draw the line at a professional level...we must be careful not to get too close...(I'm) aware of (the) trap before I get sucked into it. I imagine that once a staff member gets sucked in, it is difficult to get out.*" This intern had a keen awareness of the importance of boundaries; that's impressive because even seasoned professionals find themselves involved in situations that can be hard to escape at times.

Focus on THEORY

Burn Out, Compassion Fatigue, and Vicarious Traumatization

The work of helping professionals can be replete with stressors, including heavy workloads, a non-supportive workplace, and the challenges of severely complex and chronic client problems. As a future professional in this field, it is important that you are aware of the ways in which your life's work can negatively affect you. Self-care is a primary goal for the helping professionals. Without self-care, the professional is quite vulnerable to all three of these clinical reactions to the work. If you believe that you are experiencing any of these reactions to your work, inform your supervisors immediately so that you can learn how to prevent ongoing effects and engage the most effective interventions for you.

Burnout (BO) reflects general psychological stress that progresses gradually as a result of emotional exhaustion. Interns tend to key into their co-workers who appear to be experiencing burnout, almost as if interns had an intuitive sense of when helping professionals are in trouble with their work. Symptoms occur in any profession and are related to client problems of complexity and chronicity. There are emotional, physical, and behavioral issues, work-related matters, and interpersonal problems (Trippany, Kress, & Wilcoxon, 2004).

Compassion Fatigue (CF) is a form of burnout. It can develop when the worker is exposed to a secondary stressful event(s) and in turn becomes emotionally depleted and unable to provide the empathy needed for effective client care. Physical and spiritual exhaustion result as well. Physicians experiencing compassion fatigue have described it as being pulled into a vortex and unable to stop the downward spin (Pfiffering & Gilley, 2000). Symptoms, which develop quickly, are usually associated with a specific event and include generalized fear, sleeping problems, flashbacks of the event, and avoidance of event-related matters (Dubi, Webber, & Mascaari 2005). Other symptoms include (Pfifferling & Gilley, April, 2000):

- Abusing drugs, alcohol, or food
- Anger
- Blaming
- Chronic lateness
- Depression
- Diminished sense of personal accomplishment
- Exhaustion (physical or emotional
- Frequent headaches
- Gastrointestinal complaints

- High self-expectations
- Hopelessness
- Hypertension
- Inability to maintain balance of empathy and objectivity
- Increased irritability
- Less ability to feel joy
- Low self-esteem
- Sleep disturbances
- Workaholism

(continued)

Focus on THEORY *(continued)*

Vicarious Traumatization (VT) occurs only among those helping professionals who work specifically with traumatized clients (Trippany et al., 2004, p. 32) because of their personal reactions to their clients' trauma. If you are working with victims of physical or sexual assaults or domestic violence, victims of natural disasters such as Hurricanes Katrina and Sandy, victims of catastrophes such as the Minneapolis bridge collapse, or victims of terrorist attacks such as 9/11 and the Oklahoma City bombing, or school shootings, then you are working with clients who survived the trauma, and that makes you vulnerable to VT. Not all who survive or are affected by trauma develop Post Traumatic Stress Disorder symptoms, and not all who work with people who have experienced or been exposed to trauma develop VT (Trippany et al., 2004, 32).

Trauma is described as exposure to a situation in which a person experiences an event that involves actual or threatened death, or serious injury, and evokes intense fear, helplessness, or horror (Harvard Health Publications, 2006, p. 1). The victim may experience the event directly, witness it, or be confronted with it in some other way (DSM-IV-TR, APA, 2000). Not only can the victims be dramatically affected; the helping professionals who work with the victims can be significantly affected as well, including profound changes in the core aspects of their beings (Trippany et al., 2004). It is the traumatic histories of the clients, not the demanding nature of their cases, that contribute to developing VT. The onset of symptoms is acute and often includes the symptoms of burnout in addition to changes in trust, safety issues, feelings of control, issues of intimacy, esteem needs, and intrusive imagery (Trippany et al., 2004, p. 32). There are disruptions in the cognitive schema of the helping professional, including identity, memory, and the belief systems; and as well changes in one's sense of spirituality and one's perceptions of the self, others, and the world. These effects are not limited to the work with clients, but rather permeate the professional's private life.

The first step in prevention of these reactions is basic awareness of them. The second step is a self-care plan to ensure that you are safe from the devastating effects of the work.

Negative Reactions to Clients

You may be surprised to discover that there are clients and there are behaviors you just do not like. You may even feel frustrated and refer to these clients as "difficult." The term *difficult client* implies that the difficulty rests solely with the client. While it is true that certain behaviors, behavior patterns, and issues would be troubling to anyone working with the clients, the depth and breadth of the emotional reactions to the client vary greatly from person to person. The question is not which clients are difficult, but which clients *you* find to be difficult.

Perhaps, too, you are working with clients whose race, ethnicity, age, social class, lifestyle, or any other aspect of their cultural identity is new and very different from your own. You know how you *want* to feel about your differences, but it is important to be open, at least with yourself, about the how you *actually* feel. What you probably did not expect to deal with are your reactions as you learn about people who are quite different from what you expected. For example, there are the sex-abusing clergy, the substance-abusing physicians, and the embezzling attorneys. We have met them in our work, and you will meet them in your work as well.

In Their Own Words	The Interns' Problems

Some of the problems that our interns have encountered are listed here, written in the language of their student journals. Perhaps some of them look familiar to you. Perhaps you can add some of your own:

- *The pace here is totally insane (or incredibly slow).*
- *My supervisor is too vague (or awfully blunt).*
- *My supervisor never seems to have time for me.*
- *My supervisor is always looking over my shoulder—he doesn't trust me.*
- *If I don't have any questions, my supervisor ends the supervision hour. That's not right. She's supposed to ask the questions.*
- *I don't like the way my supervisor treats the clients (or staff).*
- *The people here are cliquish—I don't belong.*
- *They are so cynical. BORING.*
- *They are trying to get me involved in their problems. I'm just an intern here. I don't need to know what's going on. It's not my problem.*
- *I'm just an intern. Why does everyone expect so much of me?*

And, of special relevance to helping and service professionals:

- *I really can't accomplish much with these people—too much damage has been done and I can't perform miracles.*
- *I've never experienced what they have—they are writing me off. I can't work with them and they know it.*
- *A client lied to me or manipulated me. I thought they trusted me.*
- *A client had a relapse. Great. Now what am I supposed to do?*
- *There are some clients I just don't like. I can't find common ground with them.*
- *They just won't respond to me; not all of them but some of them.*
- *I have a good surface relationship but I can't move beyond that to really challenge them and explore some issues.*

Issues with the Site: Values, Systems, and Philosophy

As with other aspects of your work, seeing the dimensions of organizational life unfold can be quite different from studying about them or discussing them in the abstract. We discuss organizational issues in detail in Chapter 10, but three areas that particularly challenge interns in this regard are the organization's values, the formal and informal systems, and the operational philosophy. Once you appreciate how discrepancies can develop, you are ready to be part of the solution.

Organizational Issues What an agency values is evident in how the work gets done. What you see, though, may not always reflect what the agency says it values in its mission statement, and you may find yourself having a value conflict with the agency and the system. This is often the case with the paper-versus-people issue. Because there is never enough time in the service professions to give clients all that they need and still attend to the required documentation of the work, tension develops. "People before paper" may be what is valued by the intern, and it may be what the organization wants to value, but ideals get put aside because of the paperwork they create. Value conflicts such as this can be disheartening (Sparr, Gordon, Hickham, & Girard, 1988).

Systems of Influence The site has found ways to organize its resources to get the work done, both formally and informally. The informal organizational structure, or the structure of influence, can be a source of issues for an intern because it is this structure that influences what *really* happens in the course of the agency's work (Stanton, 1981). Sometimes these informal networks function quite smoothly and support the overall goals and work of the agency. Other times they seem to undermine those purposes.

Operational Philosophy You may be starting to feel the effects of the agency's *operational* philosophy, not the one described in the policy manual but the one that has evolved to get the work done. Many interns are impressed with their organization's operational philosophy, though there is also room for philosophical and political debate. You might find yourself surprised by the pace of the system and the realities of dealing with bureaucracies, underfunded programs, or reductions in staff.

Civic Development: The Social Contract

Your internship, whether in legal issues, foreign language, or public service, provides you with an "authentic occasion" to discuss with members of the internship community issues concerning the public good (Newmann, 1989) and your connection to them at that site. Each profession has its contract with society, as well as its ethical and moral commitments. These obligations create the web that connects the work to a larger social purpose. Importantly, you get to think about the public relevance of the profession and see in practice how its social obligations to society are or are not carried out at your field site (Sullivan, 2005).

If you are interning in a government agency, the helping professions, or in a nonprofit organization, you will find these obligations evident in its mission statement and policies. But they are just as important in the private sector in a business or corporate setting. Remember, too, that formal statements of mission are only part of the picture. Just as you need to focus on how your own values are lived, as opposed to merely stated, this is a good time in the internship to assess the operational values of the site, whether it is a business, a service, or a government level agency.

THINK About It

**The Social Contract and the Private for Profit Sector:
What Did You Expect?**

If you are interning in the private sector, it's important to consider these questions about the social contract and how compatible the answers are with your philosophy.

- What responsibilities does your site and the private sector it is part have beyond maximizing profits and developing wealth?
- What responsibility should they have if any in terms of contributing to the solutions of social problems in their communities or in the communities of their line employees?
- In this green age, what do you think the responsibility should be to eco-stability and sustainable resources? What is it in actuality?

The issues of civic development that many interns tend to deal with at this point in the internships are:

- Values and differences in values between what the profession expects and your personal as well as emerging professional values
- The meaning of the concept of "public good" and what responsibility you see yourself as having to it as an intern
- The responsibility to resolve differences through discussion and reasoning
- Understanding your civic responsibility in the profession of your internship

---

Issues with Your Life Context

You have been doing the juggling act for a while now and more balls may have been thrown your way than you can manage. As you are learning, life does get in the way of your internship—illness, a lost job, an accident, or a car that doesn't start. You may find your energy level waning, or you may be bending under the strain of multiple demands. You may even feel as this intern did: *"I have no time to be myself, or to be by myself, never mind spending time with friends or family."* Finding that time to spend with them is very important. Friends and family are central to your support system and your sense of well-being. Just as you cannot really know how emotionally demanding the internship will be, you cannot know how well your support system will respond when it comes to internship issues. It is time to take stock of your support network as well and decide what changes, if any, are needed. And it may be time to engage that network for its support.

Slipping & Sliding... While Exploring

For some interns, the problems they encounter are troubling aspects of an otherwise positive experience. Other interns reach a point where the emotional tone of the internship changes for them, and not for the better. Slippage seems to have happened. For example, you are realizing that you are not able to work collegially with co-workers or supervisors or within the organization's guidelines; and, you find little compatibility of values or philosophy with the site. You may not have been aware of these developments occurring until they seemed to suddenly be there. And, instead of enthusiasm and nervous excitement, you feel anger, resentment, confusion, frustration, and even panic.

What you also may find unexpected is the end of the hopeful feelings of earlier days in placement and the onset of more negative feelings. Concerns at this time center on many of the same areas as in the Anticipation stage, except that there is a shift in concerns from *What if?* to *What's wrong?*: *"What's wrong with my internship? What's wrong with my clients? What's wrong with the organization? What's wrong with me?"*

If you are in the middle of this sort of slump, or if you hit one later on, try to remember that many interns experience a period of disillusionment, and we have found that it tends to occur during the Exploration stage more often than during the other stages. Furthermore, many authors and researchers note that a period of disillusionment is a normal part of any internship (Garvey, 1992; Lamb, Baker, Jennings, & Yarris, 1982; Suelzle & Borzak, 1981), Understanding how you got to this juncture in your internship may be useful in stopping the slide you are experiencing and turning the situation around.

What Happened?

This upsetting shift in tone can happen for many reasons, and we certainly do not know them all. For some interns it is a cumulative process; for others, it is the result of a particularly challenging situation or a life circumstance that occurred; and for others it is a combination. You will recall that we discussed engaged and disengaged responses to challenges. Sometimes an intern takes a disengaged approach to handling problems, and it works for a while. However, with each bump and each time the intern does not proactively re-set the course, the balance is upset just a little more. You may veer from the center of the road, if you will, but you are still on the road and still moving. Disengagement and disempowerment can become habits, though, and they are hard ones to break. And if you hit enough bumps, then sooner or later you will hit the shoulder of the road and begin to lose traction.

Sometimes, there can be a bump of enormous proportions—a mega-bump—that throws an intern off course quickly and forcefully. Sometimes that bump is due to major health issues, losses through death or divorce, or economic crises such as loss of one's employment. The bump can also be due to the intern's discomfort with factors such as the intensity of the work, either its nature such as working with victims of violence or veterans in physical or mental distress, or the intensity associated with meeting very challenging timelines for reports or projects. Other times, the cause of the discomfort is much harder to identify, but the hallmark feeling remains constant: it is anxiety. Initially it is experienced in subtle and then not so subtle ways. When mega-bumps such as these occur, interns who up to this point have handled bumps along the way quite well, may find themselves unable to use resources effectively because of the overwhelming nature of the circumstances. The change in the intern can be dramatic.

We refer to disillusionment in this stage as a *crisis of confidence*. And we remind you that we use "crisis" to connote both danger and opportunity.

Managing the Feelings

One unexpected consequence of this disillusionment experience, whatever the cause, is the tendency to direct your feelings at those around you, especially those who are connected to the internship experience—your campus and site supervisors, co-workers, the clientele, or even your peers. Sometimes the seminar class itself becomes the focus of your feelings. The feelings can also be directed inward, and this can be a time when an intern's self-image can take a bit of a beating. For example, depending on the types of clientele, when they do not respond as expected, refuse your services, or react to you very negatively, it can call into question whether or not you were adequately prepared for the work, whether you are cut out to do the work, or whether you should be doing the work at this time in your education (Skovholt & Ronnestad, 1995). In addition, many interns in the service professions are used to their friends telling them that they are helpful and very easy to talk with, and the interns are often filled with a sincere desire to help. However, there is a difference between wanting to help and actually being helpful, and between being an easy person for friends to talk with and being effective with challenging students, residents, clients, or customers. When the inevitable difficulties and criticism come, these interns may become resentful, perfunctory in their job performance, or critical of the supervisor (Blake & Peterman, 1985).

THINK About It

For Real: It's a Sense of Loss
In general, we have observed four sources for these feelings of disappointment and frustration that reflect the disillusionment experience, and all have to do with a sense of loss.

Loss of Focus   You are not doing what you went there to do because the focus of your placement has changed either intentionally or through neglect. Interns who lose the focus of their placements can feel it almost immediately; it is a feeling of being sidetracked or otherwise ignored or discounted.

Loss of Accomplishment   You are not doing what you went there to do, not because the focus changed, but rather because either you are not able to demonstrate the skills or competencies needed to accomplish the internship or the design of your internship does not allow you to have the experience you want. You will probably feel frustrated and angry, feelings which you may direct at yourself or at the personnel or clients at your site.

Loss of Meaning   Essentially, you are unhappy with what you are doing at the site. Either the work does not matter to you, personally or professionally, or the internship has not been designed to provide you with a worthwhile field experience. Regardless, you may find yourself going through the motions of an internship, but without the spirit that comes from being personally invested in meaningful learning and work.

Loss of Purpose   You feel as if your internship is not of value to the organization or to your supervisor. You do have assigned tasks, but they are not interwoven in such a way that you know what you are doing from day to day and are able to go about that business on your own under supervision. And, that will not change in the future. Essentially, you do not feel grounded in the work because you do not have a clearly defined role or purpose as an intern. If you answer the question, What is your role as an intern? by saying you don't really know, then you are experiencing a loss of purpose in your placement.

What Can I Do?

Understanding how you got here may or may not help you feel calmer, but it will not in and of itself get you back on the road. It is certainly true (though rare) that a mega bump, especially if due to circumstances outside the internship, can lead an intern to leave the placement or continue the internship at a much lower or barely acceptable level; that is the danger side of the crisis. However, it has also been our experience that with focus and effort, balance can be restored, problems can be resolved, and the internship can get back on track, and that is the opportunity. When that happens, the satisfaction and empowerment you feel will be in direct proportion to the size of the challenge you meet. Our Chapter 9 offers you tools for staying on track, including a problem-solving model that can be very helpful for emerging from this experience of disillusionment.

Conclusion

This time in your internship is full of opportunities. Each day you are at your internship, there is something new to learn and a skill to develop or hone. The importance of the work you are given is increasing, as is the pace of the work and the expectations of the quality of your work. Challenges are more common and by now you realize that they are part of the internship. They may seem more like a curse than an opportunity at this point in the internship, though. You might want to ignore the challenges, but chances are that will make things worse for you. However, making sense of the feelings and concerns of this stage is critical to growing through it. And, growing through it is critical to succeeding in your internship. If you take this more constructive, engaged view of challenges and problems, then you will handle them better and learn more from them.

You've learned that the first step in moving through the challenges that you face is to take stock and have a clear inventory, awareness of the obstacles, as well as an understanding of how you respond when dealing with them. "*When I am faced with an overwhelming demand, I tend to initially stress myself out with thoughts like, 'how am I gonna do this?' Then I am able to self-talk and make a plan. I typically react by over-reacting and as soon as 'the crisis' is over, I manage and become strong willed.*" This intern knew her style of taking on challenges and when in the process she would empower herself to move the issue forward. The second step, of course, is developing a plan. The tools you'll need to move that plan forward, manage the challenges, and thrive in your internship are described in the chapter that follows.

For Review & Reflection

Checking In

Use the prompts and questions below to think about the ways in which you are growing, specifically the self-awareness you are developing and the ways you have changed.

Monitoring for Growth: What were your *moments of growth* in the past week

- What are the *effects* on your professional sense of self? On the work that you do?
- In what areas are you struggling to grow?
- Why are they important?
  - What is your plan to ensure that you *continue* to grow?
  - *How* will you know that you've grown?

Monitoring for Change: Moments of Self-Awareness

You may want to use the following prompts on a weekly basis, in your logs or journal, to better understand the nature of the self-awareness you are developing.

- I *learned t*hat I...
- *I realize* that I...
- Think back...to the first week of internship
  - What feelings do you recall having?
  - What thoughts did you have at the time about the placement? Seminar class? Assignments? Civic development expectations?
  - What hopes did you carry with you?
  - With what role did you identify at the time...that of student? As an intern? In your family role? As an employee? Perhaps in another capacity?

- **Think about**...who you are today.
  - How have your feelings changed?
  - How are your thoughts different?
  - What happened to your hopes?
  - What has changed about you?
  - Do you like the changes? If not, why not.
  - What hasn't changed and needs to?

○ Prioritize and plan for action

○ What have you learned about the power of adjusting expectations?

Personal Ponderings (Select the Most Meaningful)

1. Reread your responses to the anxieties exercise in Chapter 5 (*Personal Ponderings*, #1). Have your concerns and anxieties changed over these first weeks? If so, how? What have you learned about yourself and initial anxieties in new learning situations?

2. Which of your strengths are becoming apparent as effective tools in this internship and why?

3. Go back and reread your daily journal entries of the first few weeks. What would you say are the major things you have learned about yourself and about working at the site?

4. When it is time to go to your internship, how do you usually feel? Be honest. Has that changed over time? If so, what do you make of that change?

5. Take a look at the goals in your Learning Contract. List them in your journal and indicate whether each has been met, not met, or partially met. In the latter cases, are you disappointed? How might you move beyond these goals to the next logical challenge?

6. What things have happened that you didn't expect? Include positive and negative things and think about the three areas (knowledge, skills, and self) as well as your responsibilities and your supervisor's.

7. In general, would you say the tasks you have been given so far have over-challenged you, under-challenged you, or have been about right? Explain.

8. What have you learned about your supervisor's style, and how good a match it is to your learning and supervisory needs? Be as detailed as you can.

9. Have you and your supervisor disagreed as yet? If so, how did that happen? How did it make you feel? How was it resolved?

10. Have supervisors or others given you verbal feedback on your performance? How did you respond to what they said?

11. In what ways have your co-workers met or not met your expectations? How important is it?

12. Do you have concerns about your relationships with other interns—at your site or in your seminar class?

13. How has the agency itself compared to your expectations? Does the pace approximate what you thought it would be? What are the unwritten

rules, or values, of the agency? How do you feel about them and why does it matter?

14. Now that you have been at your placement for a while, does the support system you described in an earlier entry seem adequate? If not, in what ways is it failing you? What do you plan to do about it?

Especially for Interns in the Helping Professions

1. Are there clients to whom you are beginning to feel close? Are there clients you have a hard time connecting with? Think about why that may be in both cases. What lessons are you learning from these reactions?

2. Think about the kind of relationships you have with clients. In what ways have you been challenged by the clients? Have they tested your boundaries? If so, how? What other boundary-related issues have surfaced up to this point? Discuss your answers with classmates and try to help each other clarify the problems and perhaps some solutions.

Seminar Springboard

Checking Out the Problems.   What problems have you become aware of at your internship? Make a list of them and try to put them in order of importance. How does your list compare to those of your classmates in seminar class? What similarities are evident? What differences do you notice?

For Further Exploration

As You Sow. (2011). "Corporate Social Responsibility." Retrieved from http://asyousow.org/crs.

An example of an organization that works with corporations to promote environmental and social corporate responsibility through shareholder advocacy, coalition building and innovative legal strategies...founded on the belief that that many environmental and human rights issues can be resolved by increased corporate responsibility (http://asyousow.org).

Battistoni, R. (2006). Civic engagement: A broad perspective. In K. Kecskes (Ed.), *Engaging departments: Moving faculty culture from private to public, individual to collective focus for the common good.* Bolton, MA: Anker Publications. pp. 11–26.

An excellent summary of how academic majors can be vehicles for civic development.

Benner, P. (1984). *From novice to expert.* Menlo Park, CA: Addison-Wesley.

Seminal book in clinical practice in nursing looks at five levels of competencies and the demonstration of excellence in actual practice.

Colby, A., Erlich, T., Sullivan, W. M., & Dolle, J. R. (2011). *Rethinking undergraduate business education: Liberal learning for the profession* San Francisco, CA: Jossey-Bass.

A thoughtful treatment of business education, including discussions of the social contract.

Colby, A., Erlich, T., Beaumont, E., & Stephens, J. (2003). *Educating citizens: Preparing America's undergraduates for lives of moral and civic responsibility.* San Francisco: Jossey Bass. Chapters 6 and 8.

Particularly useful for stimulating thinking about civic development.

Dreyfus, S. E., & Dreyfus, H. L. (1980). *A five-stage model of the mental activities involved in directed skill acquisition* (Unpublished Report F49620-79-C-0063). Air Force Office of Scientific Research (AFSC), University of California, Berkeley.

Brief, informative paper describes a model of skills acquisition based on a study of chess players and airline pilots.

Kuh, G. D. (2008). *High-Impact Educational Practices: What they are, who has access, and why they matter.* Washington, DC: Association of American Colleges and Universities.

Describes educational practices that have a significant impact on student success which benefit all students and especially seem to benefit underserved students even more than their more advantaged peers.

Northwestern Mutual Life Insurance and Financial Services. (2012). Community Service Award. Retrieved from www.northwesternmutual.com

Northwestern Mutual financial representative interns don't just get to test-drive the full-time career; they can also get engaged in the company's commitment to volunteerism. Through its annual Community Service Award (http://www.northwesternmutual.com/about-northwestern-mutual/our-company/northwestern-mutual-foundation.aspx# Employees-and-Field) Northwestern Mutual interns like Chad Monheim (http://www.youtube.com/watch?v=1ne4qft8gy4&feature=plcp) can get recognition—and charitable funding— for the philanthropic causes they're passionate about. Go to http://www.north-westernmutual.com/career-opportunitiesfinancial-representative-internship/default.aspx to learn more about Northwestern Mutual financial representative internships.

Siemens. (2011). "The Siemens Caring Hands Program." *Siemens Corporation.* Retrieved from www.usa.siemens.com/answers/en.

A not-for-profit organization of this global company established to receive, maintain, and disburse funds for sponsoring and encouraging charitable activities in which employees donate their time and talents to worthy causes to carry out the company's commitment to impact the communities in which

they live and work through volunteerism in a variety of community service activities, including giving campaigns, blood drives, holiday food and gift collections, and disaster relief fund raising when their employees, operations, or business partners are affected.

The Washington Center for Internships and Academic Seminars. (2012). www.twc.edu.

Students at The Washington Center (www.twc.edu) intern at government agencies, non-profits, think tanks and businesses such as the Department of Justice, Amnesty International, Merrill Lynch, No Labels, and Voice of America. They also have the opportunity to become positive change agents by participating in civic engagement projects on important domestic and international issues, including domestic violence, homelessness, veterans, torture abolition and the environment. Students learn about these issues by interacting with national and local community leaders, then become involved in a direct service and/or advocacy project. They can choose to join a TWC-guided project or design their own individual one. For more information about civic engagement at The Washington Center, please see http://www.twc.edu/internships/washington-dc-programs/internship-experience/ leadership-forum/civic-engagement-projects. Articles about recent civic engagement activities can be found at TWCNOW http://www.twc.edu/twcnow/news/ term/civic%20engagement.

Wilson, S. J. (1981). *Field Instruction: Techniques for supervisors.* New York: The Free Press.

This book covers a number of challenges faced by interns and supervisors. It has not been updated, as the author has passed on, but it remains a valuable resource.

References

Benner, P. (1984). *From novice to expert.* Menlo Park: Addison-Wesley Publishing Co.

Blake, R., & Peterman, P. J. (1985). *Social work field instruction: The undergraduate experience.* New York: University Press of America.

Burnham, C. (1981). Being there: A student perspective on field study. In L. Borzak (Ed.), *Field study: A sourcebook for experiential learning* (65–72). Beverly Hills: Sage Publications.

Colby, A., Erlich, T., Sullivan, W. M., & Dolle, J. R. (2011). *Rethinking undergraduate business education: Liberal learning for the profession* San Francisco, CA: Jossey-Bass.

Dewey, J. (1938). *Experience and education.* New York: MacMillan Publishers.

Dreyfus, S. E., & Dreyfus, H. L. (1980). A five-stage model of the mental activities involved in directed skill acquisition. *Unpublished report supported by the Air Force Office of Scientific Research (A.F.S.C.), U.S.A.F. (Contract F 49620-79-C-0063), U.C.-Berkeley.*

Dubi, M., Webber, J., & B., Mascari J.B. (2005). Extreme conditions test counselors. *Counseling Today,* Novermber, 2005(1), 13;27–28.

Evans, N. J., Forney, D., & Guido-Dibrito, F. (1998). Schlossberg's transition theory. *Student development in college: Theory, research, and practice* (1st ed.). San Francisco, CA: Jossey-Bass.

Garvey, D. V., A. C. (1992). From theory to practice for college student interns: A stage theory approach. *The Journal of Experiential Education, 15,* (2)(August), 40–43.

Godfrey, P. C. (2000). A moral argument for service-learning in management education. In P. C. Godfrey & E. T. Grasso (Eds.), *Working for the common good: Concepts and models for service-learning in management* (pp. 21–42). Sterling, VA: Stylus Publications.

Gordon, G. R., & McBride, B. R. (2011). *Criminal justice internships: Theory into practice* (7th ed.). Cinncinatti, Ohio: Anderson Publishing.

Harvard Health Publications. (2006, March). First aid for emotional trauma: When and how must we act to prevent lasting damage? *Harvard Mental Health Letter 22,*(9), 1-3. Boston, MA: Author.

Kegan, R. (1982). *The evolving self: Problem and process in human development.* Cambridge, MA: Harvard University Press.

King, M. A., & Uzan, S. L. (1990, October). *The field experience: An integrative model.* Paper presented at the Annual Confernce of the National Organization for Human Service Education, Boston, MA.

Kolb, D. A. (1984). *Experiential learning: Experience as the source of learning and development.* New York: Prentice Hall.

Lamb, D. H., Baker, J. M., Jennings, M. L., & Yarris, E. (1982). Passages of an internship in professional psychology. *Professional Psychology, 13,* (5) (October), 661–669.

Levine, J. (2012). *Working with people: The helping process* (9th ed.). New York: Longman.

Newmann, F. (1989). *Reflective civic participation.* Madison, WI: National Center for Effective Secondary Schools, University of Wisconsin.

Perry, W. G. (1970). *Forms of intellectual and ethical development.* New York: Holt, Rinehart & Winston, Inc.

Pfiffering, J. H., & Gilley, K. (2000). Overcoming compassion fatigue. *Family Practice Management, 7*(4), 39–44.

Royse, D., Dhooper, S. S., & Rompf, E. L. (2011). *Field instruction : A guide for social work students* (5th ed.). Upper Saddle River, NJ: Prentice Hall.

Schlossberg, N. K., Waters, E. B., & Goodman, J. (1995). *Counseling adults in transition* (2nd ed). New York: Springer. Free Press.

Skovholt, T. M., & Ronnestad, M. H. (1995). *The evolving professional self: Stages and themes in therapist and counselor development.* New York: John Wiley and Sons.

Sparr, L. F., Gordon, G. H., Hickham, D. H., & Girard, D. E. (1988). The doctor-patient relationship during medical internship: The evolution of dissatisfaction. *Social Science Medicine, 26*(11), 1095–1101.

Stanton, T. K. (1981). Discovering the ecology of human organizations. In L. Borzak (Ed.), *Field study : A sourcebook for experiential learning*. Beverly Hills: Sage Publications.

Suelzle, M., & Borzak, L. (1981). Stages of field work. In L. Borzak (Ed.), *Field Study: A sourcebook for experiential learning*. Beverly Hills: Sage Publications.

Sullivan, W. M. (2005). *Work and integrity: The crisis and promise of professionalism in America* (2nd ed.). San Francisco: Jossey-Bass.

Swaner, L. E. (2012). The theories, contexts, and multiple pedagogies of engaged learning: What succeeeds and why? In D. W. Harward (Ed.), *Transforming undergraduate education: Theories that compel and practices that succeed* (pp. 73–90). New York: Rowman and Littlefield.

Trippany, R. L., Kress, V., & Wilcoxon, S. A. (2004). Preventing vicarious trauma: What counselors should know when working with trauma survivors. *Journal of Counseling and Development, 82*(1), 32–37.

Voegele, J. D., & Lieberman, D. (2005). Failure with the best intentions: When things go wrong. In C. M. Cress, P. J. Collier & V. L. Reitenauer (Eds.), *Learning through serving: A student guidebook for service-learning across the disciplines*. Sterling, VA: Stylus.

Waddock, S., & Post, J. (2000). Transforming management education: The role of service learning. In P. C. Godfrey & E. T. Grasso (Eds.), *Working for the common good: Concepts and models for service-learning in management* (pp. 43–54). Sterling, VA: Stylus Publications.

Wagner, J. (1981). Field study as a state of mind. In L. Borzak (Ed.), *Field Study: A sourcebook for experiential learning* (pp. 18–49). Beverly Hills: Sage Publications.

Watzlawick, P., Weaklund, J. H., & Fish, R. (1974). *Change: Principles of problem formation and problem resolution*. New York: Norton.

Wilson, S. J. (1981). *Field instruction: Techniques for supervisors*. New York: The Free Press.

Yuen, H. K. (1990). Fieldwork students under stress. *American Journal of Occupational Therapy, 44*(1), 80–81.

 CHAPTER *9*

Advanced Tools for Staying Engaged and Moving Forward

> *I chose to confront this situation and figure out just what was happening to me that made me feel unsatisfied.... Once I confronted these anxieties, though, everything worked out. I am beginning to confront this situation by making something good of this.*
>
> *I recognized, sensed and felt the great struggle exploring change can initiate within others as well (as me).*
> STUDENT REFLECTIONS

Throughout this book we have tried to help you learn how to adopt an engaged stance toward your internship and your learning as you move through the weeks and the stages of your internship. We have looked at what it means to take a more—or less—engaged approach to the challenges you face, and what it might mean to find yourself slipping toward the experience of disillusionment. In the last chapter, we focused on several areas in which interns often experience challenges as the Anticipation stage recedes and the Exploration stage unfolds. Now seems like a good time to introduce some tools that will help you meet challenges effectively and get back on track if you find yourself slipping and sliding. You may already have a number of skills and beliefs that

250

help you do so; what we offer you here are some resources for times when those approaches do not seem to be working or to be enough.

We also want to remind you that facing challenges is an important and necessary part of a successful internship. The ability to solve problems, and the confidence that you can do so, will serve you well in your internship, your career, and your life. Not many people look forward to challenges of the kind mentioned in Chapter 8, but you can learn to see the opportunity for growth just as clearly as the potential for danger.

In this chapter, we offer perspectives as well as tools to help you meet challenges and emerge feeling better about your internship, more empowered in terms of your ability to solve problems, and perhaps even a little wiser.

Sources of Power: Start with Your Heart

Perhaps you have had this experience either in or outside of your internship. You identify a problem and think you know clearly what is involved. You really want to do something about it. But every time you try to think clearly, you become confused and feel drained, until you finally start to think about something else just to get away from those feelings. Or you may find that you begin making a little progress, only to drown in the frustration and anger that you feel about the problem and your (perceived) inability to do something about it. This sort of emotional paralysis can be extremely frustrating.

You may feel like you don't know what to do, but knowing is not always the problem. We have found that no technique is going to work without three important ingredients—belief, will, and effort—which, interestingly, are driven not by your head, but rather by your heart. The lessons from the heart inform us that the affective domain must be an equal partner with the cognitive domain if you are to be successful in meeting challenges.

The Power of Belief

If you do not believe that a situation can change, most likely it will not. Let's rephrase that. If you wait for a situation to change on its own, it might. However, the point of this chapter is to help you be an active agent of change, as this intern was: "*I have learned that, after you have worked to put all the pieces into place…what you really want to accomplish finds its way into existence. I just knew it (would happen).*"

If you do not believe that you can change a situation, then it may become a self-fulfilling prophecy. Gerald and Marianne Corey have written a marvelous book on self-change and effectiveness entitled *I Never Knew I Had a Choice* (2010). Perhaps *I Never Believed I Had a Choice* more accurately

| Voices of *EXperience* | When the Going Gets Tough… From Those Who Know |

- *Remember, this is part of life, not life itself.*
- *Be mindful of the internship as a <u>learning</u> experience.*
- *Laugh at yourself. It puts things into perspective.*
- *Develop thick skin when dealing with peoples' negative comments.*
- *Don't take ANYTHING personally.*
- *Be passionate about your field. If you're not, then consider another field.*
- *Don't fear the unknown. Embrace it.*
- *Question <u>what</u> you are doing, not <u>who</u> you are. Live one day of the internship at a time!*
- *Find patience. Everything else will fall into place.*
- *Challenge your supervisor to challenge you.*
- *You can't possibly learn everything about your placement during the year, so just open up and drink in as much as you can.*
- *Get lost—finding your way is the fun-nest part and how you'll learn the most.*
- *Remain optimally uncomfortable for positive change/growth.*
- *This too shall pass.*
- *It is what you make it.*
- *It is what it is.*
- *Sometimes things will not be perfect or how you would like them to be.*
- *Every mistake is an opportunity to learn.*
- *Don't try to be perfect; just make sure you do enough to be proud of.*
- *No matter your experience, good or bad, you'll have changed for the better.*
- *Learn from your experience, positive, negative, somewhere in between.*
- *Approach with humility. There is no way you can know everything. Always be open to learn. You will grow even if you don't believe it in the beginning.*
- *Just when you think you can't do it anymore or have nothing left, reach deep down inside.*

reflects the reader's experience. You need to *believe* that things can be different and that you can create the change that needs to take place. "*I had an incident where I didn't agree with the director's decision about a student.… I stood up for what I believed in and what I believed should happen.*" This intern believed in the potential of the client. It was the intern's commitment to his own beliefs—which were coming from an intuitive feeling about what this client was capable of—that resulted in the client being allowed to take the upcoming GED—and the client did and passed the examination!

Of course, you can work on feeling confident, and that will help you succeed, or, you can work on solving the problem, and that will help you feel more confident. For some people in some situations, it is the belief that is the foundation for the action. For other people, the converse is true. You know best what works for you.

The Power of Will

Will is the intention and determination to make change occur. Intention and determination are powerful motivators, and the frustration and anger you may feel are normal parts of the emotional experience of trying to change something. You need to allow yourself to feel those feelings and not run from them or hide them. As with most feelings, that is how you begin to work through them and allow determination to come to the fore. As one intern recalled, "*I...just pondered...and just wondered whether I was actually helping the client or hindering the client. I was determined to know so I stayed with my decision of saying "no" and having the client stick it out. She did and I am happy that I did not give in. It would have been so much easier for me to say yes. But, it would not have helped either the client or me.*"

Another barrier to this determination is the resentment some interns feel about having to work on creating the change. You may feel that this problem is not of your making, and you may be right. So if you didn't "do it," then why should it be you who has to fix it? Accepting responsibility can feel like you are accepting blame for the situation. But you are *not* accepting blame; you are empowering yourself, as is evident in this intern's comment: "*I need to speak up if I am being given too much work. Otherwise I will burn out. Once I start saying 'hold off, I have too much to do,' I know it will affect my internship (in a negative way). But I have to do it.*" The overriding goal is for you to empower yourself so that you feel able to take charge of your life and your situation when you want to, as this student learned to do.

As long as you wait for others to act differently, you are taking a disengaged approach. "*I learned that I have to take responsibility for my internship...to take the bull by the horns as you said...I feel that it's an important step in the process...I did not ask for an agency in crisis for an internship, but whether it is or not, I have got to get my internship on track.*"

The Power of Effort

Finally, you have to do something; effort refers to what you actually *do*. It also refers to persistence and perseverance. Some of your difficulties will be relatively easy to analyze and resolve; others will take more time. "*What my heart really wanted to do (in this internship) is now going to happen.... I am so excited about this great boost (to morale). I feel so motivated now and thrilled to be able to share from my heart something I know people will enjoy and benefit from. It was a lot of work but worth the effort.*" Another intern said it more succinctly: "*At this point in my internship, it is reassuring to see that persistence does pay off.*"

Considering Character

Use Those Strengths!

Everyone has strengths, including character strengths, and it is very important that we know just what our character strengths are and how to make the most of them. Why? Because character strengths, along with positive emotions and meaning, are considered one of the three "positive" routes to having a "full life." It is easy in times of great demands and stress during the internship to forget to focus on our strengths, forgetting that they are *the* tools for getting us through the tough times.

Character strengths are found in every culture and across cultures as well. The most commonly valued *character* strengths across cultures will be no surprise to you: *kindness, fairness, authenticity, gratitude, and open-mindedness.* There are 24 in all, identified by the field of positive psychology, including character strengths "of the heart," such as *zest, gratitude, hope,* and *love,* which happen to foreshadow lifetime satisfaction more so than *curiosity* and *love of learning.*

You're probably curious about the remaining character strengths, and you should be. It's very important to know ourselves. They are: *creativity, curiosity, love of learning, having perspective; bravery, persistence; social intelligence, leadership, teamwork; forgiveness, modesty, prudence, self-regulation;* and, *an appreciation of beauty and excellence, humor,* and *religiousness.*

Being mindful of this list can prove useful so long as you are consciously aware of your strengths and make a point of intentionally using them in your daily life. Why? Because people derive satisfaction out of life if they make the most of knowing and using these strengths frequently. In other words, live your strengths to the fullest!

Where to begin?

- First, take stock of your strengths by listing the ones that are an important part of your life—now or in the past and indicate accordingly.

- Then, think about specific situations in which those strengths have helped to bring about *pleasure, engagement* and *meaning* in your life.

- Finally, plan ways to use those character strengths more often and in new ways in your internship, first on a weekly basis and then gradually on a daily basis until living your strengths become a way of living.

Sources: www.ppc.sas.upenn.edu, www.ppc.sas.upenn.edu/articleseligman.pdf.
Rashid, T., & Seligman, M.E.P. (in press). *Positive Psychotherapy: A Treatment Manual.* New York: Oxford University Press; Seligman, M.E.P. (2011). *Flourish.* New York: Simon & Schuster.

Sources of Support

By now you realize that your sources of support related to the internship have grown substantially. You have support both at the site, on campus, in your seminar class, and most likely online. In our experience, some interns naturally reach out for all the support they can get. Others have the initial instinct to try and go it alone, and yet others fall somewhere in between. You may need a little support or a lot, but if you need it and do not seek it, or do not know where to find it, your struggles will be compounded and perhaps complicated in many ways.

Remember Your Support System

In Chapter 3 we invited you to consider the various people in your life and the various kinds of support they give you. Go back now to Chapter 3 and take another look, thinking now about the context of the internship. Who are your companions, your consultants, your sounding boards, and so on? Are you making the best use of these resources when you face challenges at your internship? Perhaps some of them are not available to you given the change in your life and schedule, or perhaps this experience is so different from theirs that they simply cannot help. The point here is to determine what assets you have in your support system, be sure you are using them, and think about where else you can get support in areas where that system may be a bit thin.

Support from Your Supervisors

Supervisory support includes both sources on campus and sources at the site; peer support can come from the site (other interns) and from the seminar class and/or support teams, on campus and/or online. The supervisors on campus include the person who oversees your seminar class and your internship (campus instructor), the person who coordinated your field placement, and your academic/program advisor. For the sake of this discussion, we use only the term campus instructor, although any of your supervisors could certainly be a source of support and guidance.

The campus instructor is a valuable resource for support and consultation, and one that interns sometimes overlook. Perhaps this is because they have faculty or staff roles on campus and interns feel uncomfortable approaching them, or because the interns are just getting to know them in the class, which often meets only weekly. However, some interns have realized that the campus supervisor/instructor has an intuitive sense of what is happening during this stage of the field experience and how to help interns empower themselves and move beyond the situation. Sometimes it's a matter of taking the time to talk about the situation in a quiet space and in reflective ways. Sometimes it is a matter of learning about resources on campus that would be helpful to you at this point. Sometimes it's a technical matter, such as revisiting the initial agreement with the site or the Learning Contract to enhance your awareness of why the situation is so troubling. Regardless, all your campus supervisors are there for you when the going gets tough. Your success, though, is your campus instructor's work—a priority, in fact.

When it comes to your field site, you have sources of support there as well. You have your site supervisor, although it probably comes as no surprise to you that it is often a lot easier to discuss feelings and issues with the campus instructor than it is to raise the issues with a site supervisor. In fact, it is the site supervisor who many interns find most intimidating when it comes to dealing with these feelings in particular. Sometimes the issue is one of not wanting to disappoint the site supervisor. Other times it is a fear of the supervisor ending the placement. In reality, site supervisors are often a wellspring of resources and support for their students and welcome the opportunity to work with students in resolving such challenges.

In addition to the site supervisors, you have co-workers that may have supervisory responsibilities to you, such as the person who is responsible for your on-the-job direction and support, and office personnel who make a point to have a hand on the pulse of how the internships are going (such as the program director or office manager) or make a point to "take care" of the needs of the interns. There may also be other interns at the site, some at the same academic level you are, some before or beyond you. All of the people in the roles just mentioned can be sources of support for you, but not all offer the same type of support. So, do give thought to what it is you need so that you go to the "right" person and not expect a type of support the person can't give you.

Support from Your Seminar Class

Not all interns have seminar classes for support, but if you do be sure to take advantage of it, or at least try. As we have said before, it is difficult for people in your life who are not doing an internship, or have never done one, to appreciate your trials and tribulations. Your peers are in the same boat and even though they may have vastly different life experiences or internship experiences than you, they can listen, offer feedback if you ask for it, and empathize. It is important to remember the confidential nature of the discussion when you are talking about problems in the field. *Respect* is the operative word, and respect is evidenced in a number of ways. For example, it is very important to protect identities when discussing details of the situation in the seminar class. Our rule of thumb is to advise our interns to assume that a family relative of the individual(s) involved in the problem is in the class and to protect all identities so that even family members wouldn't recognize their relatives from the intern's discussion.

   If you are not feeling comfortable in discussing an issue with your site supervisor and it becomes apparent that you really need to, perhaps your seminar classmates can help. Maybe you and your peers could develop scenarios in seminar class that would be helpful. Or it could be that a peer has experienced a similar reluctance and can share how it was resolved.

Focus on THEORY

Feeling the Stress

There is nothing pleasant about stress in an internship—even if you know it will pass. Regardless of the sources, its effects are the same. In the context of this chapter, it is important to know that stress can actually undermine your ability to think creatively and solve problems.

   Exactly what *happens* when you feel stressed? In a nutshell, a situation you *perceive* to be a threat of some sort triggers a flushing of stress hormones throughout your body that in turn produce physiological changes. You feel the effects of the perceived danger when your heart pounds, your breathing changes, and you have an over all feeling of tenseness. It's the instantaneous activation of the survival mechanism that you are feeling, the effects of hormonal changes throughout your body's systems; you know it as the fight-or-flight response. Experiencing the effects of distress signals to your body does take a toll on your physical and emotional health after a while. Sleep is affected, as is exercise, and it may contribute to anxiety, depression and addiction.

   Exactly what *causes* these reactions? Read on IF you are curious about what takes place *in* the body. A complex and orchestrated system of responses occurs instantaneously. Key to that system is the *autonomic nervous system* (ANS) (which has two subsystems), and it controls your body's involuntary functions (this is what you experience when your blood pressure, breathing, and heart beat change). The *sympathetic nervous system* of the ANS produces the hormone epinephrine—you know it as adrenaline—which in turn triggers the release of blood sugar, fats from temporary storage sites, and nutrients that saturate your blood— all energizing and focusing you so you can respond to the threat. When the surge of epinephrine subsides, yet another system kicks in (the HPA) and more hormones are released, eventually triggering cortisol which keeps the body revved up until the threat passes. Once the perceived threat passes, cortisol levels drop and the second subsystem of the ANS is activated—the *parasympathetic nervous system*; this is the braking system in the body and it calms you down. Finding ways to put on the brakes and keep them on is the big challenge because even chronic high levels of stress keep the HPA axis operating and this contributes to

(continued)

Focus on THEORY (continued)

health problems (damage to blood vessels and arteries, increased blood pressure, heart attacks, strokes). And, what about the effects of those elevated levels of cortisol that so energized you when you needed it? Well, they build up fat tissue and cause weight gain because cortisol increases your appetite to replace the energy lost to the stress response. So, a pizza or a pasta meal or a box of cookies feels SO good after the crisis has passed.

Exactly *what* **can you do about it?** Experts advise to use *relaxation techniques* (such as the Relaxation Response, diaphragmatic breathing, repetitive prayer, visualizations) and e*xercise* (such as walking to improve breathing and relieve muscle tension) *and movement therapies* (such as yoga, tai chi, qi dong to induce calm). Importantly, they also advise to use your support system. People who have close relationships with friends and family receive emotional support that sustains them in times of crisis for reasons still not clear.

Source: Harvard Heath Publications, March, 2011.

Framing and Reframing

We introduced the concept of *reframing* in Chapter 3 and want to revisit it here. Sometimes simply the way you talk about a problem, both to others and to yourself, can make it easier or harder to solve, can affect how much you learn, and can affect how empowered you do or do not feel.

What Do You Mean, Failure?

In the book, *Learning through Serving*, Janelle Voegele and Devorah Lieberman discuss the importance of how you use the word *failure* (2005). You may say to yourself at some point, *My supervisor has failed me*, or *I have failed to meet one of my goals*, or even, with a project or client, *I am just a failure*. In our experience, the word *failure* connotes something final. When you fail a test, that's it—you failed. But many situations in your internship that you may label a failure are or can be seen as opportunities to learn, grow, and improve. It may be that your goals were unrealistic, or that you needed a different set of knowledge and skills than you thought. It may also be that a supervisor or co-worker was simply unable to help you with a task, or to provide an experience that you had hoped for. You didn't fail, unless you are convinced that you did and can't find a way to grow from the experience. We encourage you to use the word *failure* sparingly at best.

THINK About It

The Power of Metacognition

Metacognition means thinking about or reflecting on your own thinking. Students who use a metacognitive lens on situations respond to a "failure" situation by wondering what they can do differently next time to get a better ending to the story. "Failure" to them is feedback about what needs to change, not a game stopper. For the intern who doesn't use metacognitive ways of thinking about "failure" in a situation, it is a game stopper. Pointing the finger is more important than figuring out how to use themselves and their strengths to create a better ending. We use the following questions with students after major assignments as a metacognitive tool—a way for them to understand what they *now* know about what they knew at the time they did the assignment, and what they can learn from their grade to improve their performance. In the instance of the internship, if the "failure" is not connected to project work, you will need to substitute accordingly in the questions below for the situation. Some students created a checklist to use in future situations.

- How satisfied are you with how things turned out with this project/situation (1–10 with 10 the highest)?
- Did you think about such a project/situation beforehand in terms of what you expected as the end goal? If so, did you have a plan? How well did you think it through? Was it based on past experience or on your logical conclusions?
- How satisfied are you with how you approached the project/situation?
- Did you invest time in preparing for it or did you just dive in?
- Did you invest time while in the situation trying to think through it?
- When you were experiencing it, did you consider *how* you wanted it to turn out? If so, did that slow you down and let you think?
- *Or* were you experiencing it without a vision of a successful ending?
- Why didn't you know what you needed to know *or* what you needed to do?
- What was the standard you sought: Just get through it and survive? Good enough? Mastery? Excellence? Did perfection play a role? Did apathy?
- What do you now know about yourself that you didn't know before the project/situation? Before this exercise?

Matches and Mismatches

Sometimes what seems like a failure is simply a mismatch. If you have not yet done so, now is a good time to think about what you have learned about yourself and the areas of matches and mismatches in the internship. In Chapter 8, we described what can happen when your learning style is not well matched to the supervision style employed at your site, but there are other arenas where mismatches can occur as well. In one situation, an intern worked at a family planning clinic regularly fielding telephone calls asking for information on abortion. Over time, she became impatient with the phone calls, the clients, and even the agency. She said that many of the questions she answered were "stupid" and wondered why the people were so poorly informed. She chafed at the bureaucratic procedures employed by the agency. Upon further exploration, it appeared that a values issue was at work. The intern was firmly opposed to abortion for herself and had been open about that belief with her campus instructor and the agency. However, she insisted that she believed with equal fervor that everyone should make their own choices and that she was not opposed to abortion in general. However, as the semester progressed, she grew more and more uncomfortable with the fact that abortions were performed at her placement site. It seems that she had much stronger antiabortion beliefs than she thought. It took her some time to admit this, even to herself, because she believed she should be tolerant. Once she did so, she was able to direct her anxiety away from the agency and its clientele and view it as a poor match, given what she had learned about herself. The site arranged for her to work in their community education division, and the internship was ultimately successful. This intern and her supervisors worked to reframe her frustration and resentment as a mismatch, and one that could not have been foreseen.

Reframing Complaints

Complaining is a common phenomenon at most workplaces, and it is often counterproductive. Robert Kegan and Lisa Lahey have written extensively about the power of language and its ability to affect how you function at work (2001). They write, for example, about the contrasting "languages" of complaint and commitment. Complaining to or with others about troubling aspects of the work may offer some temporary relief and a sense of camaraderie, but it rarely leads to change. In fact, sometimes it can inhibit change because energy that could be put into change is put into complaining. At the heart of your more serious complaints about your internship is perhaps a commitment that is important to affirm and pursue. Furthermore, if part of your complaint is that some of the staff do nothing but complain, considering the commitment that underlies their complaints may be helpful. Complaints about high work loads, for example, may signal a commitment to quality services. And your complaint about departures from your Learning Contract signals a deep commitment to your own learning.

Owning Up!
Taking Your Share of Responsibility

Oh, how easy it is to blame others for all the ways in which life—or internships—fail to meet our expectations. And sometimes other people deserve a large helping of blame. But one of the themes of this book is the importance of self-understanding and nowhere is that more important than when you are wrestling with a problem or challenge. For one thing, your willingness to look at your own role in the problem may well facilitate others in doing the same. For another, while you cannot control what others do, you can always work on making change in yourself when you need to. Finally, as we emphasized in Chapter 4, if you do not look at yourself, you may assess a situation as somehow wrong when it is merely different.

Remember Cultural Competence

Now that you have had some time at your placement, we invite you to return to the section on Cultural Profile on p. 101 and give thought once again to your own. Think about the cultural profiles of the people you work with, the people you work for, and the people served by your organization. Some of the tension you feel may be in response to being exposed to behaviors that are culturally driven. In some cultures, for example, interrupting is a common and acceptable phenomenon. In others, raised voices are not a cause for concern. In some cultures, neither of these behaviors is acceptable. If you come from one sort of culture with regard to these behaviors and are being exposed to another with the staff, the clients, and/or your supervisor, you most likely will experience some discomfort. What's important here is to understand which differences are causing the concerns and flag what must change to make a positive difference.

Patterns Come Around and Around

Perhaps you find yourself reacting to situations at your internship in ways that you don't like and you wish you could make a change. And although will is a powerful force, willpower alone is not always enough. Suppose, for example, you have a hard time speaking up in staff meetings, even when you know you have something valuable to add. You promise yourself you will do better, and then you don't. In some cases you may be dealing with a response pattern such as we described in Chapter 4. Here are some fairly common response patterns that might manifest themselves in an internship:

- You have a hard time accepting criticism.
- You say "yes" when you want to say "no."
- You are extremely upset by confrontation.

And there are many others. If you think you might be dealing, at least in part, with a response pattern, we invite you to revisit the THINK About It box on patterns (p. 93) and try to clarify the pattern. We suggested some resources or working on patterns in Chapter 4 and we list them again at the end of this chapter.

Has Your Family of Origin Come with You?

Not many interns actually have their families come to see the site, but sometimes part of a problem can be that you are "following" a rule from your family of origin that is not appropriate in your internship, or you are playing a role that works for you in your family but not in the internship. If you had a family rule, for example, that children never question their elders, are you now reluctant to ask questions about "the givens"? or reluctant to ask about decisions or strategies made by those more experienced than you? Are you uncomfortable when others do? In another example, you may have been a jokester in your family, and throughout your life used humor and joking to gain acceptance. In an internship, however, sometimes a lot of joking is not welcomed or appropriate. Go back and look at the box on p. 95 and think about how your family's ways are helping you or may be contributing to the problems you are having.

Consider Your Style of Dealing with Conflict

For some interns, a conflict with a client can be upsetting. While few people actually enjoy confrontations or conflicts with clients, they can be necessary, especially with certain client populations where limit setting is especially important. If you are finding that sort of conflict difficult, go back and review the styles of conflict discussed in Chapter 4. Also, consider whether your response to conflict is part of a larger response pattern, as discussed earlier.

Even more troubling are conflicts with peers or supervisors. The conflict itself is often upsetting, but if you try to resolve it and it gets worse, that can be very unsettling. If that has happened, consider whether the other person is using a style of conflict that is unfamiliar or uncomfortable for you. If that is so and talking about it is not something you are ready to do, then think about why that is and what needs to be done for that to change. For the time being, though, this type of situation is something you might benefit from talking about with your peers.

What's the Difference?

We want to remind you again of the importance of viewing your reaction to people and situations as just one of many possible reactions. As we explained in Chapter 4, if you are not able to do so, you may end up putting a negative label on a behavior or person when such a response is neither warranted nor helpful. If you find yourself thinking that *"no normal person would..."* or *"anyone would object to..."* then a word of caution is warranted. Your version of normal is undoubtedly influenced by a great many factors, including your personality, your family background, and various aspects of your cultural profile. Framing issues as differences sets the stage for reaching a mutual understanding and compromise, as opposed to each of you thinking the other one is somehow wrong, both about the issue at hand and the way you are dealing with it.

A Metamodel for Breaking through Barriers: Eight Steps to Creating Change[1]

The mere formulation of a problem is often more important than the solution.

ALBERT EINSTEIN

I knew that my internship might be a little more difficult to manage than most others because I was employed at my internship site. I also knew that it would be difficult to work full time, complete all my internship and seminar requirements, baby-sit part-time, and to maintain all of my relationships and my new marriage. I knew these things but I didn't know how I would get through them.

STUDENT REFLECTION

As an intern, you will encounter all sorts of challenges. The material offered so far should help you resolve some challenges and get re-engaged when you find yourself slipping. For times when you need more, we offer a more comprehensive problem-solving model. This model should be especially useful if you are having the pervasive feeling that somehow your internship has gone wrong or you are experiencing feelings of disengagement.

We hope this model will be both easy to understand and helpful to use. Before we explain the steps, though, a few introductory comments are in order. First, the model is designed to slow you down a bit. Yes, you read that correctly. If you have a problem that is really troubling, you are probably focused on what to do about it, what action to take. Anything, you figure, has got to be better than the ways things are. Or you choose a strategy because it worked in another situation or because a friend used it successfully. Thinking intelligently about a problem involves some careful analysis, some goal setting, and only then deciding

---

[1]This model is derived from a model for interdisciplinary integration and problem solving (see Sweitzer, 1989) that in turn has its roots in the Behavior-Person-Environment (BPE) principle advanced by Kurt Lewin (1954).

on and implementing a plan of action. So, yes, this model slows you down somewhat so you can solve the problem.

Focus on THEORY

Problem-Solving Models

The literature is replete with techniques for solving various kinds of problems and with generic problem-solving models (see, for example, McClam & Woodside, 1994). The generic framework is hardly a new idea; in fact, it has its roots in the ancient teachings of Buddha, who wrote about the Noble Truths, including suffering, the origins of suffering, and the path leading to cessation of suffering (Watzlawick, Weaklund, & Fish, 1974). Most models include steps that answer the following five basic questions:

1. What is the problem?
2. What do I know about the problem so far?
3. What is my goal?
4. What are my alternatives for reaching that goal?
5. What is my plan of action?

This model is also designed to broaden the way you engage your problem; referring back to Kolb's model we are encouraging more time on reflection before you draw conclusions or formulate action. We want you to move beyond the initial stage, where you hold one individual responsible, and look at the perspectives and contributions of many other people. This model asks you to use *integrative* engagement (Swaner, 2012), that is, to think and feel your way through a process, not simply to plug in answers to questions and in turn get a score that instructs you what to do next. We also want you to use *contextual* engagement (Swaner, 2012), considering the wider context in which the problem is occurring, including the systems that are involved, such as work groups, the office as a whole, and other agencies with whom you work. It also involves a process of engaging others, namely your system of support and resources. For example, you may want to work on the problem with other interns at your site, discuss the issue(s) in seminar class or over the Internet with your support teams, work with your campus and site supervisors to find a resolution, or do all of the above.

A Word to the Wise…   Collaboration with others does not mean dependence on them. When faced with a problem on new turf such as the internship, you may find yourself slipping into dependent habits. When engaging your support systems for help, do so with a mental list of what you already have done to move the problem along so that you can continue an engaged stance when collaborating.

We suggest that you first read through the model in its entirety and follow the example we included. Next, return to the first step and, using an issue that is problematic to your internship, work your way through all eight steps to create a Change Statement. To help you move through the steps, we identify potential pitfalls and provide guiding questions for reflection before moving on. We designed the reflection components so you can write about them in your journal if you wish. You will also be advised to make reality checks before moving on, to ensure that you benefit from the support and perspectives of your resource systems.

One Step at a Time
In the middle of difficulty lies opportunity.
ALBERT EINSTEIN

Step 1: Say It Out Loud

Quick—in one sentence—what is the problem? Don't think too much about it; just say what comes to you (in the next step, you will name the problem more carefully). Then, write down what you said. You can learn something important about how you are thinking from the way you first state the problem. Consider an intern who is struggling with someone else at the internship, perhaps a co-worker, supervisor, or client. Here are some ways the intern might initially describe it:

- I can't get anywhere with her.
- She is not interested in changing.
- This co-worker needs a different unit to work in.
- No one told me how to handle this.
- Their suggestions aren't working with this co-worker.

Do you see how each of these statements frames it differently, locating the problem in a different place? Where did you initially locate your problem? Where are some other places it might be located? Is there an opportunity here for you to reframe your initial statement?

Reality Check! Does it make sense to involve anyone from your support system in the process at this early point? Your campus supervisor or instructor? Your site supervisor? Peers? Co-workers? Someone else?

Step 2: Name the Problem

Now it is time to let go of the feelings, the thoughts, and the impulses that you've been having and think clearly about the behaviors that are causing problems for you. Don't think about the people; think about the specific behaviors. Behaviors are actions that anyone can see if they are able to

Focus on PRACTICE

Using the Three-Column Method for Step 2

Here is one way to work on Step 2 using a process that will be somewhat familiar to you if you have been using the "three-column" method in your journal entries:

- Divide a piece of paper into three columns. In one, list the behaviors.
- In the next column, list your thoughts that result from those behaviors.
- Finally, list the feelings that result from each behavior.
- Now, look at what you have written and try to state the problem in one sentence. It can be a long sentence, but try to use just one sentence.
- Think about when and in response to what this problem usually occurs and see if you can express that in a sentence. If it is not a recurring problem, this sentence will not be important.

observe. You also need to think about the thoughts and feelings you are having as a result of these behaviors.

To begin your Change Statement, see if you can describe exactly how this problem affects you. Use the following questions to guide your thinking. Once you have worked on this group of questions, you will have a concise, clear statement of the problem in paragraph form to add to your worksheet.

- In what specific way does it create unpleasantness?
- Does it embarrass you?
- Does it make you angry?
- Does it hamper your ability to work?

Here is an example of what such a paragraph could look like:

When she runs away from the program, it angers me because I think about all the work I have done with her, and she just blows it every time. It's just like when I used to work with kids on the streets. They would do the same thing, not listen to me. It makes me feel helpless, sort of powerless, and useless in my work.

A Word to the Wise... Keep in mind that what first appears to be the problem may only be a surface issue. In our experience, many of the problems that interns choose to confront are issues that have come up again and again, although they may appear in different forms because the specific problems they create can vary. For example, an intern who has issues with acceptance may choose to eat lunch alone and thus not be included in the energy and discussion that go on in organizations where working lunches are part of the normal routine. Problems can develop if others interpret the intern's choice to eat alone as an indicator of a lack of investment or interest in the issues discussed at lunchtime. Spending time sorting out the *real* problem now will save time and energy later.

Reality Check! You need to ask yourself again, does it make sense to involve anyone this early in the process? Your campus supervisor or instructor? Your site supervisor? Peers? Co-workers? Someone else?

Step 3: Expand Your Thinking

Although examining a problem from a variety of perspectives is an insightful experience, it can also be frustrating for reasons we have already discussed. But someone has to do the investigative legwork, and in this case, the only one who can is you. Begin with some scrap paper. As you work your way through the following four guiding questions, try to keep in mind the role of politics and other subtle dynamics. Be patient and persistent with yourself; some of these steps may not come easily to you, but they all can be mastered with a little perseverance.

- *Who are the players involved in this issue?* Identify as many people as possible who are related to the problem. Including them does not mean that they are causing or contributing to the problem, only that they are part of the scene. Don't forget to include yourself.

- *How does each of them frame the issue?* Work your way through each person on your list and try to put yourself in that person's shoes. Try to see the situation from their points of view by imagining what it is like to have their positions on the staff, to have been there as long as they have, to work with an intern, or to have you in their work life.

- *What are the major systems involved in the situation?* Each of these individuals is part of and influenced by a number of systems.

THINK About It

Where You Sit Is How You Stand

Some of our students have found it helpful to imagine all the individuals involved with the problem forming a circle, which in turn forms the hub of a wheel. Each person on the wheel sees the issue differently depending on where they sit. Their perspective, in turn, influences their stand on the issue at hand. Try to imagine how each of these individuals views and feels about the problem. From your perspective, how might each of them be contributing to the problem? How would they perceive it from their perspectives?

(You may want to take a look at Chapter 10 if the notion of systems is not familiar to you.)

- Your first task is to make a list of all the systems that are somehow connected to the problem. You may find it easier to approach this task by brainstorming. Begin with those directly connected to the situation and move to those with indirect connections. Or you might find it easier to be more organized and work your way from the inside out or the outside in. The inside group of systems includes the staff (and any subgroups of the staff you can identify), your work group, the clients, and anyone else who connects directly with the agency.
- Now think about the agencies that work with yours, either collaboratively, competitively, or in adversarial ways.
- Next identify the sources of funding and other resources.
- Don't forget your systems of support and resources both on campus and related to the placement site. This includes your campus supervisor or instructor, professors, administrators, and if you have one, the seminar class or web-based support team. There may be more; if so, keep going!
- Now go back to your list and eliminate any systems that, on further consideration, seem to have very little to do with the problem.
- Finally, think about how each system—as opposed to the individuals in it—contributes to the problem. To do this, you need to focus on the rules and roles that drive individual behavior (again, you may want to look at this section of Chapter 10).

- *What about you?* In all likelihood, you are contributing to the problem in some way. Your initial statement of the problem (Step 1) may be a clue here. Go back to pp. 261–263 of this chapter and consider your own contribution.
- *How did it get so complex?* Now go back over all the ideas and insights you have had during this portion of the problem analysis. Write a paragraph discussing them. Notice how you have begun to create a new picture of the problem that is more complex and takes into consideration a number of perspectives on the issue.

Reality Check! At this point you have done a considerable amount of deliberate thinking about this problem. If you have not as yet engaged those in your support system, it is time to do so. Ask yourself again, who makes sense to involve at this point? Your campus supervisor or instructor are valuable resources, as is your site supervisor. If you are not involving them and they *should* know about this problem, you need to ask yourself why you are not involving them and work with someone to resolve this new issue.

The Eight Steps to Creating Change

1. Say It Out Loud
2. Name the Problem
3. Expand Your Thinking
4. Consider the Causes
5. Focus Your Attention
6. Determine Your Goals
7. Identify the Strategies
8. Create the Change

© Cengage Learning

Step 4: Consider the Causes

Now go back over your notes and summary from the last step and try to come up with a list of the potential major causes of the problem. Notice we did not say the major cause; there is always more than one potential cause. Be sure to consider all the individuals and systems you examined before concluding your list.

Reality Check! It's time to find out how you are doing seeing possible causes for this problem. It's best to check this out with others before moving on.

Step 5: Focus Your Attention

You have identified a number of dimensions of the problem. You will need to make changes to and alter the dynamics discussed in the list of causes, but where do you begin? There are really two issues here. You need to review the causes you identified and think about which ones are most important; and you will need to identify priorities in terms of affecting change in the situation. However, you also need to direct your energy towards aspects of the problem that can actually change, and over which you actually have some control or influence. This may seem like a simple enough proposition, but in our experience it is easy to get confused and it's important to be as clear-eyed as you can. Sometimes the things about which you feel most passionate are the very ones that you cannot change. And yet you keep trying. Going down that road, particularly when you are trying to address a problem in a time-limited situation such as an internship, is a real waste of time and energy. If this sounds familiar to you—and even if it doesn't—we encourage you to use the exercise in the box below.

THINK About It

Doing What You CAN Do

One approach to managing your responses to problems and setting priorities appropriately is a version of the well regarded **CAN & CANNOT** exercise. If you are not familiar with it, we will review it here for you. Divide your paper/screen in half and head the columns accordingly. In the CAN column, list everything that *can* be done in the situation; nothing is too small—list them all. In the CANNOT column, list everything you know that *cannot* be done about the situation. Nothing is too big to put in that column.

- Now go back and look again at the items in the "can" column. Are you sure that the items here are things over which you have some influence and control? They may be things that you very much want to be different (and you might be right!) but if you really have no control over them, they belong in the other column

- Now before you think about strategies, remember that when you invest your energy and time trying to change what's in the CANNOT column, you are spinning your wheels and going nowhere because *nothing* can be done about the realities in that column. They are there for you to accept and then move on. Not easy but that's what you need to do. So, why the problem to begin with? Because the tendency is to crisscross the lines of responsibility.

- Take one last look at the "can" column and try to prioritize the tasks. Which of the changes will have the biggest impact? Are you prepared to make those changes?

- Remember, too, that when you do *not* invest in the CAN column, all that can be done is not getting done. So, you are spending time doing what cannot be done rather than doing what can be done. Rather obvious, but if it were easy, you'd be doing it.

BOTTOM LINE: Doing all you *can do* is *all* that you can do. Knowing that and then accepting it will let you sleep at night.

Once you have done the exercise, go back and add a new paragraph to the Change Statement you are developing. This paragraph needs to answer two questions:

- What is blocking me from resolving this problem?
- What are the most important causes of the problem that I am able to change?

Reality Check! If you find issues on your list of things you do have control over but privately doubt you will do anything about, it is important that you consult with someone who you know will help you sort out this issue before moving on.

Step 6: Determine Your Goals

For the next step in your Change Statement, each cause of the problem that is on the "changeable" side of your list, now needs goals for change. State as specifically and behaviorally as you can what you want to be different about the situation. There is no room for fuzzy thinking here. Just wanting the issue to go away may be your most pressing emotion, but now is the time for clear and concrete statements. Be sure that you have covered all the individuals and systems that you believe have a substantial role in causing the problem.

One possible pitfall here is to confuse a goal with an action step. For example, suppose that one major cause you identified is that your duties are radically different from what you expected and were told. Changing placements is not a goal; it is an action step. The goal would be to bring your daily actions more in line with your stated learning goals. One reason this distinction is important is that there are usually multiple ways of achieving any goal. If one approach fails, take some time to be disappointed, but then think of another. In the example just given, if you made switching placements your action step and then were told you could not make the change, you might feel the problem is now unsolvable. If you keep your real goal in mind, though, you are more likely to come up with another approach.

Reality Check! Goals are a big deal. Before moving on, it is best to review them with someone who you know is able to have a clear perspective on the problem and provide you with useful feedback.

Step 7: Identify the Strategies

Your task for this portion of your Change Statement is to develop a list of possible concrete actions you could take to meet each of the goals you identified previously. In developing the action steps for each goal, be sure the interventions reflect your knowledge about the nature, cause, and context of the problem. The principle of effectiveness is operating here; that is, each action must be effective for each goal. The picture you have created of the problem in this step is one that is action packed and future focused. When you write your paragraph about this step, let yourself be guided by the question: *What must I do to make my goals happen?*

You may remember that we called this model a meta-model because it allows you to incorporate many other approaches and strategies. You have probably studied many approaches to resolving interpersonal problems and conflicts, making constructive changes in yourself, and making change at the system or community level. But if you have not or if you want to do some more exploration before deciding on a strategy, we have included some resources at the end of the chapter.

Reality Check! Strategizing is an art and a science unto itself. You may want to involve someone at this point you know will be a faithful sounding board for you, who can stay with you and the process until the goals are clear and realistic.

Step 8: Create the Change

This is the step you have been waiting for. Ever since the issue became a barrier to your internship, you have focused on this step. This is the time when change happens. If you've done your work with determination and effort, with genuineness and a willingness to be reflective and introspective in your thinking, and if you asked yourself the tough questions that needed to be asked, then you are ready to move on. It is time to do the achievable and implement your plan with commitment and perseverance. Yes, you've given it your best, and your best is good enough. The potential pitfall of this step is to fail to do what you can do about the issue, resigning yourself to the status quo when you have the capacity to make change occur.

The guiding question for this step in your Change Statement is: *How will I know when things are different?* Let this question guide you as you implement your plan and move through the barrier that has prevented you from moving ahead in your internship. Let it also guide you as you write the last paragraph of your Change Statement. Congratulations. You did it!

Reality Check! This is no time to forget those who have supported you throughout the process. Check back in with each of them and let them know how you worked things out. And, of course, thank them. What's much appreciated is to tell them how they made a difference in your decisions and the outcome of what once was a challenging problem for you.

Not Being Taken Seriously: The Case of the Rebuffed Intern

Some of you may be comfortable with this model and can work with it right away. But for others, it may be too abstract to be useful. So, we are going to work through an example for you. We will not detail all aspects of the steps, but we will show you how the eight steps work in this particular situation.

We are using the case of Kim, an intern at an adoption agency. Because of the nature of the problem, this case could apply to a wide variety of contexts for an internship. So, as you read about Kim's situation and watch her Change Statement evolve, substitute the focus of your internship in place of that of the adoption agency.

Kim is conducting an advanced internship at an adoption agency and is very excited about this opportunity. He thinks this is the kind of work he would like to do for a career, and he really liked the agency and his supervisor. He was told, and it was written into his Learning Contract, that he would eventually have his own cases and would write case notes and reports.

(continued)

Not Being Taken Seriously:
The Case of the Rebuffed Intern *(continued)*

A few weeks into the internship, things are going smoothly. Kim has been working alongside his supervisor and the supervisor thinks things are going well. The supervisor wants to give Kim two cases, one of which is a new case and one of which is a post-adoption situation where the family needs only routine follow-up visits. In both cases, the supervisor is going to monitor Kim very closely. However, about that time a new executive director comes to the agency.

The new director decides that it is not appropriate for an intern to have any cases, since the intern is not fully qualified, and directs the supervisor to take the cases away from Kim. The director understands that having cases was in Kim's Learning Contract, and he is apologetic to Kim and to the college supervisor; but the director is firm about the change. The supervisor feels badly about doing it, but explains that it is the director's decision. Kim is very disappointed and considers leaving the placement. Once able to talk about the problem, Kim had the following self-dialogue.

Step 1: Say It Out Loud

"My internship is ruined! This new director is too rigid and doesn't know me. He doesn't trust me!"

Reality Check! I need to contact my campus supervisor right away. Or, maybe the placement coordinator on campus. They won't let the director get away with this. After all, an agreement is an agreement.

Step 2: Name the Problem

I know I can handle these responsibilities, and it doesn't seem fair that I am not at least going to get a chance to prove myself. I am afraid I will lose a major learning opportunity as a result of this change in my internship. I am bored because I don't have anything to do that is worthwhile and makes me think; and I am frustrated because there is important work going on around me and I am not included in it. I'm beginning to think they don't trust me. I spend a lot of time trying to think of what I could have done differently.

Reality Check! I don't know. I left a message for my campus supervisor about this and he's not called back and it's been a whole hour. I realize he's not on campus today but you'd think he'd call in for messages every hour!

(continued)

Not Being Taken Seriously: The Case of the Rebuffed Intern *(continued)*

Step 3: Expand Your Thinking

My Supervisor…is probably embarrassed. But there is a new director, and my supervisor has to be careful not to be seen as disruptive or disrespectful.

The Director…is under pressure as the new boss. I remember that there have been some concerns about the agency and its performance, and a couple of adoptions last year did not go well and the children had to be removed.

The Clients…probably would benefit from consistency in case work, and I am not going to be here for their whole time with the agency.

My Issues…I have to admit that I have always resented not being taken seriously. It goes back to being the youngest in my family and having all my brothers make fun of me and not listen to my opinions. Maybe I am overreacting in thinking that these people don't take me seriously.

Reality Check! Okay, it's time to get focused about doing something about my situation. My campus supervisor did return the call, although it was 2 hours after I called him. I was a bit taken back when he observed that I seemed quite agitated and that this behavior seemed so unlike me. Maybe he was right in saying that usually I dig in to solve problems, and this time I seemed to be stewing over the situation instead of grabbing hold of the problem and trying to solve it. Maybe he's right. I don't know. I think I'll send the interns in the seminar class an e-mail and tell them that I'm dealing with a nightmare of a situation and could use some of their ideas. That's what Jo did last week, and she solved the mess she was dealing with.

Step 4: Consider the Causes

1. The director has legitimate ethical concerns about confidentiality and privacy.
2. The agency is worried about its performance record.
3. I have been unwilling to compromise my position.

Reality Check! Okay, I'm not in this alone. That feels good. My peers had some good ideas and dished out some blatant reality for me to face. My campus supervisor surprised me as well. He asked me what I had done up to this point to move the problem along, and I told him about e-mailing the seminar class and working with a small support group some of us in the class set up over the Internet. He liked that. It seemed like he was more willing to work with me when he saw that I was working on the problem.

(continued)

> ## Not Being Taken Seriously:
> ## The Case of the Rebuffed Intern (continued)

Step 5: Focus Your Attention

I cannot change much about this situation, except my reactions to it. But maybe I can still have some valuable learning anyway.

Reality Check! Okay. I'm onto this one. There is no magic wand being used by the campus supervisor or placement coordinator to solve my problem. This is life, and life happens. I guess this is what is meant by "learning lessons in life." I can't change the director's decision, only she can. I don't want to throw away this internship because there is so much more about it that makes it quite valuable to my career. So if I want this internship, cases or no cases, I guess I've got to find some other value in staying at this site.

Step 6: Determine Your Goals

I can think of three primary things that have to happen in order to make this internship work for me. I must:

1. Stop being so upset about this situation.
2. Identify some other responsibilities that can help me meet my learning goals.
3. Determine if this change in my learning contract meets with the approval of the academic program, my campus instructor, and the licensing board's requirements for direct service before graduation.

Reality Check! I better talk with the site supervisor. My peers were right in suggesting that if I work with my site supervisor, together we'll find other ways to keep me at the site and get me involved in work that has value for my future career.

Step 7: Identify the Strategies

-   Develop some alternative suggestions for responsibilities before meeting with my site supervisor to discuss the future of my internship.
-   Review my learning goals with my site supervisor and determine if there are other options for working directly with clients, e.g., at a sister agency.
-   Ask my site supervisor for suggestions as now I have few of my own!
-   Identify some new learning goals as a result of this situation, such as understanding some of the legal and ethical issues of internships.
-   Clarify with my campus supervisor whether this change in the Learning Contract is acceptable to the standards of the academic program and to the licensing requirements for adoption work.
-   Consider moving the internship to a sister agency for part of the time.

(continued)

Not Being Taken Seriously: The Case of the Rebuffed Intern *(continued)*

Reality Check! I can't believe how much my campus supervisor knows about this internship stuff. I should have known. He's been doing it for some time. I bet I'm not the first student who wanted to throw away the internship just because of a change in circumstances. And, boy, does he know how to brainstorm real good ideas. Guess the guy really knows how to get me to rescue my own placement from an early demise. I got some great ideas from a couple of interns in the seminar class, too. I'm going to have a great internship after all.

Step 8: Create the Change

1. I believe that I can bring about the changes that are needed. But, I must persevere until I see changes that satisfy the need for a challenging and meaningful internship for me.

2. I will e-mail the professors in my academic major and the placement coordinator about this matter immediately to determine if my academic program allows me to be in an internship that does not allow students to be involved in direct service work.

3. I will meet with the site supervisor and discuss the ideas I've developed, thanks to my peers and my campus supervisor. I just have to be sure that it is possible to take on some of these other responsibilities; they are challenging, and they are meaningful to me and to the site.

4. I will explore with the site supervisor the possibility of conducting part of the internship at a "sister" or contracting agency.

Reality Check! I am a bit surprised at what I can do to move this situation forward. It really is not all that difficult—but it sure felt that way before I started working on it as a problem to be solved, not a rejection by the site. I just have to accept that until I hear back from my professor or the placement coordinator. I don't know if I can stay at this site. If I can't, then I will have to be placed at a new site, maybe a sister site to this one. If I can stay, then I have to work closely with the site supervisor to create a new learning contract that ensures that I am satisfied with the new responsibilities. It's going to be okay. Makes me wonder why I have been stuck for so long.

Conclusion

This chapter has prepared you to confront the issues that challenge you and has given you some tools by which to do it. However, sometimes it is also helpful to pay attention to how others approach and resolve problems. Look around you. You probably won't have to look far until you find someone—a friend or a peer—who thinks very differently than you about problems and their solutions. What are those differences? How might others have dealt with the issue(s) you are facing? Can you learn anything from their approaches?

We close this chapter with this student's thoughts:

> *I never thought the crisis I experienced and the change I brought about could be used as a selling point on my cover letter.... The whole thing has made me realize how beneficial this crisis has been for me down the road.*
> STUDENT REFLECTION

For Review & Reflection

Seminar Springboards

Now It's Your Turn. Before you begin working with the 8-Step Model, think about the strengths you bring to that model—the ones you perhaps thought about as you read your way through the description and example provided in the chapter—the ones that you thought to yourself, "Oh, I knew that" or "I can do that."

Perspective Taking. In seminar class or in your support teams, select a problem—a most challenging one—that you are or a peer is having or could have. This time when you work your way through the 8-Step Model, have peers select their parts in a role-play of the problem solving. *If you do not have a seminar class nor are part of a support team, then this exercise can be done with you playing more than one role until such time that the process becomes familiar. At that point, you need only identify the key points in the perspectives of each remaining role.*

For Further Exploration

On Problem Solving

McClam, T., & Woodside, M. (1994). *Problem solving in the helping professions.* Pacific Grove, CA: Brooks/Cole.

Clearly written and concise with a problem-solving model and lots of cases to study.

On Interpersonal Effectiveness

Corey, G., & Corey, M. (2010). *I never knew I had a choice* (9th ed.). Pacific Grove, CA: Brooks/Cole.

Excellent book with a good chapter on relationships.

Johnson, D. W. (2008). *Reaching out: Interpersonal effectiveness and self-actualization* (10th ed.). Boston: Allyn & Bacon.

Clear and incisive with lots of examples and exercises—a classic.

On Making Intrapersonal Change

Ellis, A., & Harper, R. A. (1975). *A new guide to rational living*. Upper Saddle River, NJ: Prentice Hall.

Ellis's ABC model is explained with helpful self-help guides to work through your own situations.

Gibbs, L., & Grambrill, E. (1996). *Critical thinking for social workers: A workbook*. Thousand Oaks, CA: Pine Forge Press.

Provides a comprehensive skills-based approach to develop reasoning skills for effective decision making.

Grobman, L. M. (Ed.) (2002). *The field placement survival guide*. Harrisburg, PA: White Hat Communications.

Offers social workers an edited volume of writings that includes skills and advice in confronting challenges.

Levine, J. (2012). *Working with people: The helping process* (9th ed.). New York: Longman.

Self-understanding and self-modification are discussed throughout this book.

Magolda, M. B. (1992). *Knowing and reasoning in college: Gender-related patterns in undergraduates' intellectual development*. San Francisco: Jossey-Bass.

Classic book, describes how ways of knowing change over the course of the college years and how gender influences one's ways of reasoning.

Schlossberg, N. K., Waters, E. B., & Goodman, J. (1995). *Counseling adults in transition* (2nd ed.). New York: Springer. Free Press.

A classic text with a useful and informative framework for working with adult students who are in life transitions.

Weinstein, G. (1981). Self science education. In J. Fried (Ed.), *New directions for student services: Education for student development* (pp. 73–78). San Francisco: Jossey-Bass.

An easy-to-follow model for interrupting reaction patterns that may be troubling you.

On Systems Change

Garthwait, C. (2010). *The social work practicum: A guide and workbook for students*. (5th ed.). Upper Saddle River, NJ: Prentice Hall.

Offers a comprehensive approach across contexts of practice for skills development in bringing about change.

Homan, M. (2011). *Promoting community change: Making it happen in the real world*. (6th ed.). Pacific Grove, CA: Brooks/Cole.

Excellent, well-written text on community organizing with interesting thoughts on the change process.

Kotter, J. P., & Cohen, D. S. (2002). *The heart of change: Real-life stories of how people change their organizations.* Cambridge: Harvard Business School Press.

Offers an approach that affirms the importance of teamwork, visions, and communication in bringing about short-term wins that create change.

Watzlawick, P., Weaklund, J. H., & Fisch, R. (1974). *Change: Principles of problem formation and problem resolution.* New York: Norton.

Somewhat theoretical but very interesting. A groundbreaking book on why change can be so stubborn and what to do about it.

On Communication

Godwin, A. (2011). *How to solve your people problems: Dealing with your difficult relationships.* Brentwood, TN: Rosenbaum & Associates Literary Agency.

Provides tools and a framework for handling differences and draws upon practical principles to avoid conflict.

Kegan, R. & Lahey, L. (2001). *How the way we talk can change the way we work.* San Francisco: Jossey-Bass.

An extended look at the power of language to shape experience. The authors guide you step by step through exercises to refram your thinking.

References

Corey, G., & Corey, M. (2010). *I never knew I had a choice* (9th ed.). Belmont, CA: Brooks/Cole.

Harvard Health Publications. (2011, March). Understanding the stress response. *Harvard mental health letter, 27,* (9) 4–5. Boston, MA: Author.

Kegan, R., & Lahey, L. L. (2001). *How the way we talk can change the way we work.* San Francisco, CA: Jossey-Bass.

Lewin, K. (1954). Behavior and development as a function of the total situation. In L. Charmichael (Ed.), *Manual of child psychology* (2nd ed.). New York: John Wiley & Sons.

McClam, T., & Woodside, M. (1994). *Problem solving in the helping professions.* Pacific Grove, CA: Brooks Cole.

Swaner, L. E. (2012). The theories, contexts, and multiple pedagogies of engaged learning: What succeeds and why? In D. W. Harward (Ed.), *Transforming undergraduate education: Theories that compel and practices that succeed.* (pp. 73–90). New York: Rowman and Littlefield.

Sweitzer, H. F. (1989). The BPE framework: A tool for analysis and interdisciplinary integration in human service education. *Human Service Education, 9*(1), 11–19.

Voegele, J. D., & Lieberman, D. (2005). Failure with the best intentions: When things go wrong. In C. M. Cress, P. J. Collier, & V. L. Reitenauer (Eds.), *Learning through serving: A student guidebook for service-learning acorss the disciplines*. Sterling, VA: Stylus.

Watzlawick, P., Weaklund, J. H., & Fish, R. (1974). *Change: Principles of problem formation and problem resolution*. New York: Norton.

CHAPTER *10*

Navigating the Internship Site

Don't Skip This Chapter!

In our experience, interns react pretty differently to the prospect of learning about organizational dynamics. Some are very excited about it and some, particularly those who have studied business, already know a good deal about this topic. Others are not interested at all; they want to do the work of the site, not learn about the site itself. Well, we are going to encourage you to make understanding the placement site part of the exploration and expansion that happens during this stage.

Throughout this book we have emphasized the importance of engagement, and of *contextual* engagement (Swaner, 2012), which is the ability to consider issues in their larger context. This chapter is an opportunity to do just that. Additionally, if you are not prepared for the organizational dynamics and issues discussed in this chapter, you could be in for some unpleasant surprises. Suppose, for example, that your supervisor, who meets with you once a week, is not around when you are working, but another supervisor is there most of the time that you are. This person seems unusually hostile and appears to resent the time you spend with your supervisor. She also occasionally contradicts something your supervisor has told you. While this situation would not be easy under any circumstances, it will be less of an enigma if you understand that your supervisor was recently promoted to that job for which the other supervisor was passed over, even though she has been there longer. Whether you know it or not, organizational dynamics are bound to affect you, your colleagues, your supervisor, and even the people you work with if you are in direct service.

281

Lenses on Your Placement Site

We are going to encourage you to look at your placement site through the related lenses of systems theory and organizational theory, each of which you may have studied in some of your classes. Even if you haven't, we think you'll find the chapter informative and useful to your understanding of the context of your work. We have chosen a few concepts from both theories that we think are particularly relevant to your work as an intern. We will not be covering all, or even most, of the major concepts, so you may want to investigate them further or even take a course in organizational behavior. In fact, after you read this chapter, you may want to make some sort of organizational analysis part of your learning goals and activities.

Systems Concepts

A system is a group of people with a common purpose who are interconnected such that no one person's actions or reactions can be fully understood without also understanding the influence on that person of everyone else in the system (Egan & Cowan, 1979). Even if you understand each person in the system as an individual, to understand the system you must understand the way everyone interacts; the whole, in this case, is greater than the sum of its parts (Berger, McBreen, & Rifkin, 1996).

Systems can be analyzed internally or externally (Berger et al., 1996). Internal analysis involves studying the inner workings and components of the system and the way human and material resources are arranged and expended. However, it is important to realize that all systems are hierarchical. This does not mean that they use a hierarchical authority structure; some do not. But each system is part of a larger system, and most can be broken down into smaller systems. This hierarchical dimension is easy to see in large organizations, but it is present in smaller ones as well. A family services center, for example, can be broken down into the various programs it runs. However, it is also part of a system of service providers in its city or town. An external analysis is used to examine the relationships between a system and other related systems.

Organizational Concepts

In both internal and external analyses, organizational theory can help you look at, make sense of, and navigate your placement site. There are many different organizational theories, but for this chapter we rely heavily on the integrative work of Lee Bolman and Terrence Deal, whose book *Reframing Organizations* is an excellent resource if you want to explore organizational theory further (2008). We begin with some of the components of an internal analysis. First, we outline some basic and important background information that is important to know about any organization. The mission, goals, and values of an organization tell you in an official and explicit way what that organization is all about and what makes it distinctive. Funding, of course, provides the underpinning for the work and sets crucial parameters. Then we use Bolman and Deal's four lenses or frames—structural, human resources, political, and symbolic—to help you examine how your placement site is organized to carry out its mission and goals, and whether it practices the values it espouses.

Background Information

In presenting the concepts in this chapter, we will refer to a variety of kinds of placements, including a fictional agency in the community called the Beacon Youth Shelter, an amalgamation of several sites with which we have worked.

History

Every organization has a history that provides an important context for the way things are currently being done. Older organizations often have gone through a number of changes. For example, the Beacon Youth Shelter always had a rule that it would not accept residents with a history of physical violence. However, it has recently agreed to change and admit children with violent histories. That means this population is relatively new to the organization, and there is no doubt an adjustment period will go on for everyone. Also, organizations are sometimes bought by or merged with other organizations. Can you see why this information would be useful for an intern to know?

Mission

It is easy to gain a general sense of the purpose of your placement; sometimes the name is all you need, as in the Environmental Collaborative or Child and Family Services. But each organization has an overall mission, and that is a good place to start to see what is distinctive about it. It is likely that your placement site has a written statement of its goals. Here are some examples:

- From a media outlet: "NPR is committed to the presentation of fair, accurate and comprehensive information and selected cultural expressions for the benefit of, and at the service of our democracy. NPR is pledged to abide scrupulously by the highest artistic, editorial, and journalistic standards and practices of broadcast programming" (NPR, 2012).

- From a social service agency: "To provide quality and multicultural services to those whose lives have been affected by sexual assault."
- From a large insurance company: "HSB's technical knowledge and expertise is the key that has helped our customers avoid losses, and recover promptly from losses that do occur, for over 140 years. The Hartford Steam Boiler Inspection and Insurance Company benefits businesses and industries worldwide by providing Equipment breakdown insurance and reinsurance; Other specialty insurance and reinsurance; Risk management consultation; Engineering and inspection services" (HSB, 2012).
- From a criminal justice agency: "To serve adults and youth who exhibit or are at risk of criminal or delinquent behavior, substance abuse, or mental illness, as well as other socially disadvantaged persons."

These formal mission statements may be well known to the people who work at the organization and/or to its clients or customers, or they may be unknown to either group. A lot depends on who determined the mission and through what process (Caine, 1994). There will most likely be greater investment in and awareness of mission statements that were arrived at collaboratively, as opposed to those created by one or two people and then handed down.

Goals and Objectives

Mission statements are a good place to start, but you will find that agencies that do similar work often have similar mission statements. There is, however, a wide variety in the specific goals and objectives among organizations, even among those with very similar titles. Goals are rather broad and hard to measure. Putting an end to violence against women is an example of a goal statement. It is a goal to work toward, but not one that is likely to be reached any time soon and therefore it is hard for the agency to measure progress toward it. That same agency might have several objectives, including a 25% reduction in the incidence of teen dating violence, the adoption of certain laws and procedures, and the provision of some sort of service and assistance to every woman who requests it. These objectives are much easier to measure.

Values

Organizations also have values or principles that they try to adhere to in doing their work. Again, there are often formal written statements of these values, which may include general principles, rights, and responsibilities. Here are two brief examples:

- "Our coverage must be fair, unbiased, accurate, complete and honest. At NPR we are expected to conduct ourselves in a manner that leaves no question about our independence and fairness. We must treat the people we cover and our audience with respect" (NPR, 2012).

- "The values of Interval House define who we are, what we stand for, and how we do our work. They are principles against which we measure the importance and worth of our decisions and actions. The staff and the board of Interval House hold ourselves accountable for transforming these values into action: empowerment, support, collaboration, safety, diversity, integrity, compassion, confidentiality, dedication, and equity" (House, 2007).

Value Statements from Two Organizations

Interval House

Empowerment

We believe in power with, not power over. We help one another acquire power over their own lives and we respect the decisions that others make. People have the right to make their own decisions and to accept responsibility for them.

Support

We listen well and we provide support to one another in reaching our goals.

Diversity

We seek and welcome cultural diversity of all kinds as well as diversity of opinion, perspective, talents and gifts. There can be no excellence without diversity.

Compassion

Although we may disagree, we do not judge. We accept people for who they are, regardless of whether we agree with their behavior, and we believe in the capacity for growth and change.

Dedication

We believe in what we do and we work hard. We hold one another accountable for putting forth the effort and commitment that our work demands.

Collaboration

We believe that the greatest results come from the combined efforts of diverse organizations and individuals. We reach out to work cooperatively with others to promote meaningful and mutually beneficial change.

Safety

We believe that each person has a right to live and work free from fear and safe from harm.

Integrity

We are honest and reliable, open and sincere. We hold ourselves accountable for doing what we say we will do. We evaluate our effectiveness and change as necessary.

(continued)

Value Statements from Two Organizations *(continued)*

Confidentiality

We protect the privacy and confidentiality of those with whom we work, unless doing so would lead to harm to others.

Equity

We believe that all people deserve fairness, impartiality, justice and opportunity.

National Public Radio

Fair means that we present all important views on a subject. This range of views may be encompassed in a single story on a controversial topic, or it may play out over a body of coverage or series of commentaries. But at all times the commitment to presenting all important views must be conscious and affirmative, and it must be timely if it is being accomplished over the course of more than one story. We also assure that every possible effort is made to reach an individual (or a spokesperson for an entity) that is the subject of criticism, unfavorable allegations or other negative assertions in a story in order to allow them to respond to those assertions.

Unbiased means that we separate our personal opinions—such as an individual's religious beliefs or political ideology—from the subjects we are covering. We do not approach any coverage with overt or hidden agendas.

Accurate means that each day we make rigorous efforts at all levels of the newsgathering and programming process to ensure our facts are not only accurate but also presented in the correct context. We make every possible effort to ensure assertions of fact in commentaries, including facts implied as the basis for an opinion, are correct. We attempt to verify what our sources and the officials we interview tell us when the material involved is argumentative or open to different interpretations. We are skeptical of all facts gathered and report them only when we are reasonably satisfied of their accuracy. We guard against errors of omission that cause a story to misinform our listeners by failing to be complete. We make sure that our language accurately describes the facts and does not imply a fact we have not confirmed, and quotations are both accurate and placed properly in context.

Honest means we do not deceive the people or institutions we cover about our identity or intentions, and we do not deceive our listeners. We do not deceive our listeners by presenting the work of others as our own (plagiarism), by cutting interviews in ways that distort their meaning, or by manipulating audio in a way that distorts its meaning, how it was obtained or when it was obtained. The same applies to text and photographs or other visual material used on NPR Online. Honesty also means owning up publicly and quickly to mistakes we make on air or online.

(continued)

Value Statements from Two Organizations *(continued)*
Respect means treating the people we cover and our audience with respect by approaching subjects in an open-minded, sensitive and civil way and by recognizing the diversity of the country and world on which we report, and the diversity of interests, attitudes and experiences of our audience.

© Cengage Learning

Funding

The organization's budget may not seem very important to you; after all, you may are not be getting paid! However, the financial as well as human resources of an organization do determine what it can do for clients or customers and often affect the general tone of the workplace. Your placement site certainly has an annual budget. The budget may be part of a larger budget, as in the case of a senior services center that is funded by the city; its budget is part of the social services budget, which is in turn part of the city budget. The budget is usually broken down into categories, such as salaries, benefits, supplies, food, transportation, and so on. It is also important to know how the budget is set, who must approve it, and how changes are negotiated. Of course, budgets can change suddenly; in nonprofit organizations, which many community and service agencies are, that usually means they get cut, but occasionally an agency will be awarded a grant or get a new program approved and funded, which may in turn affect the agency's goals.

Of Special Relevance to The Helping and Service Professions

Sources of Funding

One aspect of human service agencies that is often overlooked by interns is where the money to operate actually comes from. Even if your placement site does not charge for its services, they are not free. The money to pay the staff and administer the agency that provides the "free" services has to come from somewhere. Generally speaking, agencies can be divided into three groups: public, private nonprofit, and for profit.

Public Agencies are funded through local, state, and federal tax revenues. They may have some other sources as well, such as foundation support or grant money, but taxes provide the majority of their funds. Sometimes the clients are charged a fee on a set scale or one that varies according to their income, but those fees do not begin to cover the cost of operation. What that all means is that these agencies are ultimately accountable to the legislative bodies. That is usually how they were created, and it is typically the body that decides on the agency's budget. Berg-Weger and Birkenmaier (2010) note that public agencies tend to be large, with a vertical authority structure (we will be discussing structural issues more later on). They further note that public agencies tend to have complex rules and procedures and a lot of accompanying paperwork. These agencies, or their parent organizations, lobby legislators to ensure that their funding is maintained or increased. Examples of public agencies include departments of child welfare, aging, corrections, and education.

Private Nonprofit Agencies get their funding from a variety of sources, including charitable organizations such as the United Way, foundations, corporations, and individual donors. This diversity of funds is a challenge, because it means maintaining developing relationships with and being accountable to a number of groups, but it is also a strength in that the organization may be able to avoid being overly dependent on any one source (Berg-Weger & Birkenmaier, 2010). Nonprofit agencies may receive some public funds, usually through contracts with state agencies or though Medicare. Beacon Youth Shelter, for example, contracts with the state's child welfare agency for a certain number of beds each year. Nonprofit agencies are accountable to a board of directors, who are typically volunteers. If your agency has such a board, you may find it interesting to attend a board meeting, if the agency will permit it. Nonprofit agencies generally have a lot of paperwork, but unlike public agencies they tend to go to a variety of funding sources, and sometimes to regulatory agencies as well. Fund-raising also takes up a great deal of time and energy, and as an intern you may have the opportunity to be involved in some fund-raising activities.

For-Profit Agencies get funds to start their work from individuals or private organizations, but they expect to be self-supporting. In fact, they expect more than that—they expect to make a profit. Kiser (2012) notes that this gives these agencies a dual mission to serve their clients and to make a profit. Birkenmaier and Berg-Weger (2010) assert that profit is always the driving factor in these agencies. In any case, they have to pay attention to making money, and they are held accountable for that goal, usually by a board of directors. Keeping client interests primary while still fulfilling their obligations to the board is one of the key challenges for these agencies. Like nonprofit agencies, for-profit agencies may access public funds indirectly, by contracting with state agencies. A for-profit agency might, for example, monitor children who are in court-mandated placements and services, and there have been some well-publicized experiments where public schools are operated by private organizations. The growth in for-profit human service organizations is also referred to as privatization and extends into many areas, including nursing homes, group homes for people with mental health issues, and rehabilitation.

This issue of funding is especially important because the source of funding has great power over and influence on the organization. Insurance companies, for example, usually insist on a formal diagnosis before authorizing treatment. Some human service providers are troubled by these diagnostic labels, which tend to stay in clients' files and follow them around. Nevertheless, a diagnosis must be made, and it must be one the insurance company will pay for, or the client will have to pay cash or be referred to another agency. Organizations funded by tax dollars are vulnerable to the opinions (informed or not) of the taxpayers and must worry about public relations and influencing the local political process.

Organizational Structure

An organization's structure is the way in which it is set up to accomplish its goals (Bolman & Deal, 2008). There are two basic elements of structure—division of responsibilities and coordination of work—and there are also many subcategories within each of those two elements. Each feature of an organization's structure can be tight or loose, clearly defined or ambiguous. There are endless configurations and no one of them is the best in all situations. However, some may work more effectively in certain situations than others. Let us look more closely at your placement site using a structural lens.

Division of Responsibilities and Tasks

Who will do what, and in what configurations, is one of the basic questions for any organization. Most have developed a formal structure to answer this question, although as we will see later on sometimes the real operation of the agency differs from the formal structure.

Roles

Roles describe the positions in the organization and the duties or functions that each one performs on a regular basis. The formal roles in an organization can be found in two places: job descriptions and the organizational chart. Understanding these two contexts for roles is important to understanding the integrity of the organization's structure. So, too, is understanding the degree of specialization and the purpose for grouping roles.

- *Job Descriptions*　Most organizations have job descriptions for each position that state, at least in general terms, the responsibilities of the

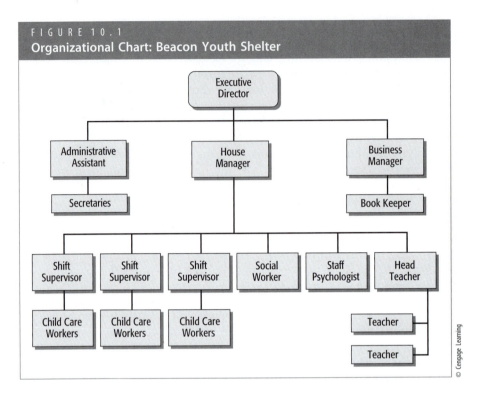

FIGURE 10.1
Organizational Chart: Beacon Youth Shelter

© Cengage Learning

positions. These descriptions may be found in the policy manual, or they may be found in the human resources department. They are often generated, or reviewed, when someone is hired. A prospective employee will usually want to see the description, and in a large organization, a higher-level administrator needs to see a job description to approve the hiring. Many placement sites do have written job descriptions for interns.

- *Organizational Chart* An organizational chart is also valuable when examining formal roles. An organizational chart does not describe the duties of each position, but it shows their position relative to one another. It shows who is responsible for whom and to whom each person is accountable. A chart for the Beacon Youth Shelter appears in Figure 10.1.

Specialization vs. Generalization

Another issue in examining roles is the degree of specialization assigned to each role. Some organizations have highly specialized jobs, each assigned to one person (or one category of people). For example, at a child welfare agency there may be one "role" that investigates reports of child abuse, often called an intake worker. Workers who assist victims of abuse could be in a different role—they may be called treatment workers.

If the family needs economic assistance, it is referred to yet another worker. In other agencies there is less specialization. For example, at a small advertising agency there are several people who handle accounts and try to handle all the needs of their clients.

Both of these models have advantages and disadvantages. It is daunting to be expected to know all of the functions and information; and if someone's role is too generalized, people may get inadequate advice or miss out on important options because of the organization's structure. On the other hand, when services are highly specialized, people can get caught in a maze of different staff roles and may feel overwhelmed or confused. It is frustrating to ask a question and be told "that is not my area." It might have already happened to you!

A Word to the Wise...   If you look at the organizational chart above, you will notice that there is no position for the intern. If an organization does not have an intern each semester or year, it may not include one. However, it is important to find out where you fit on the chart. Most interns tell us they fit "on the bottom." While that may or may not be true, it does not help us or the interns not to know to whom they are responsible or that person's place in the organization.

Groupings or Teams

Roles are often organized into teams. Sometimes those groupings are functional, and people who have the same or similar tasks are placed in the same department. Other ways to organize groups are according to time (day shift vs. nights shift), clients (as in the case of intake vs. treatment workers), and place (such as at a satellite office) (Bolman & Deal, 2008).

Coordination and Control of the Work

Regardless of the particular roles and teams in an agency, there has to be a way to coordinate what everyone does and guide or control their efforts. How much coordination is necessary depends on the interdependence of the roles involved. In a hospital or a school, for example, the people in various roles are very dependent on one another in order to accomplish their goals. According to Bolman and Deal (2008) there are two dimensions of coordination and control: vertical and lateral. Every organization has them both, but some emphasize one more than the other. And of course, there is considerable variation within each of these dimensions.

Vertical Coordination and Control

Vertical coordination and control refers to the ways in which the work of the agency is assigned, controlled, and supported. This dimension consists of structures for authority (the chain of command), communication, decision making, policies, rules, and evaluation procedures.

Authority Structure The authority structure of an organization defines the chain of command, and this chain is visible in the organizational chart. You can see from the chart of the Beacon Youth Shelter that the top administrator is the executive director. The executive director of Beacon Youth Shelter is responsible for all phases of the operation. However, other people do most of the actual running of the shelter. The executive director is most closely linked to the house manager, who is in turn responsible for several professional positions. The executive director may or may not attend staff meetings or know any of the child care workers by name.

An organizational chart also gives you an easy view of the length and complexity of the chain of command. Organizations that have many layers and a long chain of command are referred to as *tall* organizations, while organizations with fewer layers and a shorter chain of command are referred to as *flat* organizations (Queralt, 1996). Each has its advantages and disadvantages in carrying out the work and in putting the organization's values into action. Consider collaboration, for example. In a flat organization, more people have input into decisions, but the decision-making process can become quite cumbersome. In tall organizations, fewer people at each level are involved in collaboration, which makes it easier to make decisions but is not as inclusive.

Of Special Relevance to
The Helping and Service Professions

Where Are the Clients?

If you look at the organizational chart on p. 290. you will also notice that there is no place for clients on the chart. While that is not unusual, it is important to know the position occupied by clients in your agency (Queralt, 1996). In medical settings, especially hospitals, there often are patient advocates and that is one way to bring client/patient voices to the organizational table. In some agencies, clients are included in decision making. There may be an advisory board, for example, or a client council. This arrangement is more likely in organizations where clients can choose to use (and pay for) the service or go elsewhere. This arrangement gives clients some power and leverage. However, in agencies where the clients are not there voluntarily or have no other alternatives, clients may have very little influence on decisions.

Communication Flow Successful communication is, of course, the critical factor in making any organization run smoothly. In a small organization, communication may be very informal. Everyone sees and works with everyone else all day, and important information can be shared relatively easily, although it sometimes is not. Communication in larger organizations is more complicated and often happens through memoranda (memos), or increasingly through e-mails or other forms of electronic or social media. They can be sent to just one person or to a whole group of people.

It is possible to look at patterns of communication in an agency. In some organizations, communication flows from the top levels down, with each level playing a role, and from the bottom levels up, with relatively little communication among people or departments at the same level. This arrangement is called a *chain pattern*. In a *wheel pattern*, on the other hand, one person or department is at the hub. Most communication flows directly from various departments to this hub and from the hub to the departments. Again, there is relatively little communication among the departments themselves. In an *all-channel network*, everyone communicates with everyone else (Queralt, 1996). There are many other configurations, and it can take some time to determine what pattern a given agency actually uses.

Decision-Making Mechanisms Decision making is another important structural consideration. Suppose you are working at Beacon Youth Shelter and you have an idea for a field trip. There is an interesting exhibit at the science museum, and some of the residents are interested in attending. But the exhibit is not going to be there much longer. Who must you ask about this trip and how long will it take to get approval? Your supervisor thinks it is a great idea, but you are surprised to learn that the executive director, who is out of town, must approve all such trips. You also have to obtain permission from each child's parent, guardian, or case-worker, and teachers must approve the absence from school for a day. This story is a bit exaggerated (although in some cases, not by much), but it illustrates the importance of understanding the decision-making mechanisms at your placement and how they affect your work. What projects or interventions have you planned that may require multiple approvals?

Policies and Procedures Every organization has at least some formal written rules that everyone is supposed to follow. These rules can be found in an employee handbook or policy manual, which you may already have seen. Some organizations have very thick manuals; some have more than one. You may need to ask for help in finding the most relevant material. It is also important to know how these rules and procedures were established and by whom (Caine, 1998). Sometimes it is done by a committee; other times it is delegated to one person or written by a top administrator or manager. The change process is also significant. Some policies have not been changed, or even reviewed, in many years. Other organizations have scheduled reviews and updates in place for their policies and procedures.

It is also interesting to see how many people at the agency are really familiar with the rules. Both of our universities have a student handbook; chances are that yours does, too. In our experience, students read that book when they have a question or when they think they may have broken a rule or been treated unfairly. Otherwise, they may never look at it! At some placements, you will find that this is true for the policy manual as well.

Evaluations   Evaluations are an important structural consideration. Remember that part of the vertical dimension is control, influence, and guidance. Evaluation is one mechanism that can serve those purposes. In most organizations, there is some mechanism for evaluating or assessing performance. Sometimes a new employee is given an initial evaluation after a few weeks. Periodic evaluations may be done annually, semiannually, or even monthly. Evaluations are usually conducted by the person to whom you are accountable (see the organizational chart on p. 290).

- *Evaluation Methods*   There are many approaches to evaluation. Some methods allow the employee to set goals and then be evaluated on how successfully these goals have been reached. Other times, the criteria and goals are set by the supervisor, and there are many approaches between these two extremes. Some organizations use self-evaluation or peer evaluation in addition to or in conjunction with the supervisor's evaluation. The format of the report can also vary from a simple checklist or set of numerical ratings to more complex narrative formats.

  An important consideration is the frequency of the evaluations. Some sites prefer to evaluate according to the schedule used for their employees, which is often twice yearly. Other sites prefer a more frequent schedule, especially during the first six weeks of placement. Your campus program has an expectation about the frequency of evaluations, which you will need to know as early as possible.

- *Evaluation Criteria*   The ways in which staff members are evaluated says a lot about what is important to the organization. One important issue here is the aspects of the job that are included in the evaluation. Think about your classes in college. You have readings assigned, and many of your instructors may tell you that class participation is important. However, if you are never tested on your understanding of the reading and participation is not a part of your grade, you may be tempted to let these areas go, especially if they are difficult. By evaluating you on certain components of your performance but not on others, the instructor is sending a message, perhaps unintentionally, about what is important in the class.

  Returning to your internship site, employees are supposed to be evaluated based on how well they perform the duties in their job descriptions. In practice, however, some of these duties may not be reviewed, commented upon, or counted toward promotion, pay increases, and other organizational rewards. In addition, a supervisor may comment on aspects of the job that are not in the job description. One professional we know was criticized in writing for not being sociable with others at the site, even though his work was handled quite well.

- *Performance vs. Outcomes*   Some organizations evaluate performance (also called input), some evaluate outcomes, and some evaluate both (Caine, 1998). Again, an analogy to school may help illustrate this distinction. Teachers of a U.S. history class may be evaluated based on whether they cover all the material by the end of the year and whether students seem to enjoy and be stimulated by the class. Another approach would be to evaluate teachers based on what students are learning. Some schools believe that it is the teacher's responsibility to lay out the information and the students' responsibility to learn it. Others believe that the school and its teachers have a responsibility to see to it that every child learns. The evaluation methods chosen reflect these two approaches. Similarly, an organization may evaluate employees based on the number of clients seen or the number of hours worked, or it may choose to find ways to measure the change in clients and evaluate its workers in that manner. In private organizations and for-profit human service agencies, the emphasis is almost always on outcomes.

Lateral Coordination and Control

Although every organization has a vertical structure, it is not always the most effective way of getting the work done. Policies and chains of command are not always followed, nor are they always efficient. Lateral methods of coordination and control are those less formal and more flexible structures that emerge, over time (Bolman & Deal, 2008). Don't confuse lateral control with a "flat" authority structure. Flat structures are less hierarchical, but lateral methods are still needed. Lateral methods may include meetings, which can occur monthly, weekly, or daily. Some organizations begin each work shift with a meeting. Various staff groups then meet once a week for more in-depth discussions, and all of the groups meet together once a month.

Task forces or temporary work groups are another form of lateral coordination, and ones that you may have the opportunity to be involved in as an intern. For example, most service agencies have to be evaluated by local, state, or even federal licensing organizations from time to time. Generally, the organization has to gather data and write a self-study for the visiting team to review. In addition, the team visit, which can last for several days, needs to be organized and coordinated, and large numbers of documents have to be made available. Some of this work can be done on an ongoing basis and built into someone's role. But some of it is temporary and occurs only once every few years. In these cases, a temporary team may be organized to prepare the self-study, arrange documents, and handle the logistics of the team visit.

Task forces usually arise out of some particular problem or challenge. For example, an organization may be concerned at the lack of racial and ethnic diversity in their workforce, feeling that it does not represent the diversity in the surrounding community. A task force may be appointed, consisting of individuals from many segments of the organization, and asked to study this problem and propose a solution.

Human Resources

Another way to look at your placement site is to focus on the people rather than the structure. In doing so, you will see a whole other, very rich, dimension of life in that organization (Bolman & Deal, 2008). Here we are concerned not so much with what people are supposed to do, but with what they *actually* do as they try to match their own needs to the needs and features of the organization. As with the structural lens, there are a number of components to consider when assessing the human dimension.

Communication Skills in Organizations

Using the structural lens, you looked at patterns of communication. These patterns emerge over time and it is almost as if you must observe from a distance to see them.

One aspect of communication is its clarity. Consider how well people in the organization listen to one another and how they are helped to do so. Another aspect of communication is its openness. In some organizations, people generally say what they think about issues and about one another, in honest and clear ways. There is no prescription for how to accomplish that goal; specific approaches vary depending on the constellation of individuals and cultural profiles in the organization. Still, every organization needs to find a way to accomplish this goal, and some do so far more effectively than others.

Conflict resolution is another set of communication skills that is visible in various ways and to various degrees in organizations. If there are not some structured, accepted, and supported ways of dealing with conflict, more informal and often less effective methods will be used. One of us had an intern whose co-workers complained about her in formal and informal meetings, yet no one spoke directly with her, nor did anyone try to stop the conversations that were taking place behind her back. When someone finally took her aside and told her of these conversations, she spoke with her supervisor. The supervisor then went to the individuals involved, but not to the group, and not in the presence of the intern. Needless to say, the situation did not get better.

Norms

As you read the section on the rules of organizations, you may have occasionally wondered how to answer some of the questions about your placement. In some cases, you don't know the answer at all, but in others, you may be wondering, *"Do you want to know what's written down or do you want to know what's really going on?"* Often, the formal rules, roles, goals, and values only tell part of the story. Consider the speed limit on the interstate. The formal limit may be 55 or 65 mph, but how fast you can go before risking a ticket is the informal rule that most people concern themselves with. If you live in a college dormitory, there is probably a time after which you are supposed to be quiet. But the unwritten rule usually is that you can make all the noise you want as long as no one complains. These are examples of informal rules or norms that contradict or modify the formal rules. Other norms are not products of the formal rules. Most schoolchildren will tell you that it's wrong to "rat" or "tattle," often to the chagrin of adults. This is a very powerful rule for many children, yet it is not written anywhere.

Every organization has unwritten rules. Think back on jobs you have had and on your current internship placement. There is an official start time, but there is usually a grace period of at least a few minutes. In some places, you are expected to work late—for no extra pay—and in other places, you are not supposed to do that even if you want to. Many organizations have an informal dress code to go along with the formal one. Sometimes the informal rules come about because the formal rules have eroded or became obsolete but have never been changed (Gordon & McBride, 2011).

Here are some other examples of unwritten rules that interns have encountered:

- Go to lunch with your co-workers or they will think you are a snob.
- Never go out to lunch; work at your desk.
- Don't miss the boss's holiday open house.
- You can date another employee, but keep it quiet.

And here is another one observed by one of our interns: *"the more time out of the office the better. The assumption is that when you are out in the field you are visiting directly with clients...which is how the work is billed."*

The written rules are usually easy to find, but how do you go about finding the unwritten ones? First of all, you will usually know when you have broken one. Just like in high school, when you wore the "wrong" thing, used an expression that was suddenly "out," or sat at

A Case in Point

Unwritten Rules *Rule!*

At one family services agency, for example, the hours were 9 A.M. to 7:30 P.M., Monday through Friday. However, in recent years, the agency had been doing more and more outreach programming, working with groups in the community, meeting during the evening in church halls or people's homes, and developing a lot of weekend programs. Fridays are an especially busy day, as the finishing touches are put on the week-end activities. As a result, Mondays have become an unofficial day of "downtime." Few appointments or meetings are scheduled, and many staff members do not come in until noon. If you were an intern there and wanted to work on Monday or tried to launch a new group that meets on Mondays, you would receive a chilly reception, and you might wonder why.

the "wrong" table in the cafeteria, people will react strangely; perhaps someone will even take you aside and clue you in when you violate an unwritten rule. Another way to discover the unwritten rules is to know the written rules and observe what happens when they are broken (Caine, 1998). In some cases, nothing happens. Either the rule has been bent in a way you don't know about, or it has been ignored altogether. As one of our students said: *"I learned the rules by observing my surroundings and conversations."*

Informal Roles

It will not take you long to notice that some people perform duties that are not in their job descriptions. It may be that they were asked to do it and they didn't feel that they could say "no." Consider, for example, the employee whose supervisor discovers that her route home takes her past where his son has piano lessons. One day, the supervisor asks if she would mind picking up his son, since he will be in a meeting. The employee is happy to help, but it soon becomes an expectation, and that may become problematic.

Another way job descriptions expand is when a task is not in anyone's job description but needs to be done; someone just decides to do it, and after a while everyone relies on this person. There may be someone, for instance, who remembers everyone's birthday, takes up a collection for a gift when someone gets married or retires, or makes the arrangements for parties. Other roles in organizations might include the jester, who makes everyone laugh, and the go-between, who seems to have access to just about everyone.

Sometimes people expand their job description to make their jobs more interesting, especially if they have had the same job for a while (Gordon & McBride, 2011). At the Beacon Youth Shelter, a child care worker became interested in grant writing, volunteered to work on some grants, and eventually became quite good at it. Of course, his job description expanded and his paycheck did not, but it did make him a more influential person in the organization, and he developed a new skills set which are reasons people sometimes try to expand their roles.

Cliques

There are informal subgroups or cliques in many organizations. At Beacon Youth Shelter, for example, there is a group of child care workers who have been there a long time; they are referred to as *the veterans*. Few of them had any formal education to prepare them for their work; they learned on the job. Some of them rose in the organization to become shift supervisors, and until recently, the house manager position was held by a former member of this group. In recent years, though, there have been more and more workers hired who were graduates of programs in social services. They are often younger and full of new ideas. Some of these ideas work, but others don't. The veterans sometimes shake their heads at these "whiz kids," and sometimes the whiz kids get impatient with the veterans.

Of course, both groups have a valuable perspective to offer the organization, and most of the time they work well together. Still, there is some tension and competitiveness between them. It doesn't help that the new house manager did not come from within, is a graduate of a human services program, and has a master's degree. The veterans miss the easy relationship and open access they had with the old house manager; they could just approach her privately and accomplish a lot. Not only do the veterans not have this relationship with the new manager but some of the whiz kids seem to be developing this informal access.

In Their Own Words   **Interns Encounter Cliques**

I have noticed that there are several subgroups: a sports one, a political one, and a friendship outside work one.

I've noticed that there is a core group that seems closest to the director.... I would term this as the "in crowd." I don't believe the director purposely treats anyone with less respect professionally, but this core group is afforded more personal time, where lingering, joking, and story telling take place.

Management Style

Just as your supervisor has a style of managing you, people in supervisory positions have their own styles with the people or groups they supervise. There is considerable variation among individuals, of course, but there is often a prevailing style in the organization itself. If you go back and reread the section in Chapter 6 on supervisory styles, you can think about how this applies to your placement site.

Staff Development

Most organizations put some time and energy into helping their employees grow as professionals and do their jobs better. These activities are often referred as *staff development*, and they may take place during a portion of regular staff meetings, or special meetings may be held exclusively for that purpose. Some organizations or units take their staff away on retreats. Others send them to workshops offered by colleges or training organizations. Still others provide in-kind or financial support for their employees' ongoing education.

The connection to values is not just in how much time and money are put into staff development, but rather in the purpose of the activities (Caine, 1998). Some activities are directed at helping people learn to do their current jobs better. Others are directed at learning a new set of regulations, policies, or procedures. Still others are aimed at helping the employees learn more about theories so that they can come up with and share applications to the organization's work. Finally, some staff development activities are directed at career development. These activities are undertaken with the understanding that most of the employees will not and should not stay in their current jobs. They may move up in the organization or they may move on to a parallel, or better, position somewhere else. Staff development of this kind is an investment in an individual and the profession as well as in the current functioning of the organization.

Organizational Politics

Your age, your interests, and your work experience affect the associations you have with the word *politics*. In our experience, many interns associate that word only with elections for political office. Others apply the term more broadly but find that the word has negative connotations of wheeling, dealing, and backbiting. In fact, we have had a number of students tell us that they want to go into the helping professions to "get away from all the politics" that they imagine exist in the corporate world.

In reality, you cannot escape politics. Politics are everywhere, and they affect you whether you see them or not. The politics of an organization can be difficult for an intern to see, especially if you are not there for very many hours each week or for very many weeks. Still, the more you can learn how to look at your placement and organizations in this way, the more successful you are likely to be.

Bolman and Deal (2008) call this lens the *political frame*. When you look through this lens, you see organizations as coalitions, made up of different individuals and groups. They may be formal groups or the informal groups and cliques described earlier. Those individuals and groups have substantial differences in perspective, beliefs, and priorities. That diversity can and should lead to a richness in the decisions made in the organization, but it is a mistake to think that the differences can be resolved or integrated all the time.

According to the political frame, when resources are scarce, decisions about them are critical. Many organizations of every kind experience periods of shrinking budgets, due to factors such as funding cuts, differing corporate priorities, or poor profits. In most nonprofit organizations, resources are *always* scarce; scrambling to raise money and adjusting to reduced allocations from legislatures, shortfalls in fund-raising goals, and escalating costs are just facts of life. In these circumstances, conflict, bargaining, and negotiation are inevitable, and not necessarily bad things to do. So, the first thing to look for when you examine your placement site from this perspective is the various groups and the resource issues over which they compete. The next thing to look at is how each group accesses and uses power and influence.

Power and Influence in an Organization

Formal authority and power are not the same thing. Sometimes in organizations individuals or groups who are "lower" on the organizational chart have more power and influence than some of those above them. Secretaries often appear at the lower levels of the chart; in theory, they have little power and influence. However, it is often secretaries who know where everything is and how to navigate the paperwork maze. They overhear a lot of conversations, and they often control important resources, such as supplies. If you want to know how to get something done, you are better off consulting an experienced secretary than the policies and procedures manual.

There are a number of ways to have power and influence in an organization (Bolman & Deal, 2008; Homan, 2011). Position power is one of them, but power also comes from information and expertise. If your organization is using a new set of software programs or a new computer system or mobile device, and you have someone who is very knowledgeable about them, that person has enormous power and influence. Pity the people who have insulted, slighted, or ignored the tech expert; their devices will have a problem sooner or later, and they will get fixed later, not sooner. Control of rewards, resources, and "inside" information is another source of power, as we noted in the case of secretaries. People who are well connected to those who control these aspects of agency life generally have what they need and know what they need to know, often in advance of everyone else. Finally, individuals can have personal power. This power may come from high levels of charisma, from natural charm and sociability, or from carefully building networks of friends and attending to each of them.

Power in organizations can also be found in the informal networks of influence. In some instances, people on the staff are related (or were related) to each other. In other instances, they have shared decades of connection through schooling, neighborhoods, or churches. In still others, outside activities such as softball may unite them. Determining these informal influence networks takes time and reflection. You might try arranging everyone in the order of influence they seem to have on decisions or on daily operations. Russo suggests that you interview people in the organization and ask them to whom they give advice, to whom they give directives and from whom they accept advice and directives (Russo, 1993). This may be impossible in practice, and not necessarily well received by the staff, but if you could do it, you could then construct a web of connections that would tell you a lot about how power and influence flow in the organization.

Organizations as Cultures

A final way to look at your placement site is as a culture that, like any culture, is understood at least in part by its symbols and rituals (Bolman & Deal, 2008). When you look at behavior in organizations this way, you are looking not so much at the function of certain practices, but at their *meaning*. Commencement exercises are really not very functional if you think about them. Diplomas could be mailed to graduates and it would be a lot easier and cheaper. But the ceremony means a great deal to the graduates, faculty, staff, administrators and parents. It is a celebration of the values of the institution, a way to say, "Look, here is what we do well." It is an important annual reminder, in the face of the day-to-day frustrations and aggravations, of why faculty members, administrators, advisors, coaches, and others on campuses do what they do. Ceremonies marking thirty days—or five years—of sobriety in Alcoholics Anonymous are another example.

Less dramatic events, such as staff meetings, can be looked at in this way as well. They may not be the most efficient way of communicating critical information, or even the most productive forum to discuss difficult issues, but they are a way for people in the organization to see each other visibly interacting and inhabiting the same space.

Like other organizations, your placement site may have smaller, less formal rituals as well. Going for pizza on Thursday, or stopping for coffee on the way back from an off-site meeting, or gathering in the lounge during a break may seem silly to you, but you ignore them or decline an invitation to participate at your peril. These rituals symbolize collegiality, concern, and cohesion.

Of Special Relevance to The Helping and Service Professions

The Importance of Language

Language is another aspect to the symbolic or cultural frame and another key to operational values. For example, consider the way staff members talk about clients and the labels they apply to them. Sometimes staff members will refer to their clients in openly derogatory terms, like *druggie*, *emo*, *SpEddy*, or *geezer*. Other times, the labeling is more subtle. For example, sometimes certain clients are referred to as *difficult* or *hard cases*. When we hear this, we are always tempted to ask: Difficult for whom? This comment is made using the experience of the staff as a base for description, not the experience of the client. The same client could be described as "having a hard time accepting authority" or "unwilling to accept personal responsibility." These phrases are much more descriptive of the *client's* experience. Of course, people are going to become frustrated and say negative things from time to time, although preferably not to the client. A pattern, however, of one sort of comment or label can be a clue to an unconscious organizational value.

Organizations also use rituals or symbols to make their values and mission visible, both to the public and to the employees, so that they do not become irrelevant to daily work. Posters, letterhead, pens, even e-mail signatures can reflect the mission and values.

The internal workings of any organization are complex and fascinating. It will take you some time and energy to learn to look at organizations in the ways discussed here. We hope that you learn to do it better than you do now and that you come to see the value of taking time for these analyses. Still, the picture is not complete. Just as it is impossible to understand a system without understanding all its component parts and relationships, it is impossible to understand that system without understanding its relationship to other systems and to larger societal forces and pressures.

The External Environment

You may already be beginning to see how the world outside the site has an impact on its functioning. The number and nature of the organization's relationships with other organizations and agencies, sometimes called the *task environment* (Queralt, 1996), will affect your work and the ability of your placement site to do its job. In a more indirect but no less important way, the surrounding community, the economic climate, and some political issues affect you, your clients, and your agency (Queralt, 1996).

The Task Environment

Most organizations cannot function without smooth relationships with other organizations; this is certainly true in human services. For example, the Beacon Youth Shelter works with two state agencies that refer and pay for residents. A local food bank helps supply their food. A large clinic in the area contracts with Beacon for certain psychological services. Some of the students attend school off-grounds. Occasionally, a resident runs away or causes some problem in the community; good relationships with neighbors and the local police are essential. In addition, residents do not stay at Beacon forever; sooner or later, a new placement must be found for those who cannot return to their families or live on their own. This aftercare placement process involves detailed knowledge of and good relationships with a variety of other placement sites.

Of course, other organizations can be sources of competition as well as collaboration. That is particularly true in the world of for-profit organizations, but it is not limited to them. If you are interning with a nonprofit, other organizations in your area that provide similar services may compete for grants, state contracts, corporate sponsors, and so on.

Furthermore, many organizations (and most human service agencies) have one or more agencies that monitor them in some way. Some agencies are part of a larger parent organization. That is not the case with Beacon Youth Shelter, but other shelters in the city are funded by organizations like Catholic Charities or the Salvation Army.

The Sociopolitical Environment

There are a number of political issues that affect placement sites; we will just mention a few of them that seem to be common across many different placements. Local politics can have a considerable impact. For example, in some states, there is a cap on property taxes, which make up a large portion of each town's budget. Towns can vote to exceed the limit, and these votes are often taken to provide a special service, such as a new gymnasium or road repairs. If your placement is heavily funded by the town, these are important issues for your agency.

The attitudes of the people in the town toward your organization and the work it does are important. Some social service agencies, like group homes, seem to inspire at least apprehension, if not fear, among neighbors. People agree on the need for them but want them located somewhere else—the "not in my backyard" way of thinking. The same is often true of certain kinds of manufacturing plants, such as biodiesel, or large installations such as cell phone towers or wind turbines. If your organization has just opened up or launched a venture in a hostile or unwelcoming community, it has a large public relations task ahead of it. Some agencies make deliberate efforts to involve people in the community in their work. This is especially important when the clients are residents of that community, as they are in a Planned Parenthood Center, for example. If these residents are suspicious of and hostile toward the agency, chances are that fewer people will come for services.

State and federal politics can have an impact as well. A new governor or mayor often means new people in charge of government agencies and sometimes means a change in philosophy, along with new regulations and changes in the way funds are distributed. Federal issues such as health care reform, abortion rights, or changes in the tax code may also have a large impact on the mission, goals, and tactics of your placement site.

The Organization and Your Civic Development

In Chapter 1, we made the case that an internship can be a vehicle for civic development and provided stage-related examples in Chapter 2 for developing civic opportunities throughout your internship. In Chapter 6, where we discussed the Learning Contract, we invited you to develop learning goals and activities that were aimed at this aspect of your development. Using the lenses of organization and systems theories we have discussed in this chapter, you can further explore the civic dimension of your internship by examining how your placement site recognizes and fulfills its civic mission. See the For Further Exploration section of this chapter for examples of what organizations are doing about contributing to or investing in their communities.

You may recall that in Chapter 1 we pointed out that some private organizations embrace a commitment to the identification and solution of societal problems. Does your placement site make that commitment? Consider this statement from an organization we discussed briefly earlier in the chapter, Hartford Steam Boiler (HSB, 2012):

The Hartford Steam Boiler Inspection and Insurance Company is committed to improving the quality of life in our communities. We support this effort through:

- *Direct corporate donations;*
- *Our employees' participation in fundraisers and volunteer opportunities;*
- *Matching our employees' charitable gifts.*

HSB places a high value on improving the quality of life in Hartford, its home for more than 140 years, and the cities where we have branch offices. HSB believes businesses that are dedicated to the community reap more benefits than simply their own prosperity and growth. They help to improve the ultimate bottom-line—life—by creating a place where everyone wants to live, work, and grow.

Corporations also execute their civic mission by inviting community leaders to serve on their boards of directors or on subcommittees of their boards. They may also invite community leaders in for more informal conversations about the activities of the company.

Of Special Relevance to The Helping and Service Professions

The Civic Dimension

If you are in a social service setting, you may think that the civic dimension of your agency's work is obvious, and parts of it might be, but we invite you to think about it more deeply. Remember that part of the concept of a civic professional is one who employs the art of *practical reasoning* (Sullivan, 2005), which goes far beyond the acquisition and application of skills. Does your agency promote this sort of development in its staff?

Another question to ask is how well your agency recognizes the community context of the work it does. We will be discussing the community context in depth in Chapter 11, but the question here is whether your agency chooses to be aware of and work with the constellation of assets and challenges in the surrounding community, or whether it provides services to individuals and families in ignorance of these assets and challenges.

Finally, just as in a corporate setting, you can ask whether your agency actively seeks input and involvement from the community in determining its priorities and assessing its effectiveness. And again, this can be done by inviting community leaders or community members to sit on committees or participate in agency evaluations.

Conclusion

In this chapter, we have tried to introduce you to some ways of thinking about organizations. Although you may feel somewhat overwhelmed by the information, we have really only scratched the surface. Systems theory, organizational dynamics, organizational development, values, and staff development are enormous topics. In our experience, many interns find these topics fascinating. Others may not. However, we believe that the concepts we have introduced here, at a minimum, will help you understand both your placement site and what is happening to you there. At the very least, this chapter should stimulate some thought now and be a resource if you begin to have problems at the placement.

For Review & Reflection

Checking In

Dynamics & Systems Matter.   This time around, when you evaluate your internship experience on a scale of 1–10, think about how the site's organizational dynamics and systems affect the rating the experience earned for the week. Then ask yourself the tougher question: *So What?* And, by now, you know the sequence of questions to work your way through: *Now What?* And, finally, *Then What?* The dynamics and systems of any and all places of your employment—even if you own your own company—matter significantly to the quality of the organization's culture and climate.

Experience Matters (For the EXperienced Intern)

If you have worked in one or more organizations, how many of these ideas seemed familiar to you? How many were new? If this is your second internship, it may be useful to compare the features of that organization to the one you are in now. Again, look for similarities and differences between your workplace(s), where you interned previously, and your current internship site

Personal Ponderings

As you can see, there are many aspects to an organization. Whether reading about them appealed to you or not, you do have strengths that you bring to your experience of those aspects of you field site. The questions below are broken into categories. Some of them will require that you ask questions and do some research. Choose the questions within each category that seem most meaningful to you and think about *why* that particular question has meaning for you.

Background Information

1. How many years has your agency existed? Have there been any major changes during that time? Describe them.
2. What is the agency's overall goal and philosophy concerning the clients? Working with clients? Managing employees and interns?
3. What is your agency's annual budget? What categories is it broken into?

4. What are the sources of the money? Do clients have to pay? If not, does some other agency pay? If the services are "free," where does the agency get the money for payroll and operating expenses?

Structure

5. Obtain or create an organizational chart of the agency. Are interns on the chart? If not, place them in appropriate locations.
6. Where are the formal rules of the agency located? Have you looked at this resource?
7. How are decisions supposed to be made at your site? What evidence do you have that decisions are or are not actually made that way?

Human Resources

8. Have you observed any informal rules? Do any of them conflict with the formal rules?
9. Have you observed instances of people performing tasks and functions that are not in their job description? What do you know about how this happened? How about informal roles such as "office clown," which aren't in anyone's job description? Do there seem to be informal or unwritten expectations of you as an intern? How do you feel about those expectations?
10. Are you aware of the cliques or informal subgroups that exist at your agency?

Politics

11. What are the various subgroups within your agency that may be vying for power or resources?
12. Which of the sources of power identified in this chapter can you see being used in your agency?

Organizational Culture

13. What are some important symbols and rituals at your agency?

The External Environment

14. What are some of the agencies with which your site has important relationships? What do they do?
15. Are there any agencies that monitor or control your agency? Are there periodic site visits?
16. What have you been able to learn about the relationship between your agency and the surrounding community?
17. What local, state, or national political issues are affecting your agency and its clients?

Civically Speaking

To what extent and in what ways does your internship site fulfill its civic mission? Does it foster a culture of civic professionalism? Can you see additional ways in which its civic mission can be fulfilled?

Seminar Springboards

Analyzing Home Base. Analyzing organizations is not easy, and some of the concepts in this chapter may be abstract or difficult to understand. If so, one thing you could do as a class is to analyze the campus you attend, using the various frames discussed in the chapter. If you are business students, you might want to share with your peers outside of your major other organizational and systems lenses you've worked with in the past.

For Further Exploration

As You Sow. (2011). "Corporate Social Responsibility." Retrieved from http://asyousow.org/crs.

An example of an organization that works with corporations to "promote environmental and social corporate responsibility through shareholder advocacy, coalition building and innovative legal strategies...founded on the belief that that many environmental and human rights issues can be resolved by increased corporate responsibility" (http://asyousow.org).

Berger, R. L., & Federico, R. C. (1985). *Human behavior: A perspective for the helping professions* (2nd ed.). New York: Longman.

A thorough discussion of systems theory, including many concepts not discussed in this chapter.

Egan, G. (1984). People in systems: A comprehensive model for psychosocial education and training. In D. Larson (Ed.), *Teaching psychological skills: Models for giving psychology away* (pp. 21–43). Pacific Grove, CA: Brooks/Cole.

Egan, G., & Cowan, M. A. (1979). *People in systems: A comprehensive model for psychosocial education and training.* Pacific Grove, CA: Brooks/Cole.

A wonderful book that integrates a systems and individual development approach to human services. Excellent chapters on systems. Unfortunately, the book is out of print. If your library does not have a copy, try the preceding Egan reference.

Gordon, G. R., & McBride, R. B. (2011). *Criminal justice internships: Theory into practice* (6th ed.). Cincinnati, OH: Anderson.

An excellent chapter on organizations from a criminal justice perspective. Covers both public and private settings.

Northwestern Mutual Life Insurance and Financial Services (2012). Community Service Award. Retrieved from www.northwesternmutual.com.

Northwestern Mutual financial representative interns don't just get to test-drive the full-time career; they can also get engaged in the company's commitment to volunteerism. Through its annual Community Service Award (http://www.northwesternmutual.com/about-northwestern-mutual/our-company/ northwestern-mutual-foundation.aspx#Employees-and-Field) Northwestern Mutual interns like Chad Monheim (http://www.youtube.com/watch? v=1ne4qft8gy4&feature=plcp) can get recognition—and charitable funding—for the philanthropic causes they're passionate about. Go to http:// www.northwesternmutual.com/career-opportunities/financial-representative-internship/default.aspx to learn more about Northwestern Mutual financial representative internships.

Queralt, M. (1996). *The social environment and human behavior: A diversity perspective.* Needham Heights, MA: Allyn & Bacon.

A social work text with an excellent chapter on organizations.

Russo, J. R. (1993). *Serving and surviving as a human service worker* (2nd ed.). Prospect Heights, IL: Waveland Press.

Interesting chapter on organizations, with examples from criminal justice and mental health settings.

Senge, P. M. (1990). *The fifth discipline: The art and practice of the learning organization.* New York: Doubleday/Currency.

Very interesting and accessible discussion of systems theory and organizations. The applications are business oriented, but the concepts are very clearly explained.

Siemens. (2011). "The Siemens Caring Hands Program." Siemens Corporation. Retrieved from www.usa.siemens.com/answers/en.

A not-for-profit organization of this global company established to receive, maintain, and disburse funds for sponsoring and encouraging charitable activities in which employees donate their time and talents to worthy causes to carry out the company's commitment to impact the communities in which they live and work through volunteerism in a variety of community service activities, including giving campaigns, blood drives, holiday food and gift collections, and disaster relief fund raising when their employees, operations, or business partners are affected.

The Washington Center for Internships and Academic Seminars. (2012). Retrieved from www.twc.edu.

Students at the Washington Center (www.twc.edu) intern at government agencies, non-profits, think tanks and businesses such as the Department of Justice, Amnesty International, Merrill Lynch, No Labels and Voice of America. They also have the opportunity to become positive change agents by participating in civic engagement projects on important domestic and international issues, including domestic violence, homelessness, veterans, torture abolition, and the environment. Students learn about these issues by interacting with national and local community leaders, and then become involved in a direct service and/or advocacy project. They can choose to join a TWC-guided project or design their own individual one. For more information about civic engagement at the Washington Center, please see http://www.twc.edu/internships/washington-dc-programs/internship-experience/ leadership-forum/civic-engagement-projects. Articles about recent civic engagement activities can be found at TWCNOW http://www.twc.edu/twcnow/news/ term/civic%20engagement.

References

Berger, R. L., McBreen, J. T., & Rifkin, M. T. (1996). *Human behavior: A perspective for the helping professions* (4th ed.). Boston: Allyn & Bacon.

Birkenmaier, J., & Berg-Weger, M. (2010). *The practical companion for social work: Integrating class and field work* (3rd ed.). Boston: Pearson Allyn & Bacon.

Bolman, L. G., & Deal, T. E. (2008). *Reframing organizations: Artistry, choice and leadership* (4th ed.). San Francisco: Jossey-Bass.

Caine, B. (1994). What can I learn from doing gruntwork? *NSEE Quarterly* (Winter), 6–7; 22–23.

Caine, B. T. (1998). Using Bolman and Deal's four frames as diagnostic tools: Key concepts and sample questions. In B. T. Caine (Ed.), *Understanding organizations: A ClassPak workbook*. Nashville, TN: Vanderbilt University Copy Center.

Egan, G., & Cowan, M. A. (1979). *People in systems: A comprehensive model for psychosocial education and training*. Pacific Grove, CA: Brooks/Cole.

Gordon, G. R., & McBride, B. R. (2011). *Criminal justice internships: Theory into practice* (7th ed.): Anderson Publishing.

Homan, M. S. (2011). *Promoting community change: Making it happen in the real world* (6th ed.). Belmont, CA: Wadsworth.

House, I. (2007). *Proudly facing the future: Strategic plan 2007–2010*. Hartford, CT: Interval House.

HSB. (2012). Retrieved from www.hsb.com.

Kiser, P. M. (2012). *Getting the most from your human service internship: Learning from experience* (3rd ed.). Belmont, CA: Brooks/Cole.

NPR. (2012). National Public Radio, from www.npr.org.

Queralt, M. (1996). *The social environment and human behavior: A diversity perspective*. Needham Heights, MA: Allyn & Bacon.

Russo, J. R. (1993). *Serving and surviving as a human service worker* (2nd ed.). Prospect Heights, IL: Waveland Press.

Sullivan, W. M. (2005). *Work and integrity: The crisis and promise of professionalism in America* (2nd ed.). San Francisco: Jossey-Bass.

Swaner, L. E. (2012). The theories, contexts, and multiple pedagogies of engaged learning: What succeeds and why? In D. W. Harward (Ed.), *Transforming undergraduate education: Theories that compel and practices that succeed* (pp. 73–90). New York: Rowman and Littlefield.

CHAPTER *11*

Finding the Beat of
the Community

*I think that every community in someone's life plays a
role in who they are and how they interact with others.
Because there are so many different communities in my
agency, it is hard to try and understand every community
that every client belongs to or is from. Trying to
understand is helping me learn how diverse the agency
really is.*

STUDENT REFLECTION

Introduction

In Chapter 10, we encouraged you to widen the scope of your engagement, to stretch beyond your colleagues and clients and consider the
placement site itself. In this chapter, we encourage you to widen that
scope even further and consider the community context of your work. In
our experience some interns are naturally drawn to the community context, others are led to it by their experience with clients and their struggles, and still others never really feel drawn to learning about the
community. However, all interns and all internships are affected by the
community and its dynamics.

In this chapter, we will spend some time helping you understand why
the community context is important. We will also spend a little time on
descriptions and categories of communities. Finally, we will help you consider some of the important aspects of any community, and how you
might go about learning this information in your particular situation.

The Community Context and The Civic Professional

Regardless of where you are interning or what profession you hope to enter, learning about the community is an important part of your personal, professional, and civic development. We have commented several times in this book that every profession has a social obligation and its work has public relevance. Sometimes that social obligation is enacted nationally, or even globally, but it is also enacted locally in an organization's relationship with the community where it is located. And we like to think of that obligation as not just a one-way street, but a mutual relationship where an organization and the professions are engaged with the community. Effective local engagement is not possible, however, without a deep knowledge of and respect for the community itself.

By now you are becoming more familiar with the concept of a *civic professional* (Sullivan, 2005). Part of becoming a civic professional lies in learning to understand the community context of your work. The art of *practical reasoning* calls for you to move back and forth between the theoretical and human contexts of your profession; the community forms an important part of the human context for many interns.

In Chapter 1, we discussed the internship as an opportunity for civic development, a way for you to acquire and strengthen knowledge, skills, attitudes and values, that will make you better prepared to contribute to a healthy democracy. We revisited this concept in Chapter 2 and Chapter 6 where we encouraged you to become aware and set civic development learning goals for yourself. Wherever you end up living, you will be part of a community. Being an active citizen of that community means understanding it as well as you can. For some students, the dynamics of communities become clearer when they leave their home communities and go somewhere very different, as when a student from a suburban background does a service-learning project or internship in a poor, urban area. However, even though the elements of community that those students see may look different from those they are used to, they are really the same elements, just manifested differently. Fred traveled to New Orleans early in 2007 and visited the Ninth Ward, one of the areas most devastated by Hurricane Katrina and one of the slowest to recover. In a speech at the conference he was attending, a speaker reminded the attendees that while there may not have been a hurricane to point it out and draw national attention, there are neighborhoods like the Ninth Ward in every city. Where, she asked, is your Ninth Ward (Hughes, 2007)?

Of Special Relevance to
The Helping and Service Professions

But I Don't Want to Work in the Community!

For some of you in social service settings, the community *is* your client. If your agency helps to organize community discussions of neighborhood or municipal issues, or if you are helping a group of people mobilize for political advocacy, you are, working with some individuals and groups, but the *main* target of your work is the community itself. If that describes your work, you probably don't need much convincing about studying the community; for you, this is as fundamental to your work as thinking about the clientele is to other interns.

However, even if your work does not focus on the community directly it is important to take some time to think about the community. It is all part of learning to think in systemic terms, something you have probably studied in your program, and something that we encouraged you to do in considering your family (Chapter 4) and the organization (Chapter 10). It is a hard lens for many people to acquire, and it takes practice and persistence. But once you do acquire it, it is equally hard to stop looking through it! Once you learn to see the world in this way, things will never look the same.

For one thing, your clients live their lives and struggle with their issues in a larger context. Communities shape the context of people's lives in many ways. You should not ignore some of the larger reasons why your clients have the problems that they do (Homan, 1999; Rogers, Collins, Barlow, & Grinnell, 2000), nor larger arenas in which assistance or solutions may lie. You are probably accustomed to thinking at least a little bit about your clients' families, even if they are not the main focus of your work, but the context goes beyond that. One of our students, for example, worked with adults who were developmentally disabled. After a while, she became quite frustrated at how few of them lived independently and had jobs, believing that they would feel much better about themselves and probably function at a higher level that way. She learned that attempts to build group homes or get employers to accept adults with developmental disabilities were often delayed or scuttled by the perception, whether based in reality or not, of the attitudes of the community at large. She also learned that her agency had done very little to try to educate or influence that community. Some of her clients were frustrated too, and she wondered whether her energy should be spent trying to help them feel better about and adjust to something she believed was unfair.

For another thing, your agency does not exist in a vacuum any more than do your clients (Garthwait, 2010). Many agencies grew directly out of community needs. The range of agencies and programs that have expanded to meet the needs of persons with HIV/AIDS is a perfect example. Regardless of their origins, though, all agencies are affected by the rhythms of the community. For example, even though a homeless shelter that admits clients who are actively using drugs and alcohol meets needs in the community, there are some people in the community and in the neighborhood who are adamantly opposed to it. Some of them are philosophically opposed to the approach that these agencies use; more are concerned with the safety and general appearance of the surrounding neighborhood. People concerned about promoting tourism worry about the effect of a homeless population. When factions like these are able to gain power and influence, or when a high-profile crime or death occurs, efforts are renewed to close these shelters. In other communities, efforts are made to close programs for the homeless and conduct sweeps of people living on the streets, in hopes that the homeless will simply go away.

Finally, communities are powerful. They can be powerful allies or powerful opponents. One of us once worked with a local school to create a series of parent workshops. The topics were chosen based on input from school counselors and social workers, who in turn were listening to the parents. Flyers were sent home to parents, transportation was arranged, and child care was provided. However, almost no one came. What we had not done was to find the people who were influential in the parent network in the community and involve them from the beginning. They did not oppose our work, but they did not support it, either. We later learned that our university, like many others, had a long history of offering programs to the community, and a long history of not delivering what was promised, or pursuing its own as opposed to the community's agenda. The parents did not know us, and their general instinct was not to trust us.

Communities are important factors in your work for all these and many other reasons. So let us help you identify a community to study and suggest some of the things you may want to find out.

What Is a Community?

The first step in "finding the beat" is to identify the community, and this is sometimes not easy. For one thing, we use the word *community* to refer to different kinds of groups (Garthwait, 2010; Homan, 2011). As common a term as it is, it is also elusive because of the variety of ground it covers (Kempers, 2002). Sometimes we are referring to a geographical area—a city, a town, a neighborhood, or even a block. Other times we are referring to an age group, as when we refer to the community of senior citizens; or an interest group, such as local environmentalists; or a cultural group, such as the Caribbean American community. Members of these groups may live in different parts of the geographical community. To muddy the waters even further, most communities of every sort can be further divided into smaller communities.

Furthermore, the community where your organization or agency is located may be a different community from the one where its clients or employees live. In that case, both communities are probably important to know about. In some social service settings, the clients come from a wide range of geographical locations and are members of many different communities of interest, age, or culture as well. So, if you work in a clinic that provides services to persons with HIV/AIDS, the community where the clinic is located, the communities where your clients live, and the community of people with HIV/AIDS are all communities that have an impact on your work and your clients.

We say all this not to confuse you or to undertake an abstract theoretical exercise. The point is to get you to think about all the communities there are for you to consider; we acknowledge the difficulty in choosing one or two. Chances are, though, that one or two is all you will have time for. We encourage you to consult your instructor and site supervisor in identifying the community or communities that are most important in helping you and your organization achieve your goals.

A Community Inventory

Communities are complex and fascinating and it takes time, experience, and some expertise to get to know them. We encourage you to invest your time in some kind of community inventory and to make it part of your learning goals. We cannot begin to cover all of the important aspects of communities or working in them in this short chapter, but we will suggest some aspects of communities for you to consider and explore. We chose facets of the community that seem especially relevant to you as an intern, with ideas drawn from our experience and from the work of Mark Homan, author of two seminal texts on community development (1999, 2011), as well as Garthwait (2010) and Rogers et al. (2000). We begin with basic information that you need to know, leading off with the concepts of *assets* and *needs* and moving on to ideas about analyzing communities. Just as we did in Chapter 10, we have organized this discussion into Bolman and Deal's (2008) categories of structural considerations, human resources, symbols, and political considerations.

Basic Information About Communities

If you have chosen a geographic community to explore, begin by learning its dimensions. How large is it, and what are its recognized boundaries? Boundaries can be a town line, a street, a river, or anything that people generally accept as marking the edge of a community. Next, consider the general appearance of the community. What sort of impression does it give at first glance? What catches your eye? Are there visible local landmarks? How about landmarks you might miss at first, but that every resident knows about (like a controversial land fill or high power transition lines)?

Next, consider some basic information about the people in the community. How many people live there? Is the population dense or spread out? Is it stable or does it fluctuate? Communities with seasonal workers, summer residents, or colleges are examples of communities where the population changes over time. Finally, what is the average income of the people who live there? How does it compare to state or national averages and to the cost of living in the community?

Assets and Needs

When considering any community, it is easy to think in terms of needs. And to some extent, that makes sense; all communities do have needs and it is important to look at how well and in what way the needs are met. It is easy, though, to move from "needs" to "needy," to look only at what the community does not have or does not do well. That is especially true if your interest in the community is stimulated by the desire to understand problems being experienced by individual clients. Working with youth in a Boys and Girls Club in the inner city, for example, it is easy to look at crime rates, poverty, and poor transportation as factors contributing to the problems your clients have.

When you think of communities this way, you are equating needs with deficits. It is all too easy, then, to miss the strengths or assets of the community. And that is especially easy if the community meets its needs in ways that are unfamiliar to you. Some writers in the field of community development have been reframing discussions of communities to focus on assets, or capital (Hodgson, 1999; Kretzmann & McKnight, 1997). As we proceed with our discussion of community inventories, we are going to try to use the language of strengths and assets, and we encourage you to do the same.

Structural Considerations

This category of assets concerns the formal and informal structures that help people in the community to meet their needs and how well those structures work. We begin with basic human needs such as food, shelter, clothing, and medical care, and the issue here is *access* to these assets across a variety of incomes and circumstances. What provisions has the community made to assure that these needs are met? How successful are they? For example, how does the community handle those who are temporarily or permanently displaced from their homes? Some people will have insurance for such circumstances, but some will not. What then? Are there shelters? Perhaps the community relies on networks of family and friends to handle situations like this. Does that work? Another example is grocery and other retail stores. How easy are they to find and get to?

Education

What is the quality of the schools in the community? What about adult education and training opportunities? Is accessible and affordable post-secondary education available? And don't forget about the informal learning, mentoring, and educational networks that exist in many communities.

Employment

Employment is another basic need, and opportunities for employment are a community asset. Some communities have a variety of employment opportunities available while others have almost none. Sometimes people in the community have to travel a long way to get work. Also, sometimes the jobs that are available require special skills but no training is available. As you learn about these features of the community you are studying, the unemployment statistics (which you should also determine) will have more meaning.

Communication

Communication is also a basic need; people need to know what is going on in their community and in the world around them. Most communities have both formal and informal means of communicating necessary information, and these are powerful community assets. Identify the major newspapers and TV and radio stations available in the community. Do those media outlets provide adequate coverage of news and events in the community you are studying? Large metropolitan newspapers often overlook issues in outlying areas or in certain segments of the city. Most communities have access to the internet, but not all, and that can be a real challenge. Also, some communities have their own means of communication; some have newspapers, web sites, blogs or newsletters, and many have gathering places where people go to catch up on what is going on. Finding these local resources is crucial.

Basic Systems

Basic systems in communities meet needs or provide access to the services described earlier. Adequate public transportation to, from, and within the community makes an enormous difference in people's abilities to meet their needs, as do the quality of roads, sidewalks, crosswalks, and walk lights. In some communities, clean air and water are taken for granted. Others struggle with pollution or even with a lack of potable water. Similarly, you probably don't think very much or very often about waste removal and drainage systems unless you have been in places where sewers back up, garbage overflows, and water sits in stagnant pools. Not only do these problems detract from the general attractiveness and desirability of the community, they each pose serious health risks.

Human Resources

People, of course, are an enormous asset, and the idea here is to get to know something about the people in the community, their social and emotional needs, and their strengths as well as their liabilities. In some communities, it is easy to see that people feel safe and secure. They move about freely and easily; they congregate in a variety of places. In other communities, people are afraid to go outside or to gather in certain places. And of course there is lots of room in between these two extremes. Pride is another social and emotional issue. Find out what people who live in the community are proud of.

In the chapter on organizational issues, we mentioned operational values and informed you that you can learn a lot about an organization by how it spends its time and money. The same is true for people in the community you are studying. When people have free time, where do they go and what do they do? There will, of course, be a range of answers to that question in any community, but you can probably figure out some patterns. Some communities, for example, have a strong recreational focus; people gather in large and small groups to play sports, play cards, bowl, and so on. In others, there are a large number of volunteers and a large number of volunteer opportunities. In still others, though, people tend to come home from work, lock their doors, and spend time only with themselves and their immediate family.

As you learn about the people in the community, remember to look for their strengths. In social services, since the focus is often on problems, it is easy to look at people and communities in terms of what they do not have and what they do not offer. This can also be true of corporations as they consider the relevant issues in a community. And it is an important part of the picture. Communities also have strengths, however, and those strengths are often found in the skills, talents, and values of the people who live there. They are to be embraced for what they are and how they can contribute to the changes people want in their communities.

Community Symbols

Communities, just like organizations, can also be looked at as cultures. Among the assets of any culture are its symbols and traditions. Earlier we mentioned landmarks. Some community landmarks have symbolic importance beyond what is immediately obvious. A church, for example, can be a center of social life in a community. As such, it is a symbol of unity in the community. Communities also often have special celebrations, like town fairs, carnivals, dances, or parades. They also sometimes have their own traditions around holidays.

Political Considerations

The issues here are who has power and how decisions are made, and there are both formal and informal aspects to these dynamics. Find out the status and strength of the major political parties in the community. In some communities, there are parties in addition to Democrats and Republicans, such as Libertarians, that have a strong following. Another part of the formal structure in a community is its local decision-making structure. There are many of these, ranging from small towns that are governed by small committees and a town meeting to large municipal governance structures.

There is more to the political life of a community than decision making, and many more sources of political power and influence than the formal mechanisms. At issue is the control of vital resources, such as money, goods and services, and information. In some cases, the institutions that control money, energy, natural resources, and information are located outside the community and operate without any input from or regard for community interests. And don't forget the informal means of access; there are both legal and illegal avenues to loans and credit, for example, and to goods and services.

Sometimes power is conferred formally, by election or appointment or by occupying a position of authority in a major institution. Other sources of power in the community are less obvious from the outside. People in communities derive power and influence from their families, for example; in some communities, there are families who have lived there and been powerful for generations. Other people derive power from their connections, their personality, or both.

The following chart depicts the dimensions of a community that we have discussed and should serve as a summary as well as a guide for your investigation:

Elements of a Community Inventory

Basic Characteristics

Geographic

Size and boundaries

Landmarks

General appearance

People

Number

Density

Stability

Income

Cultural Issues

Symbols

Landmarks

Gathering places

Rituals

Fairs

Celebrations

Parades

Structural Issues

Basic Needs

Food

Shelter

Clothing

Medical care

Employment

Opportunities

Training needs

Location

Unemployment %

Education

Schools

Adult education and training

Communication

Media

Informal communication

Transportation

Air and Water

Waste and Drainage

Human Resources

People

Sense of safety

Sense of belonging

Sources of pride

Core values

Skills

Talents

Other strengths

Political Considerations

Formal Power

Major political parties

Other parties

Governance structures

Formal Control of

Money

Energy

Natural resources

Goods and services

Information

Informal Sources of Power and Influence

How Do I Find All This Out?

First of all, you are probably *not* going to find out all of this information, depending on the length of your internship and the importance you, your campus instructor, and your site supervisors assign to these issues. However, everything you learn about a particular community and about how to do this sort of research will be valuable, and there are several places to look. Some of this information you can find by yourself on the Internet or by visiting local or college libraries; most libraries have reference librarians who can help you as well. Homan (2011) suggests several resources that you might find in the community and a library, including:

- Census data
- Electronic data bases

- Community web sites
- City directories
- Newspaper files
- Local magazines
- State yearbooks
- Town reports
- Political directories

Another source of information is needs assessments of the community, which are undertaken by charitable organizations such as the United Way or by local hospitals, colleges, or universities (Homan, 2011; Kiser, 2012).

For other sources of information, you will have to go out into the community itself. Local businesses, or large businesses that operate in your community, are good resources for information. So are chambers of commerce, local social service agencies, and other groups that either service or advocate for your community (Homan, 2011). If you can find someone at your organization—or anywhere else—who can connect you to some of the informal networks we have discussed, you will learn things about your community that you cannot learn anywhere else. Be prepared, though, to learn that no one in your placement site has these connections. Homan notes that too often, agencies and other organizations sit *in* not *with* the neighborhood and conduct their business in isolation of them. In that case, you begin with whomever you can find—clients, merchants, or faith community staff—and ask them who else you should be talking to in order to get a better feel for the community.

Conclusion

We hope that we have aroused your interest and curiosity about the community or communities that form the context of your work. Even if we have not, we hope you will go out and look for some of the information discussed here to see if you can connect it to the lives and struggles of your clients.

For Review & Reflection

Checking In

What Does It Matter? How has reading this chapter made a difference in the way you think about communities? How is that useful to your internship work? To your future career? Remember, there are no wrong answers to these questions:

Think About your involvement in the community through your internship. What strengthens in particular do you bring to that work?

Experience Matters (For the EXperienced Intern)

Your life experience has informed you about communities. Some of what you have learned may be very helpful to your internship work; some may actually affect your perspective in ways that do not support the position of your internship site or, if you have clients or a clientele, their goals. Think about the implications of each of these possible scenarios to a future at the internship site as an employee.

Personal Ponderings

If you have a clientele or clients (if not, then use your personal circumstances)…think about the sorts of communities that have an impact on them. Include communities of place as well as communities of interest, age, or culture. Are their web-based communities that need to be considered, such as networking groups? Which of these communities do you think is most important for you to learn about, and why?

Civically Speaking

You have spent an entire chapter immersed in a discussion about community. How has doing so affected your perspective on civic professionalism and practical reasoning?

Seminar Springboards

Plan to conduct an informal community assessment. First, discuss with your instructor the scope of the assessment and the amount of time you have to devote to it. Next, choose from the aspects of communities listed in this chapter a set that fits the scope of your assessment. Keep track of *where* you find information, not just of what you find.

Or

If your classmates are all working in the same community, or if several of you are, you may want to consider taking a quick tour of it, by car or public transportation. Plan ahead for what you will be observing, and divide up responsibilities. One person might keep track of the number and nature of small businesses, another of churches, another of bars, and so on.[1]

If actual tours are not possible, use resources available via the Internet. Your local chamber of commerce may be helpful in this regard for suggested websites for businesses considering a move to the community. Local real estate offices may also be helpful in the same way from a residential perspective. Then conduct the same observations being done by those in actual tours and compare the results. *What* do those results tell you? The next question you already know: *So What?* Spend some time with this question. Community leaders have to understand the

---

[1]Thanks to Mark Homan at Pima Community College for suggesting this exercise.

implications of this answer. Continue with *Now What?* which is followed by *Then What?*

For Further Exploration

As You SOW. 2011. "Corporate Social Responsibility." http://asyousow .org/crs.

An example of an organization that works with corporations to "promote environmental and social corporate responsibility through shareholder advocacy, coalition building and innovative legal strategies...founded on the belief that that many environmental and human rights issues can be resolved by increased corporate responsibility."

Homan, M. S. (2011). *Promoting community change: Making it happen in the real world* (6th ed.). Belmont, CA: Wadsworth.

Clear, cogent, and thorough introduction to principles of community development.

Homan, M. S. (1999). *Rules of the game: Lessons from the field of community change.* Belmont, CA: Wadsworth.

Packed with practical suggestions and accumulated wisdom.

Kretzmann, J. P., & McKnight, J. L. (1997). *Building communities from the inside out* (2nd ed.). Evanston, IL: ACTA Publications.

Good resource for adopting an assets-based approach to communities

Northwestern Mutual Life Insurance and Financial Services. (2012). Community Service Award. Retrieved from www.northwesternmutual.com.

Northwestern Mutual financial representative interns don't just get to test-drive the full-time career; they can also get engaged in the company's commitment to volunteerism. Through its annual Community Service Award (http://www.northwesternmutual.com/about-northwestern-mutual/our-company/northwestern-mutual-foundation.aspx#Employees-and-Field) Northwestern Mutual interns like Chad Monheim (http://www.youtube.com/watch?v=1ne4qft8 gy4&feature=plcp) can get recognition—and charitable funding–for the philanthropic causes they're passionate about. Go to http://www.northwest ernmutual.com/career-opportunities/financial-representative-internship/default. aspx to learn more about Northwestern Mutual financial representative internships.

Siemens. (2011). "The Siemens Caring Hands Program." Siemens Corporation. Retrieved from www.usa.siemens.com/en/about_us/corporate_ responsibility/caring_hands.htm.

A not-for-profit organization of this global company established to receive, maintain, and disburse funds for sponsoring and encouraging charitable activities in which employees donate their time and talents to worthy causes to carry out the company's commitment to impact the communities in which they live and work through volunteerism in a variety of community service activities, including giving campaigns, blood drives, holiday food and gift collections, and disaster relief fund raising when their employees, operations, or business partners are affected.

Shumer, R. (2009). New directions in research and evaluation: Participation and use are key. In J.R. Strait and M. Lima (Eds), *The future of service-learning.* New York, NY: Stylus Publishing.

This chapter is about the evolution of evaluation practice and the use of participatory approaches in the service-learning field.

Shumer, R. (2012). Engaging youth in the evaluation process. In Velure Roholt, R; Baizerman, M., and Hildreth, R. (Eds), *Civic youthwork: Co-creating democratic youth spaces.* Lyceum.

This chapter covers the use of youth participatory evaluation as part of the civic engagement process. It includes several examples of young people evaluating their own civic involvement programs.

Shumer, R., & Shumer, S. S. (2005). Using principles of research to discover multiple levels of connection and engagement: A civic engagement audit. In Root, Callahan, and Billig (Eds). *Improving service-learning practice: Research on models to enhance impacts.* Greenwich, CT: Information Age Publishing.

This chapter describes a civic engagement audit at Metropolitan State University and provides a template for institutions of higher education to examine their civic involvement efforts.

The Washington Center for Internships and Academic Seminars. (2012). Retrieved from www.twc.edu.

Students at the Washington Center (www.twc.edu) intern at government agencies, non-profits, think tanks and businesses such as the Department of Justice, Amnesty International, Merrill Lynch, No Labels and Voice of America. They also have the opportunity to become positive change agents by participating in civic engagement projects on important domestic and international issues, including domestic violence, homelessness, veterans, torture abolition, and the environment. Students learn about these issues by interacting with national and local community leaders, then become involved in a direct service and/or advocacy project. They can choose to join a TWC-guided project or design their own individual one. For more information about civic engagement at the Washington Center, please see http://www.twc.edu/internships/washington-dc-programs/internship-experience/leadership-forum/civic-engagement-projects. Articles about recent civic engagement activities can be found at TWCNOW http://www.twc.edu/twcnow/news/term/civic%20engagement.

References

Bolman, L. G. , & Deal, T. E. (2008). *Reframing organizations: Artistry, choice and leadership* (4th ed.). San Francisco: Jossey-Bass.

Garthwait, C. L. (2010). *The social work practicum: A guide and workbook for students.* (5th ed.). Upper Saddle River, NJ: Prentice Hall.

Hodgson, P. (1999). Making internships well worth the work. *Techniques,* September, 38–39.

Homan, M. S. (1999). *Rules of the game: Lessons from the field of community change.* Belmont, CA: Brooks/Cole.

Homan, M. S. (2011). *Promoting community change: Making it happen in the real world* (6th ed.). Belmont, CA: Wadsworth.

Hughes, M. (2007). *Fulfilling the promise of a just democracy: New Orleans after Katrina.* Paper presented at the Association of American Colleges and Universities, New Orleans, LA.

Kempers, M. (2002). *Community matters.* Chicago: Burnhams, Inc.

Kiser, P. M. (2012). *Getting the most from your human service internship: Learning from experience* (3rd ed.). Belmont, CA: Brooks/Cole.

Kretzmann, J. P., & McKnight, J. L. (1997). *Building communities from the inside out.* Evanston, IL: ACTA Publications.

Rogers, G., Collins, D., Barlow, C. A., & Grinnell, R. M. (2000). *Guide to the social work practicum.* Itasca, IL: F. E. Peacock.

Sullivan, W. M. (2005). *Work and integrity: The crisis and promise of professionalism in America* (2nd ed.). San Francisco: Jossey-Bass.

CRESCENDOS

You have come quite a distance in your internship. You have dealt with the concerns of beginning the internship and you have launched and nurtured a number of relationships. In all likelihood, you have experienced a period of awareness, knowledge, understanding, exposure to values, and skills development. You also have come to understand the advantages of taking engaged approaches to your internship and its challenges. Importantly, you have learned how to identify, cope with, and move toward resolving problems using the basic and advanced tools available to you in the Essentials chapters (Chapter 3, 6, and 9).

You are about to embark on a time of great productivity and enjoyment, although they will undoubtedly have their challenges. Chapter 12 examines in greater depth the satisfaction and challenges of the Competence stage. Along with the exhilaration that most interns experience during this stage, you also begin to take notice of the complexity of the demands of the work and the workplace, especially in the areas of legal, ethical, and professional matters. Chapter 13 provides you with an overview of these issues, which you may become more aware of now that you are grounded firmly in the work. You may already have explored this Essentials chapter along the way; now, though, you have an opportunity to consider the issues from your evolving perspective at this point in the internship. Another aspect of reality that creeps into consciousness at this time are thoughts of the field placement ending; this is yet another critical time in the internship. Ending well is not easy, but it can be done and in many different ways. Chapter 14 will help you deal with the concerns and challenges of the final stage—that of Culmination.

CHAPTER 12

Riding High:
The Competence Stage

The price of greatness is responsibility.
WINSTON CHURCHILL

T here comes a time in nearly every internship when interns tell us that they are enjoying their internships in a whole new way. For many interns, this is the time in field placement when their experiences most closely match their "dream" internships—the ideals and images they carried of what an internship can be. Some describe it as the "emotional moment." The initial anxieties have subsided, problems big and small are being managed, and considerable confidence has been gained. There is a sense of feeling grounded—finally! If this sounds like what you are experiencing, then you have entered the Competence stage. For many interns and supervisors alike, this is the most exciting and rewarding time in the field experience. With confidence and with pride, each day in placement the intern demonstrates the knowledge, skills, values and attitudes expected by the site, along with the interpersonal skills, motivation, and cognitive abilities needed to attain success (Klemp, 1977). Interns arrive at this stage at different times during their internships. Some of your peers may not be here yet, while others may have been here for some time. What is important is not how fast you or your peers reach this juncture but rather having *confidence* in yourself that you will get here in due time.

In this chapter, we discuss the pleasures that most interns experience—and have earned—during this stage. As we have for other stages, we also identify concerns and tasks that await you and encourage you to take an engaged approach to them. And while it is less likely that an internship will go off track at this stage than in earlier ones, it can happen, and we end by considering the shape of the experience of disillusionment in the Competence stage.

| In Their Own Words | Voices of Competence |

Things that didn't seem to come together have finally come together.

I feel that all my years of schooling have been for this exact purpose. It's like nothing that I have ever experienced before.

This by far has been THE best experience I have ever had (in a workplace setting). For the first time in a very long time, I feel on top of what I am doing. I know my work, I am respected by my supervisor and co-workers, I get great pleasure from seeing the difference I make in the community, and I just can't get enough of the work. I finally know what I am going to do "when I grow up!"

I am confident now. I take a lot of pride in my work, and I think more independently. I am doing important work—busy work. My supervisor is great. I have developed important relationships with her and the workers, and I feel a part of the team. I have learned so much about myself too. This I didn't expect when I started out. This internship is everything I hoped for, and it happened.

Enjoying the Ride

Students often liken this time in the internship to that point in a flight when the aircraft is no longer ascending and begins to level off. When this happens, you will experience the *cruising effect*: The intensity of accelerating awareness, learning, knowledge, understanding, and responsibilities begin to decrease, and you settle into a rhythm of working that is more realistic and predictable in nature. However, to pursue the metaphor, there is plenty to see and plenty of ways to enjoy the ride.

New Perspectives

Chances are that you are seeing your internship through a set of new and different lenses. Having the knowledge to feel competent in doing the work is critical during the Competence stage. It's not as simple as what you do and don't know. Discovering knowledge you didn't know you had and discovering gaps in knowledge you thought you had can be exhilarating and/or overwhelming at times. One intern described it this way: *"I went to a training session yesterday...that made me think about my approach with all of my clients and how far I have to go. I found myself thinking about this new knowledge and how learning this little bit of knowledge was just the tip of the iceberg. It was all the stuff I didn't even know I didn't know!"*

This is often a time when interns are able to differentiate more clearly between what is and is not important to their work (Dreyfus & Dreyfus, 1980). Subtle dynamics that escaped you before are probably visible now. Remember when you were concerned about acceptance issues? *"Now I look back on all those feelings of inadequacy and laugh!"* was one intern's reaction to that period in her placement.

Even though you spent quite a bit of time trying to find your way around the organization while you were concerned about these issues, the organization's internal politics and stance in the community may have been invisible to you because your primary concerns were focused elsewhere. Now, however, you have a better opportunity to explore and understand the norms and rhythms of the organization.

A Time of Transformation and Empowerment

Your emotional landscape is changing as well. Many interns feel an enhanced sense of autonomy and mastery as their skills become sharper, and they need less constant supervision and direction. There is also a tremendous sense of confidence associated with this stage. The anxiety, awkwardness, and trepidation of earlier stages have gradually given way to a period of calm and inner strength. Interns often tell us that things have settled down at the site. However, life at the internship site is still hurtling along at the same or even an increased pace. It is not the world around you that has changed and become more peaceful; rather, it is you who have developed an inner sense of calm from which you will derive strength as you move through this intensely productive period in your internship.

Becoming the New Me

Another aspect of the ride is a transformation in your sense of self from that of a student to that of an aspiring professional. You think differently, you act differently, and you feel differently. How you think about yourself is continually affected by your experience and by the process of reflection and meaning making. For example, have you noticed that you no longer refer to yourself as "just an intern" or "the intern"? Rather, you identify more and more with the role of the professional and as a member of the profession. One intern reported, *"I am fitting in around the office. Everyone is helpful and giving me tips on how to do things. They even ask me what I would do in certain situations. I feel they respect what I have to say."* Another offered, *"I realize that people really do depend on me here...my advice is important."* You have a greater sense of self-awareness and self-respect. You are most likely learning that you can trust your knowledge and skills, and you are developing a good sense of your strengths and limitations. Your work might suddenly be taking on new meaning because you realize that you are making a difference in the lives of others. Your professional identity is beginning to emerge.

Redefining Relationships

When we ask our interns to reflect on the changes that have occurred in their placements, they often identify changes in relationships with their supervisors. They smile, for example, as they tell us that they no longer need to check in with their supervisors as they go about their work. *"At the beginning of my internship, I was constantly asking: 'What do I do next? What do I do now?' Now I am not asking what to do. I am just doing it."* We are often struck by the quality of the relationships our students develop with their supervisors. They are not only supportive and encouraging connections, but often become redefined into mutually satisfying relationships that last long past graduation day.

Working with supervisors and co-workers becomes easier during this stage. Communication becomes more comfortable and open, issues can be approached without concern about rejection or conflict, genuine teamwork can develop, and supervision can be a source of insight and feedback. Often, a change in the openness of the supervisory relationship becomes evident at this time as well. Most of you would not have considered challenging your supervisors on theoretical grounds earlier during the placement. However, a combination of confidence and a sense of emerging equality may make a big difference in your willingness to engage the supervisors in theoretical or philosophical discussions or disagreements (Lamb, Baker, Kennings, & Yarris, 1982).

Becoming Mentor and Protégé: To Be or Not to Be

Some site supervisors develop more than a supervisory relationship with their interns. They develop what is traditionally called a *mentoring relationship*. The mentoring relationship differs from supervision because it is interpersonal in nature and is one in which the supervisor takes an interest in the intern's professional development and career advancement (P. Collins, 1993). It can be a very empowering relationship, leaving both the mentor and the mentee (protégé) better professionals for it. The mentor not only oversees the development of knowledge and skills but provides "valuable insight into an organization's environment and culture, as well as psychosocial support for the student" (Stromei, 2000, p. 55). It may interest you to know that the mentoring relationship is often considered the most important aspect of graduate education (Mozes-Zirkes, 1993). At the undergraduate level, the mentoring relationship is a most instructive and influential factor in the quality of the internship experience.

(continued)

> ## Becoming Mentor and Protégé: To Be or Not to Be *(continued)*
>
> The best mentoring relationships occur spontaneously between supervisors and interns. The effective mentor makes sure that the intern becomes part of the organization very quickly and is given highly visible tasks; this type of mentor ensures that the intern is introduced to the profession through such resources as networking, luncheons, and conferences. Although many supervisors do make sure that all these bases are covered for the intern, the mentoring supervisor *personally* invests time and resources in coaching the protégé for success (Tentoni, 1995).
>
> Of course, this relationship, like others, takes time to develop (P. Collins, 1993). It moves from a relationship of positive role modeling, when much of the learning occurs through interactions, observations, and comparisons with the supervisor, through a time when you have grown to really like each other, and then to a point where you might be right now: valuing the relationship and recognizing how mutually rewarding it is for you and your supervisor. When this happens, the intern tends to become increasingly self-assured, competent, and autonomous, no longer needing the mentor in the same ways for guidance or support, and eventually becoming more independent of the supervisory relationship. In the best of mentoring relationships, a more equal relationship emerges over time. That becomes evident when the supervisor accepts you as a colleague, and you accept the supervisor on equal footing (P. Collins, 1993).

The Tasks You Face

As you move through this stage, along with the joys there are some challenges and, as always, you can approach them on the engaged or disengaged ends of the continuum, or somewhere in between. Many interns, even the most engaged, fall prey to the misconception that the days of difficulties or problems are behind them. Some say they expect more difficulties in their internships, but when the difficulties arise, they ignore or otherwise do not acknowledge them; other interns just want to focus on the good feelings of having resolved issues. We refer to this common experience as "freezing the moment," staying in the comfort zone of the internship, intentionally or unintentionally. And still others fall into the trap of believing that real professionals (like "really good" interns) should be beyond having difficulties.

Well, you are not beyond them—no one is. Your skills, your relationships with clients, supervisors, peers, and co-workers, and your understanding of the field site are all evolving, and you could continue to experience new difficulties in all these areas. The good news is that you are better equipped than ever to meet these challenges; in fact, your ability to handle difficulties can be a source of ongoing pride. The difficulties you face at this time may actually feel different now and not be so over-whelming or anxiety provoking. However, they still need to be managed in engaged ways or the difficulties will become problems.

In this stage, the general shape of engagement is that interns begin to *take charge of the learning experience* and develop a pace and style of learn-ing that is their own. Striving for competence often means being all that you can be in your role as intern and doing your very best. This engaged approach tends to be adopted by most interns during this stage. Those who shy away from "being all they can be" in the internship—for what-ever reasons, may be disengaging. As you encounter the concerns of further challenging yourself, or defining and experiencing success, of maintaining (or re-achieving) balance, and of deepening your identity as a professional, your level of engagement will go a long way to determin-ing the depth of learning during this stage. In the chart below, we provide some examples of engaged and disengaged approaches to the concerns of this stage. They and others are discussed in the chapter.

The Competence Stage

Associated Concerns	Critical Tasks	Response to Tasks	
		Engaged Response	Disengaged Response
High accomplishment	Raising the Bar: Accomplishment and Quality	Embraces tasks and challenges	Is content to continue with current level of challenge and activity; bored but not willing to change
Seeking quality		Sets high aspirations	Is satisfied with status quo
		Sets personal standards of excellence	Sets standards of "good enough"
Emerging view of self			
Feeling empowered	Having Feelings of Achievement and Success	Actively seeks fulfillment. Engages self and others in achieving the feeling of success	Accepts whatever sense of fulfillment is present. Relies on others to supply it. "Guts it out" if not feeling success.
Exploring professionalism	Maintaining Balances	Keeps personal and internship demands in-check	Is unable to manage conflicting demands effectively
Doing it all	Professionalism	Seeks to understand and adhere to professional and ethical guidelines	Ignores or accepts transgressions
Ethical issues			
Worthwhile tasks		Seeks to identify as a emerging member of the profession/field	Is content to identify as "just a student"

Raising the Bar

In our discussion of the Exploration stage, we encouraged you to revisit your Learning Contract and make adjustments to the goals and activities accordingly. Well, you are past that now, and you have most likely achieved some of your goals and on your way to many more. If you are in a year-long internship, the Competence stage lasts for quite a while, and in that time more and more of your goals may be met. That is a satisfying feeling, but the satisfaction can begin to give way to boredom or complacency; you can stay right there if you want to, but then the joyful feelings we described earlier will begin to fade. So, it's time to challenge yourself again. When we say this to groups of interns, sometimes the groans are audible. Remember, though, that you have a number of new tools to challenge yourself with, that you have been through this before, and that the risk of new challenges (and possibly some failures) also carries the reward of new satisfactions and new learning. We cannot say for sure what the new challenges will be, but we offer you two issues that come up over and over in our experience: commitments to quality and to the growth of personal integrity.

A Commitment to Quality

Not so long ago, in the throes of the transition from learning the work to doing the work, "good enough" may have been all you expected of yourself in your endeavors; just getting the work done seemed a lofty goal! That was especially so if you were dealing with a multitude of responsibilities in addition to your internship. Now, though, the "good enough" standard that prevailed earlier in the internship can give way to a standard of mastery and even excellence. If you want that to happen, you become more demanding of yourself, regardless of the number of responsibilities you have. This expectation of excellence extends to the full range of issues impacting your personal, professional, and civic development. Interns who are not engaging in such ways tends to be satisfied with "good enough" as the standards that guide their work.

A Commitment to Integrity

Developing integrity is a lifelong process that begins during the college years (Chickering, 1969). This was not necessarily a goal of your internship at the outset, but you may want to think about it now. You won't finish the process, but you can attend to it more consciously by paying attention to it in intentional ways. Developing integrity means clarifying your values, making them the subject of reflection, and attempting to live and act in ways that are consistent with what you truly believe. An internship does not necessarily change your core values; however, it certainly does challenge you to clarify them professionally and personally. Accepting that challenge is the first step to engaging the process of developing integrity. A second aspect of the process is taking a clear-eyed look at your actions and how they do and do not reflect the values you believe in. If they do not, then it may be your behavior or your values that need to shift, but if you continue this reflection you will bring these two aspects of yourself more and more in line with each other. And a third aspect of the process is developing a value system that is more and more internal and integrated. Many of your values—the rules you live by—initially come to you from outside yourself, from parents, family members, friends, religion, and culture. But as you work to challenge both your values and your behavior, your "rules" become more and more a part of *who* you are. This does not always mean that the values and principles that came to you from external sources are abandoned; many will remain. But they are now a deeper and more integrated part of you. The closer you are to adopting this value structure openly and without a façade, the closer you are to living with a sense of personal integrity (Chickering, 1969). This is a lifelong task, and your journey has just begun. You may or may not be ready to engage in this process, but this stage of the internship is a glittering opportunity to do so.

Focus on PRACTICE

Choose Your Companion: Excellence or Perfection?

At the beginning of the internship in the first seminar class, we ask our students, "Who here is a perfectionist?" Some interns acknowledge it immediately, with moans, grunts, or laughs. Other interns acknowledge it after thinking about it, with resignation. We take that moment to share these observations with them: Many students have learned from experience not to confuse the need to do work well (excellence) with the need to do work perfectly (perfection). Although both may guarantee success, one guarantees headaches as well—for all involved. When not giving yourself permission to make a mistake while learning, you deny yourself opportunities to learn how to recover from mistakes and how to solve problems. If perfection is your only or preferred way of doing things, you are bound to be chronically exhausted from the pressure you are putting on yourself. Our suggestion is that for the duration of the internship, if at all possible, give up the need for perfection and replace it with a need for excellence.

Feeling Success on the Ride

Some interns become concerned and challenged at this point in their internships when they realize that they appear successful to others, but they do not necessarily *feel* successful about their accomplishments. Such a situation can become a problem because your experience of the internship is based on what you *feel*, not on the perceptions of others. Being able to *feel* your success, often described as a sense of fulfillment, is what makes the success genuine, both emotionally and cognitively (King, 1988); anything less can compromise those accomplishments.

Sources of Fulfillment

The aspects of my experience which contribute to my sense of fulfillment are the feelings of satisfaction I get when I know I have connected with a client and really made their day better in some way.... In my internship, it is rare to receive gratitude from a client, but I feel pleasure just knowing that I am giving of myself in a positive, caring way.
STUDENT REFLECTION

There are three sources of fulfillment that nurture the emotional experience of feeling one's success: *worthwhile work, responsible relationships,* and *self-defined success* (King, 1988). As you read about them in the Focus on THEORY box on p. 340, think of your own internship and the ways in which it provides you with what you need to *feel* the success you have earned.

Worthwhile Work

The first source of fulfillment in success is embedded in the "outer" or social dimension of success (Huber, 1971). Whatever your work is, it must be considered *worthy* or *worthwhile* by you and by your supervisor. It must be productive, with responsible activities that are personally meaningful, and the work must allow you to accomplish clear goals. There are five aspects of the work that allow these feelings to develop (King, 1988):

- *Accomplishments* The work itself needs to provide you with opportunities to develop your skills and apply concepts in real-life situations. For your part, you need to take an active, engaged role in creating your work and your achievements. What you accomplish must be *purposeful in nature, constructive, challenging,* and be *a source of pride* for you. Additionally, both your work and that of the agency need to reflect *socially responsible goals* to feel good about the public relevance of your work. Interning at a site that is a known abuser of the environment or has a reputation for questionable practices with

Focus on THEORY

Sources of Fulfillment in Success

- **Doing Worthwhile Work** (*Social, Outer Dimension of Success*)
  - Accomplishments
  - Acknowledgment
  - Self-determination
  - Self-actualization
  - Intrinsic rewards

- **Developing Responsible Relationships** (*Personal, Inner Dimension of Success*)
  - Site Supervisors staff, & co-workers
  - Peers at the site and on campus
  - Campus supervisors
  - Community contacts
  - Family, partners & friends

- **Self-Defined Success**
  - Conscious choices
  - Freely made
  - Self-determined goals
  - Engaged involvement
  - Source of enjoyment

(King, 1988)

employees or clients would certainly call into question—by you and by others—the *quality* of your accomplishments.

- *Acknowledgment*   Being *recognized* and *respected* by your supervisors takes on paramount importance to how the achievements are experienced. Such acknowledgment is especially significant when your contributions are above and beyond what is expected of an intern.

- *Self-determination*   This aspect of the work has to do with the potential for autonomy in your internship work, which is evident in the *freedom to create* and *carry out* the tasks.

- *Self-actualization*   Opportunities for *creative expression* and *personal growth* need to be built into the work. Otherwise, the work can feel stagnant.

- *Intrinsically Rewarding*   Above all, the work has to provide you personal pleasure, regardless of the demands or stresses inherent in the work.

In Their Own Words	Student Reflections on Worthwhile Work

Accomplishments *By allowing me to take charge and work independently with a group of clients yet having the support of a professional.... I am proud of my internship now that I feel like I'm making a difference and doing some meaningful work.*

Acknowledgment *It really made a difference when my supervisor and my co-workers complimented me and told me that I had done outstanding work. Up to that point, I thought that I had done a good job on the project, but I didn't really know. I needed to hear it from someone there. Up until then, I began to question myself and my abilities.*

Self-Determination *I coordinated the event and actually made it happen.... I was given the opportunity to develop something on my own. It felt great.*

Self-Actualization *I was getting worried because everything I was expected to do was already laid out for me. There was no room for "me" in the work. I was just carrying out someone's prescription for how to do the work. I had some great ideas, really creative ideas, for how to do it better, and I felt so frustrated because I didn't think my ideas were welcomed. It's not really my work if my ideas aren't reflected in it. I hated how I felt. I was going nowhere just doing what I was told to do.*

Intrinsically Rewarding *It made me feel good to help someone, and it was nice to be appreciated for my efforts. I feel like I am making a difference at my internship. It feels good.*

Responsible Relationships

The second source of fulfillment in success concerns the "inner," or personal, dimension of success (Huber, 1971) and that is derived from the personal relationships that are integral to your internship experience. It is this sense of inner success—your success as a *person*—that renders your achievements worthy (Huber, 1971; King, 1988). So, if you want to reap such psychological rewards as the internship being "the best experience of my life," then you'll want to pay attention to the quality of your relationships. In particular, they need to be *genuine, cooperative,* and *mutually satisfying* for all involved; in other words, they need to be *responsible relationships* (King, 1988).

Real commitment on your part and on the part of co-workers, peers, supervisors, and the community is needed to make responsible relationships happen, and that calls for engaged approaches. Your co-workers must be receptive to your coming on board, and when that happens, it is an important milestone. The supervisor needs to establish an effective supervisory relationship with you and include you in work groups/teams, staff meetings, and social functions (when appropriate). You, in turn, need to be willing to engage both the supervisor and your co-workers in responsible supervisory and collegial connections. You also need to be willing to engage community groups in genuine, cooperative, and mutually satisfying ways.

Defining Success for Yourself

The third source of feeling the success you attain is being able to *define success for yourself*. Self-defined success essentially means you are doing what *you* truly want to do, not what someone else wants you to do. You made a choice, engaged fully in the process and free of the influences of other people (King, 1988).

When it comes to your internship, making *conscious choices* can be evident in your involvement in the selection of your field site and/or in the development of the goals of your Learning Contract. Preferably, you were actively and consciously involved and engaged in both decisions, the decisions were freely made, they reflect self-determined goals, and they are a source of enjoyment or pleasure. Now that's a tall order and not necessarily possible in all academic programs, which means that you may not have been given such opportunities early in the process. However, if the field coordinator determined your site, and the site supervisor determined your Learning Contract, and you were not actively involved in an informed way in either of these processes, that does not mean all is lost! Even if you went into the internship not consciously aware of what you really wanted from it, those overseeing your actual placement may have been able to create a context that allows you to eventually *own* the decision, i.e., emotionally avow the decision as your own. It could be that the placement supervisor assessed your interests and goals quite well and placed you with a supervisor who provided support for personal growth and awareness, while entrusting you with worthwhile work. If that was the case, then there is a good chance that you will eventually fully accept the choice of field site and the learning goals and, in turn, feel success in your achievements.

Wait! Wait! It's Not Success That I Feel

What about those of you who, instead of feeling the success you are achieving in your internship, are experiencing something that feels like anxiety? It may be that you are experiencing what is generically referred to as *achievement-related anxiety*. Initially, it is experienced in subtle and then not so subtle ways. This anxiety can be felt in the arena of expectations: living up to expectations of others or oneself; working in a competitive atmosphere where you are expected to compete; or, feeling a need to avoid failure at all costs. It also can be experienced as a need to avoid success at what you are doing, as is the case when being successful means you must move onto a position that you don't want, one that would turn your life into turmoil due to the relocation and increased time commitment involved. If this is what you are experiencing instead of enjoying success in your internship, then you might want to bring it to the attention of your supervisors and peers, depending on your circumstances. In that way, you can create a context within which to explore the anxiety you are feeling, as well as what might be contributing to it, and brainstorm ways to move beyond it. Your peers may be feeling similarly or your supervisors may have had similar experiences during their internships. Even if they haven't, they are there to support you through this. Before you call in the troops, spend some time with the THINK About It box below and on p. 344 to see if you might develop insight that is useful.

There are a number of reasons why an intern may not feel the success being achieved. If you are not enjoying this ride, even though from the

---

THINK About It

Are You Doing What YOU Want to Do in This Internship?

It's time for some soul-searching about this issue of *conscious, active choices* about your internship, and now might be a good time. How important is it that you own your internship? These are the thoughts of one student who did engage in such soul-searching. *"Oh, my … I'm not sure that I even want to be a counselor now! After all this work and putting off starting a family."* The following engaged approaches will guide you in your reflections.

- *Think about…*what you really wanted to achieve in this internship, not about what others think you should achieve or what others want for you in this experience.

- *Think about…*the extent to which your placement site is a good match for your personal, professional, and civic goals, including how compatible it is with your views of its civic responsibilities.

- *Think about…*the work of the internship and whether it is really what *you* want to do in this opportunity to learn.

- *Think about…*the extent to which the choice of internship was made freely by you. Just how involved and engaged were you at the time decisions were made?

The bottom line here is that this *is* your internship now, no matter how you reached this point, and it is your responsibility is to make the best of it. This is when you can learn how to get through situations in which you may not be fully committed but have the responsibility to do your best by being fully engaged in it. The tools in Chapter 9 offer guidance on how to do just that.

outside it looks like things are going great, then it's important to take the time to think about what may be amiss. For example, are you stressed to a point where you are not enjoying any or only few aspects of your life? Are you feeling down? Are you worried about a matter to the point of constant distraction? Most of us have experienced some if not all of these situations, and we know from experience that when we are, it is difficult to smile, let go and have a good time, and/or enjoy our daily lives. If any of these apply, then it might be difficult to enjoy your success *at this time*.

There is a big difference between a situation that is transient in nature and one that is pervasive in one's life and chronic over time. Regardless, an engaged approach is needed to move beyond this time in your internship. If the latter situation seems to apply more, we urge you to talk with one of your supervisors about it so that you can be pointed in the right direction for resources. If the former applies more, then the reflection questions in the previous THINK About It box (Are You Doing What YOU Want to Do in This Internship) may be useful. Most importantly, you will be doing all you can do to assess what needs shoring up in your internship and make that happen.

As you consider the following queries, keep in mind the overall need for *productive work* that allows you to accomplish *clear goals* through *responsible activities* that are *personally meaningful*. Your responsibility is to ensure that each of these aspects of worthwhile work is in place, and if perchance they aren't, your responsibility is to make that happen. Not sure how? Begin with your supervisor. During supervision, review the qualities that need to be shored up and create a plan to move each of them forward.

THINK About It

NOT Enjoying the Ride?

If you are not particularly enjoying your internship, there could be a number of things going on. For example, new challenges may have arisen, calling for skills and knowledge that you have yet to adequately develop. However, there are some common issues that can interfere with the sense of "cruising" typical of the internship at this point. So, ask yourself:

Is Deep Learning Occurring in the Internship? The learning that needs to take place in an academic internship is not intended to be training or shallow in nature, but instead meaningful and contributes to your growth. Be sure the internship is designed to allow enough time to process experiences so that the Experiential Learning Cycle (see Ch.1) can take place.

Are the Sources of Fulfillment There for You?

○ *Are You Doing Worthwhile Work?* Review what makes work worthy (*accomplishments; acknowledgements; opportunities for self-determination and self-actualization; intrinsically rewarding*) and ensure that they are in place.

(continued)

THINK About It *(continued)*

○ *Are Your Collegial Relationships "Responsible"?* Examine the relationships with people important to your field experience and be sure they are genuine, cooperative, and mutually satisfying. Are they disappointing in any way?

○ *Are You Living Your Definition of Success?* Remember that success must be self-defined or it is not experienced emotionally for all it can be.

What Affirms You in Your internship? List the sources that nurture your feelings of empowerment and competence and keep it handy. Can you identify the areas in your internship that you would like to see strengthened? If so, the time is right to move ahead and do just that. Not sure how? Talk with your site supervisors and with your peers about ways that they know that have worked and share your ideas for their feedback.

Reclaiming a Balanced Life

If you are like many interns, you may suddenly realize that you had a life before the internship—a social life, a family life, a private life—and you find yourself needing to reclaim it. If you are an EXperienced intern, you have been dealing with this issue every day as you managed to keep your priorities clear and in place. This is no easy feat if continuing on in your academic major or graduation depends on the grade earned in the internship. One intern was stunned by the realization: *"I realized I was not giving enough attention to my relationships with my friends and family. It had been nearly three months since we had spoken. The incident made me stop and think about how important my relationships are and how, even though I am busy, I must not neglect them because these are the people I count on for support."*

Perhaps that awareness will come to you on your way home from the field site, when you are no longer so preoccupied and are thinking about aspects of your life that were left behind in all the excitement of the internship. Or perhaps it will happen during a morning break as your mind wanders to more personal needs—for relationships on-site or in your personal life, for family time, for time alone. *"I can really understand the need now to slow down and pace myself and try to create a more livable, useful pace for other people in my life, including at the site,"* noted one intern. Another student's experience was a nagging feeling: *"I can feel a tug going on between family demands, work, and my academic requirements.... The holidays can also complicate my situation to the point where I am beginning to feel both pressure and stress."*

Interns usually have a need to get on with living at this point in their internship. In our experience, these changes signal a healthy shift of energies toward a more *balanced way of living* with your professional life. Although your internship still remains very important, it may no longer be the driving energy of your life. As one intern noted: *"The kids notice when I am not there. Continuity is important to them. This incident reinforced my need to balance my job and internship both for me and especially for the students I see."*

Feeling the Crunch

Up to this point in your internship, you have had to struggle with the ups and downs of learning skills, developing competencies, resolving differences between real and ideal expectations, becoming part of an organization, and contending with changing perspectives and life's big questions. Just when you thought it was clear sailing ahead...*wham!* You've run into yet another wall in your journey; and just about everyone hits this one.

This wall is different from previous ones. This one you know you can manage; you just don't know whether you can muster the energy you need to do it. You feel some anger about the amount and level of work you have and frustration over the amount of time you do not have. Something subtle is happening as well. Your perspective seems to be changing again. You have little tolerance for trivialities, and you find yourself feeling indifferent toward some of your assignments and responsibilities, especially if they are not directly related to the internship or do not measure up to the importance of the work you are doing in the field. This is *not* a crisis in growth you are having; this is a *crisis in management* of your time and tasks. The crisis takes on added importance because it is in fact a real threat to your schedule and the quality and the pace of your internship.

You are encountering what we refer to endearingly as *the crunch*. It is an early warning signal to let you know that the end of the internship is fast approaching. This crisis in management is no different from the crunches you faced at the end of every semester (usually at the end of the mid-semester slump, remember?) when papers, exams, and projects all become due at the same time. However, this time, much more is at stake academically, financially, and professionally. On top of that, you are probably carrying the greatest number of responsibilities you've had to date in your studies, whether you are a traditional age or EXperienced intern.

If any of the following questions best describes you at this time, know that you are not alone! Most of your peers are or will be going through "the crunch" about the time you do. It's important to be aware of the extent of its effects so you can manage it effectively.

Crunching It

Have you begun to...

- question the merit of each assignment?
- challenge requests for more work?
- struggle to avoid the wall of ineptitude?
- feel overwhelmed?
- feel stuck in time?
- think you have...

  too many deadlines?

      too many details?

          too few resources?

              too little social life?

                  too little family life?

                      too high standards?

                          too many responsibilities?

Managing the Crunch

To get through it, you'll need to use engaged approaches, some of which can be found in Chapter 9, and some of which are discussed here. You know yourself best in times like this. What works for you? What allows you to manage time and workload crunches? Some people actually thrive when they are up against a wall of deadlines, while others crumble under the same circumstances. Perhaps one of these engaged approaches will work for you as they have for many of our students.

- *Taking Time for You*   A day will do; an afternoon or evening could do just as well. The important thing here is to stop. Just stop everything. Take a holiday from the assignments, the responsibilities, and the timelines you are facing. Of course, it is not just the internship that is overwhelming you. Many of you have responsibilities beyond your internship, and you may be overloaded for what you can effectively handle at this time. However, the internship is still the new kid on the block, and its workload is most demanding right now. It is very important that you do something that you enjoy. Sing. Dance. Shop. Bike. Hike. Climb. Swim. Draw. Paint. Read. Run. Play. Ski. Work out. Create. Sew. Meditate. Sleep. Listen. Make music. Drink chocolate. Eat chocolate. Take in a movie. Go out to dinner. Do something, anything, whatever it takes to help you...*let go!*

- *Focusing on Task*   Things simply must get done. So, it is time to make decisions about your schedule and prioritize the tasks. You will need to decide what is doable in the time you have and what can get done only if you make significant changes in your schedule. This means prioritizing and scheduling as efficiently and effectively as you can. If this is not a strength of yours, and it isn't for many people, then get help for yourself. Use the following Focus on SKILLS box and seek out people who know how to manage organizing and consult with them. That person could be a peer, one of your supervisors, a colleague, or a friend. Also, instead of focusing on the long list of things you will need to do eventually, focus on what you need to do now. Try saying to yourself "all I need to do *now* is _____ and I can do that."

- *Eight Stepping It* What is obvious is that something has to change to put you back on track and in charge of your internship. You already have had to deal with difficult challenges when you confronted the barriers in your internship that prevented you from moving ahead. This hurdle is far less complex and demanding and actually quite manageable by you simply, if for no other reason, because you have gotten this far in your internship. So, if you are struggling with this challenge, roll up your sleeves and get to work. Use the model discussed in Chapter 9 (*Eight Steps to Creating Change*) to help you clarify the biggest hurdles you face during this time-and-workload crunch and bring this crisis in management to an end.

Focus on SKILLS

Organizing for Time Management

Here is an exercise that our students find helpful in accomplishing this task. First, list all the tasks that you have to complete. Then, next to them, indicate the time frame within which they can be finished, not necessarily must be finished. For example, if you have a final culminating statement to write for your seminar class, it obviously cannot be completed before the internship ends. However, you can work on much of it before the end of your placement. Then, when the internship is over, there is relatively little content to add. The time frame for this assignment actually begins on the date you start this exercise and ends on the due date. By listing tasks, assignments, and responsibilities in a similar manner, you will have a more informed framework for prioritizing tasks and assigning calendar dates to work on them.

Preparing for the Profession

As the headiness of the Competence stage continues to define your daily feelings about the internship, you may find yourself thinking more and more about the profession for which you are preparing. Writing your résumé, seeking out job opportunities, or deciding where you want to live after graduation perhaps come to mind immediately. And those are certainly important ways to begin your journey to your professional home.

Leaving Your Footprint

We find that interns are preparing themselves in other ways as well, ways that reflect their growing awareness of the importance their profession will have in their lives. Two of those ways are a need to "give back" to their internship site and a need to feel a continued part of the organization and profession after the internship ends. There certainly are ways to ensure that you made a difference at your site and will be remembered for it. Whether you are in public service settings, corporations, civic organizations, research laboratories, or nonprofit agencies, you may be feeling a need to make contributions to the field site in meaningful ways. One intern described this engaged stance in this way: *"I think, no, I know that the (police) station needs to have the back room cleaned out. I want to do that for the chief. He has been very supportive, and I want to give back in some way... I guess it means a lot to me to know that I somehow am still going to be part of the department in some way after I leave."*

These contributions rarely are part of the Learning Contract; rather, they tend to be given from the heart. They are contributions that benefit the organization but may not have been completed because of the workload demands, fiscal constraints, and task priorities faced by the organization. The contribution may take the shape of creating an internship manual for the site, developing databases, creating a resource library, or reorganizing workspaces. A popular contribution for our students has been reorganizing and downsizing data files that need to be re-catalogued into new data systems. For the most part, the contributions tend to be on the "wish lists" of site supervisors or others in the organization. If you have been thinking about making such a contribution, now is a good time to think about how to do it and when. You might want to use the predictable downtime during the last few weeks of the internship.

Moving Beyond the Textbooks

A second area of awareness that typically develops in students during the Competence stage has to do with an appreciation for what is beyond textbook knowledge. By now, you realize that remembering facts, working with efficiency and speed, approaching problems in logical ways, and using resources to meet the needs of the work are effective ways of going about your work. In fact, they may be exactly what is expected of you in your future profession. But you also realize by now that it takes more than these skills to handle challenging situations and resolve complex problems in the workplace. There are two invaluable resources that interns look to when they realize that what they need to thrive in their future professions is not all found in textbooks.

The first is *wisdom*. Wisdom transcends textbook knowledge. It is wisdom that makes a supervisor have an empowering effect and a remarkable role model and mentor. The differences between these two modalities reflect the differences between what can be instilled and developed in you (*intelligence/skills*) and what you must learn to develop through experience and commitment to a more visionary way of dealing with the work and the issues (*wisdom*).

THINK About It

Your Supervisor as a Source of Wisdom

If your field site or your supervisors encourage and affirm a tolerance for ambiguity, an ability to intuitively understand and accurately interpret situations, and an ability to identify and frame a problem accurately, then it is wisdom as well as intelligence that is being valued by them. If you are interning at a site where empathy, concern, insight, and efficient coping skills are valued, again wisdom as well as intelligence are being valued (Hanna & Ottens, 1995). Ideally, your supervisor possesses both these qualities and has modeled the importance of them for you.

The second invaluable resource is a combination of the *habits of reflectivity* (Schon, 1995) and *practical reasoning* (Sullivan, 2005) as ways of being. When professionals go about their work in these ways, they make a practice of reflecting deliberatively as they consider circumstances and situations as well as the decisions. They are constantly integrating and reintegrating theory and practice. Developing these engaged habits takes focus, practice, and conscious intent.

Becoming a Civic Professional

As you continue to prepare for your professional field and develop knowledge, hone skills, and assimilate into its culture, you will probably find yourself thinking more and more about your civic responsibilities as an adult and as a professional. The goal of civic development is to prepare you to participate fully, responsibly, and actively as a member of a community or communities. Active participation refers to an engaged approach to the responsibilities and being an "owner" instead of a "consumer" of government.

The internship has given you an opportunity to learn about the public relevance and social obligations of the profession in which you are interning and how these social obligations are or are not carried out at your site. It has also given you the opportunity to think about yourself in relation to the community and what that means in terms of your civic learning and preparation for civic effectiveness. Does that mean more is expected than voting in elections and obeying the laws? That is a question you will have to come to grips with yourself, for only you can determine what your place will be in your community. However, during this stage of the internship, there are three areas of awareness that tend to give interns reason to reflect while performing competently as they realize the importance of the work and the field to society.

The Public Relevance of the Work

Some of you are in organizations that serve the public and are publicly funded to do so. Some of you are in the private sector, where generating profit is required in order to exist and provide products or services. Others of you could be in private, not-for-profit organizations providing services to the public that are funded through a variety of sources of revenue. And still others of you could be in foundations, where private monies are distributed for social benefits. Regardless, and whatever the profession, your work has a purpose in the community and society at large—through a social type of contract for community betterment.

As mentioned in Chapter 1, the purpose of your site in the community tells you a lot about the relevance of your work to the community. If you are in the helping professions, the mission of the organization—be it government, private, or not-for-profit—spells out that purpose and describes its relevance to the public good. If it is government administered, most likely the mission and its relevance are part of a statute that authorizes its existence and funding by tax dollars. If, on the other hand, you are in an organization that is private and not serving the public in such a way, the company probably has a statement of responsibilities to the public or the community, beyond maximizing profits and generating wealth (see Chapter 10, for examples). If your internship is at such a site, you might be impressed with the organization's operational philosophy, although you still find room for philosophical and political debate about its commitment to the community.

The profession for which you are preparing is replete with values about how the work gets done and the intent of the work. For example, fairness could be a value of your profession, as could integrity, justice, beneficence, and so on. You know those values by now because you work with them each day at your site. You may be expected to uphold them in how you present yourself to co-workers and to the general public, to the community, and to clients, whether it is to provide services to military families, make the community greener, or build safer bridges.

The Site's Civic Stance

You may already know what is expected of the professional workforce at your site in terms of providing service time to the community each year, whether sitting on boards of directors or advisory groups, picking up trash, or being part of river cleanups. What you might not be aware of, though, is the company's position on eco-stability in the community and its commitment to a greener presence there. You also might not be aware of the company's policy about being part of the solution to social problems, even though it was not part of creating the problem. For example, it could be that the company has a commitment to employing a percentage of the immediate or neighboring community or it contributes financially and with in-kind donations to a local shelter for battered women (Sullivan, 2005). If you've not yet read about the commitments to communities being made by the four companies and organizations we feature in this book, see the *For Further Exploration* section at the end of this chapter to learn more about them.

THINK About It

The Civic Investment

As you reflect upon the public relevance of your work, that of the site and its stance on civic commitment, think about:

- The social obligations the site has to the communities it affects or serves.
- The ways you see the professional workforce carrying those obligations to the communities.
- If the site is not carrying out its social obligations, how do you know and how do you make sense of that?
- The ways in which you have been engaged by your supervisor or co-workers to carry out the organization's social obligations.
- Your reactions to being included or not being included, if that is the case.

Your Civic Readiness and Competence

When you think about being prepared through your education for responsibilities to the democracy in which you live, what comes to mind? Do you think in terms of your knowledge of civic matters, such as public problems, the causes of social problems, the challenges individuals face on a daily basis, or government laws and institutions? Perhaps, instead, you think in terms of the civic skills you are developing, such as communication competencies, organizational analysis, or advocacy for change in the workplace. Or you may be thinking in terms of the civic values that you are seeing in practice, such as being a good neighbor, involving minority opinions, or an obligation to the common good (Battistoni, 2006). Whether it is the knowledge, skills, or attitudes and values needed to develop your civic learning, they are being nurtured in you so that you can contribute in direct ways to your community.

Perhaps you wonder whether you have the basics in place to grow into a civically effective individual. Most likely, you have been laying the groundwork for that for some time now. Being politically informed is important to civic effectiveness, as is sensitivity to diversity issues, having leadership know-how and awareness of social responsibilities, taking academic coursework in basic civic knowledge, and developing personal competencies to work with others in the community. How you learned them—through studying, employment, previous career, volunteer work, or this internship—is not important. What is important is that you are aware of how your decisions and judgments can make a difference, now and in the future; it is that ability to make good choices that is the essence of civic effectiveness (McKensie, 1996; Howard, 2001).

Slipping and Sliding… in the Midst of the Ride

Remember mega-bumps? They can happen in this stage as well, for any number of reasons, including achievement anxiety. Those circumstances may or may not be related to the internship, but they can affect your ability to continue to invest at the level you are and perform in the ways you have been. You realize that you are not using your resources to ground yourself as you have in the past. Sometimes, if you are disengaged enough when approaching the issues discussed in this chapter, your whole feeling about the internship will shift, and not in a good way. It typically signifies the onset of a crisis. If the crisis is clearly internship related, then it usually is a *crisis of capability* to continue on and complete the work at the high standards that have been a source of pride.

This crisis can change you in dramatic ways. The danger is that the high levels of morale and investment that characterized the internship up to this point are replaced with apathy and incapacitation. The aspirations that once steered you with high levels of achievement are replaced by indifference. Standards of excellence that once guided the internship slip into standards of indifference or "just getting by." The commitment to the role/profession and associated identity that has developed is replaced by cynicism. The intern looks and acts disempowered by the changes that have occurred, and the intern feels that way as well. The frustration and anger that are being experienced and most often directed at the internship are complicated by the reality that "the dream," which has powered this internship since its beginning—or even before—seems to have slipped into the whirlpool, the vortex of disillusionment.

The question becomes whether or not the "dream internship" is retrievable. We know that it is, and finding that out is the opportunity that comes with this crisis. We also know that much concerted effort is needed on your part to make that happen. However, you have made it this far, so you know that you have what it takes. Now to get the dream back! The advanced tools in Chapter 9 will guide you in doing just that.

Conclusion

I was given the opportunity to prove to myself that I could do it. This alone has allowed me to feel competent. I tested out my skills and got a professional feel about them. Now I have the key in my hand. I feel ready to move on. I am still not quite sure which doors this key will open, but I am sure that whatever I face I will deal with as best I know how.
STUDENT REFLECTION

As you join others in reaching this stage, keep in mind that interns have different styles of expressing their emotions, and while you may experience this stage as a subtle, yet profound, shift in feelings, others may experience it as a positively giddy swing in emotions. Keep in mind as well that no one gets to this point all by themselves. Along the way, there have been people who have been there for you, who have believed in you, who have given you the benefit of the doubt even when you might not have earned it. Perhaps it is a professor or staff member on campus, perhaps it is your campus or site supervisors, or perhaps it is your seminar peers who made all the difference in being where you are today and feeling as you do. It could be a loved one, a colleague in a previous job or career, or someone who said something in passing that made the difference in the path you have chosen. At this point in the internship, it is important to take the time to reflect on what you've been given by others. It's also important to let them know how they made a difference in your internship experience or how you can do the same for someone else.

The rewards of this long journey in experiential learning are finally realized when you experience the Competence stage. Although hardly without its concerns and challenges, this stage is one in which you can indulge yourself and enjoy the feelings of finally reaching your goals. The transformations in development that you have experienced and the sense of empowerment that you have developed are evident not only to you in how you feel and go about your work, but to those around you as well.

Your new status as an emerging professional has a consequence: knowing your profession's social obligations and your responsibility to civic readiness and future competence. As you continue to develop competencies and a sense of professional identity, you begin to realize that you also are developing awareness of the behaviors of professionals in general and of your co-workers in particular. This awareness tends to take the form of curiosity about politics and staff actions and their effects on the quality of the work. Up to this point, they did not seem to carry the importance they do now. In addition, you have the responsibility of an awareness of the professional, ethical and legal issues common to the workplace. And, that is the focus of Chapter 13.

For Review & Reflection

Checking In

Getting it Right. On a piece of paper, jot down all the factors you can think of (including the ones mentioned in this chapter) that are important to your internship (positive and negative). Then, next to each one, indicate the levels of "too much," "too little," or "just right" for each of the factors. Chances are that what's the "just right" level leaves you with the greatest feelings of satisfaction. Once you've done your inventory, think about what you need to do to reduce or increase amounts to reach "just right" levels across the board.

Experience Matters (For the EXperienced Intern)

You've had to deal with many transitions to reach the point where you today in your education. The "crunch" and the "balancing act" are two that you've probably had plenty of experience learning to manage. However, the "cruising effect" may be something you've not experienced. If that's so, what questions did that experience raise for you and why? If that's not the case, which transitions are most challenging for you and why might that be?

Personal Ponderings

During this stage of the internship, you are navigating more and more of the work on your own and finally doing what you went there to do. You've taken charge of your internship, which means you are the one who determines the pace of your learning and to a large degree, the shape of your learning. As you think about your learning, what tasks/experiences do you want added to the learning contract that you either consider important to your knowledge base or something you want exposure to? What has been most challenging for you to manage in determining the pace of your internship, and why do you think that is?

Seminar Springboards

Being a Professional: What's It All About? Your understanding of what a professional is has been affected by your internship. In what ways has it changed and why does knowing that matter? How do you measure up to your current description of a professional and what aspects of being one do you plan to work on next? If your description of a professional includes having a civic responsibility by being in the profession, what aspects of that responsibility will be most challenging for you as you begin your career? And, if it doesn't include having that civic responsibility, why is that?

For Further Exploration

As You Sow. (2011). "Corporate Social Responsibility." Retrieved from http://asyousow.org/crs.

An example of an organization that works with corporations to promote environmental and social corporate responsibility through shareholder advocacy, coalition building and innovative legal strategies...founded on the belief that that many environmental and human rights issues can be resolved by increased corporate responsibility (http://asyousow.org).

Baird, B. N. (2011). *The internship, practicum, and field placement handbook.* Upper Saddle River, NJ: Prentice Hall.

Comprehensive and covers a wide range of topics relevant to the clinical/counseling dimension of human service work.

Colby, A., Erlich, T., Beaumont, E., & Stephens, J. (2003). *Educating citizens: Preparing America's undergraduates for lives of moral and civic responsibility.* San Francisco: Jossey-Bass. Chapters 6, 8

Particularly useful for stimulating thinking about civic development.

Jackson, R. (1997). Alive in the world: The transformative power of experience. *NSEE Quarterly, 22*(3), 1, 24–26.

Explores what it is about experiential education that "gives experience the power to transform" an individual's thinking.

Kuh, G. D. (2008). *High-impact educational practices: What they are, who has access, and why they matter.* Washington, DC: Association of American Colleges and Universities.

Describes educational practices that have a significant impact on student success which benefit all students and especially seem to benefit underserved students even more than their more advantaged peers.

National Taskforce on Civic Learning and Democratic Engagement. (2012). *A crucible moment: College learning and democracy's future.* Washington, DC: Association of American Colleges and Universities.

This report from the National Task Force on Civic Learning and Democratic Engagement is a "call to action" for higher education to act on its mission to educate students for citizenship as well as the workforce.

Northwestern Mutual Life Insurance and Financial Services. (2012). Community Service Award. Retrieved from www.northwesternmutual.com.

Northwestern Mutual financial representative interns don't just get to test-drive the full-time career; they can also get engaged in the company's commitment to volunteerism. Through its annual Community Service Award http://www.

northwesternmutual.com/about-northwestern-mutual/our-company/northwestern-mutual-foundation.aspx#Employees-and-Field) Northwestern Mutual interns like Chad Monheim (http://www.youtube.com/watch?v=1ne4qft8gy4&feature=plcp) can get recognition—and charitable funding—for the philanthropic causes they're passionate about. Go to http://www.northwesternmutual.com/career-opportunities/financial-representative-internship/default.aspx to learn more about Northwestern Mutual financial representative internships.

Siemens. (2011). "The Siemens Caring Hands Program." Siemens Corporation. Retrieved from www.usa.siemens.com/answers/en.

A not-for-profit organization of this global company established to receive, maintain, and disburse funds for sponsoring and encouraging charitable activities in which employees donate their time and talents to worthy causes to carry out the company's commitment to impact the communities in which they live and work through volunteerism in a variety of community service activities, including giving campaigns, blood drives, holiday food and gift collections, and disaster relief fund raising when their employees, operations, or business partners are affected.

The Washington Center for Internships and Academic Seminars. (2012). Retreived from www.twc.edu.

Students at the Washington Center (www.twc.edu) intern at government agencies, nonprofits, think tanks and businesses such as the Department of Justice, Amnesty International, Merrill Lynch, No Labels and Voice of America. They also have the opportunity to become positive change agents by participating in civic engagement projects on important domestic and international issues, including domestic violence, homelessness, veterans, torture abolition and the environment. Students learn about these issues by interacting with national and local community leaders, then become involved in a direct service and/or advocacy project. They can choose to join a TWC-guided project or design their own individual one. For more information about civic engagement at the Washington Center, please see http://www.twc.edu/internships/washington-dc-programs/internship-experience/leadership-forum/civic-engagement-projects. Articles about recent civic engagement activities can be found at TWCNOW http://www.twc.edu/twcnow/news/term/civic%20engagement.

References

Battistoni, R. (2006). Civic engagement: A broad perspective. In K. Kecskes (Ed.), *Engaging departments moving faculty culture from private to public, individual to collective focus for the common good,* (pp. 11–26). Boston: Anker Publishing.

Chickering, A. W. (1969). *Education and identity.* San Francisco: Jossey-Bass.

Collins, P. (1993). The interpersonal vicissitudes of mentorship: An exploratory study of the field supervisor-student relationship. *Clinical Supervisor, 11*(1), 121–136.

Dreyfus, S. E., & Dreyfus, H. L. (1980). *A five stage model of the mental activities involved in directed skill acquisition* (Unpublished Report F49620–79-C-0063). Air Force Office of Scientific Research (AFSC), University of California, Berkeley.

Hanna, F. J., & Ottens, A. J. (1995). The role of wisdom in psychotherapy. *Journal of Psychotherapy Integration, 5*, 195–219.

Howard, J. (Summer, 2001). *Service-learning course design workbook.* OCSL Press, University of Michigan: Michigan Journal of Community Service Learning, pp. 40, 42.

Huber, R. M. (1971). *The American idea of success.* New York: McGraw-Hill.

King, M. A. (1988). *Toward an understanding of the phenomenology of fulfillment in success.* Unpublished doctoral dissertation. University of Massachusetts, Amherst.

Klemp, G. (1977). Three factors of success. In D. W. Vermilye (Ed.), *Relating work and education.* San Francisco: Jossey-Bass (pp. 102–109).

Kolb, D. A. (1984). *Experiential learning: Experience as the source of learning and development.* Upper Saddle River, NJ: Prentice Hall.

Lamb, D. H., Baker, J. M., Jennings, M. L., & Yarris, E. (1982). Passages of an internship in professional psychology. *Professional Psychology, 13*(5), 661–669.

McKensie, R. J. (1996). Experiential education and civic learning. *The National Society of Experiential Education Quarterly, 22*(2), 1, 20–23.

Mozes-Zirkes, S. (July 1993). Mentoring integral to science, practice. *APA Monitor,* 34.

Schon, D. A. (1995). *The reflective practitioner: How professionals think in action* (2nd ed.). Aldershott, UK: Ashgate Publishing.

Stromei, L. K. (2000). Increasing retention and success through mentoring. *New Directions for Community Colleges, 11*(2), 55–62.

Sullivan, W. (2005). *Work and integrity: The crisis and promise of professionalism in America.* The Carnegie Foundation for the Advancement of Teaching. San Francisco: Jossey-Bass.

Tentoni, S. C. (1995). The mentoring of counseling students: A concept in search of a paradigm. *Counselor Education and Supervision, 35*(1), 32–41.

CHAPTER 13

Internship Essentials

Professional, Ethical, and Legal Issues: What's to Know and Why?

I have to make sure that I am not in the wrong place at the wrong time. I have to be careful...because I don't want to be in a bad situation. I try to stay out of situations where I could be forced to make a bad decision.
STUDENT REFLECTION

As you go about working independently now, making decisions and testing the limits of your own competence, you will encounter situations that make you feel uncomfortable, that give you cause to pause and think. You already may have found yourself feeling this way and were unsure about what conclusions to draw, what decision to make, or what action to take. Welcome to the world of professional work and its many challenging issues. For the purposes of this chapter, an *issue* refers to a point that is in question or in dispute. Internship issues, then, pertain to some aspect of an internship that can become a matter of debate among others, usually supervisors, co-workers, clientele, or members of the community. Some of these issues have ethical dimensions; some have legal dimensions; some have professional dimensions; and some have all three. This chapter will provide you with a way of thinking about these issues and identify tools and resources to help you deal with them.

We have one reminder before we move forward with this discussion. Although many of you using the text are interns, not all are. Some students are conducting practica or are in other field-based learning experiences such as service- learning. The academic levels of your field experiences vary considerably as well, from two-year undergraduate programs through doctoral programs. Consider, too, that your academic majors vary considerably and include liberal arts majors such as political science and journalism, as well as professional studies majors such as communications, nursing, and business. As a result, some of the issues we address will be of significant relevance to all readers because of the commonalities across the academic disciplines. At other times, however, you will need the resourcefulness and guidance of your campus or site supervisors for legal, ethical, or professional guidance.

Internship Issues

We have grouped issues pertaining to internships into two general areas: *internship role issues* across academic disciplines and student affairs programs; and, *internship work issues* across professions, which include *practice*, *intervention*, and *integrity* issues. Two of the categories—*internship role* and *practice issues*—tend to apply to all interns; the remaining two categories—*intervention* and *integrity issues*—are particularly relevant to the helping and service professions as discussed in this chapter. Of course, each profession addresses *integrity* issues in its own way, and if you are not in the service or helping professions, your site supervisor is best informed to discuss that category with you. We do, though, describe situations that interns may face in each of these four categories. As you read the examples, consider your own internship and situations that could occur.

Internship role issues listed in Table 13.1 apply to all interns. Although this may seem unnecessary to say, your role, first and foremost, is that of an *intern*. It is easy—very easy—to lose sight of that fact as you slip into the role for the professional work—after all you became an intern to be in that professional role!

Internship Role Issues

The issues identified in Table 13.1 deal specifically with your role as an intern and the issues of academic integrity, competence, and supervision. When you were first introduced to these issues in Chapter 3, you were a fledgling intern; now that you have been in the role for some time, we suggest that you review them in terms of how well they are being addressed in your internship and what different meanings they have taken on since you first perused them.

TABLE 13.1

Internship Issues Across Campus Programs and Academic Disciplines

Internship Role Issues (All Interns)

- Right to quality supervision
- Responsibility to confront situations in which educational instruction is of poor scholarship and nonobjective
- Disclosure of risk factors to all potentially affected parties (to site supervisor about intern; to intern about site supervisor)
- Behavior consistent with community standards and expectations
- Awareness of risk status of agency
- Active involvement in the placement process and consideration of more than one placement site
- A clearly articulated learning contract that identifies mutual rights, responsibilities, and expectations
- A service contract with the agency that defines the limitations of the intern's role
- Liability insurance
- The prior knowledge clause
- Assurance of work and field site safety
- Assumption of risk as limited to ordinary risk
- Employer-employee-independent contractor relationship

- Compensation: stipend, scholarship, taxable/tax-free
- Deportment
- Negligence
- Malpractice
- Implication of federal funds and related statutory and regulatory requirements
- Use of college work-study funds for interns
- Grievance processes
- Informed consent in accepting the internship
- Respecting the prerogative and obligations of the institution
- Responsibility to confront unethical/illegal behaviors
- Public representation of self and work
- Disclosure of status as intern
- Boundary awareness
- Boundary management
- Personal disclosures
- Criminal activities
- Political influences/corruption
- Subversion of service system
- Office politics

© Cengage Learning

- *Academic integrity* issues include a quality field site, responsible contracts, and a seminar class that ensures a "safe place" for reflective discussions (Rothman, 2000).

- *Competence* issues include knowing your limitations and finding a balance between challenging work and a realization you have exceeded your level of competency. It is important that you know the limits of your skills and seek help as needed (Gordon & McBride, 2011; Taylor, 1999, p. 99).

- *Supervision* issues include the assignment of an appropriate supervisor who knows how to supervise interns in particular and can appropriately address such issues as client abandonment, deportment,

---

A Case in Point

Not Yet an Intern and Already a Role Dilemma!

You believe that unethical and possibly illegal practices are occurring at a field site that you want for your internship. To complicate matters further, you have been offered a paid internship by that agency's director and you support yourself. The site is out of state and is not one of your campus's regularly used sites. Your family is relocating to that state because of financial reasons, and having a paid internship in that area of the state would be of great help financially by making it possible for you to live with them. You doubt that the field placement coordinator on your campus is aware of the improprieties at the agency. You are questioning whether to inform the campus about what you know or be silent and help your family. Both choices make sense to you, and that is what makes it a dilemma.

© Cengage Learning 2014

attraction in the supervisory relationship, legal and policy issues, and quality evaluations of the intern's competencies.

Internship Work Issues

Practice Issues

This set of issues has to do with how you engage your profession and includes such issues as educational preparation, diversity awareness, and dressing for the role. An example of a dilemma that falls into this category is your having information about a teacher in a residential school who is demonstrating insensitivity to the culture of a student's family, is misrepresenting qualifications to work with such families, and is now being promoted to supervisory teacher of your work group at the site. It happens that you and the student share a similar cultural identity. It also happens that you and this teacher have had difficulties working together in the past. You are wondering whether you should disclose your concerns to your site or campus supervisors, talk directly with the teacher, or say nothing.

Of Special Relevance to
The Helping and Service Professions

Internship Issues and Aspects of the Work

There has been a growing need since the late 1980s for helping and service professionals in particular to know about their rights and responsibilities in the helping process and for others to be aware of those specific to their professions. Within the helping professions,

(continued)

Of Special Relevance to
The Helping and Service Professions *(continued)*

there is a wide variety of professional roles to consider: human service and social workers; mental health, addiction, family, marriage, school guidance, school adjustment, vocational, rehabilitation, and pastoral counselors; and psychologists. In addition, many of you are in public service work, such as in criminal justice (probation, parole, corrections, and law enforcement); conservation agencies; immigration agencies; and positions in all levels of government and public administration.

If you are in the helping or service professions and have the responsibility of being a care provider to others, then accountability is demanded in all four categories of internship issues. The potential for liability in all these areas is very real. Interns may be held to the same standards and ethical documents as employees. In addition, there are a number of issues that are specific to experiential education and internships in particular. For example, as an intern, you have the multiple roles of being a student and a recipient of services by virtue of your need for supervision. As you consider the issues listed in Tables 13.1 and 13.2, think about how such situations could play out at your site.

Although a detailed discussion of the rights and responsibilities across all helping professions is beyond the scope of this chapter and text, we organized the issues by aspects of the work you do (Chiaferi & Griffin, 1997) and, with some adaptation, incorporated issues across human service professions to inform Table 13.2 that follows (American Psychological Association, 2010; Baird, 2011; Chiaferi & Griffin, 1997; Collins, Thomlison, & Grinnell, 1992; G. Corey et al., 2011; Goldstein, 1990; Gordon & McBride, 2011; Martin, 1991; Schultz, 1992; Wilson, 1981).

Intervention Issues

This set of internship issues has to do with working directly with the clientele. The work includes managing the clients' issues, making referrals, and overseeing a caseload. An example of a dilemma that would fall into this category is finding out at an Alcoholics Anonymous (AA) meeting, which you attend for personal reasons, that your supervisor's patient is planning to leave the country in the next couple of weeks. Of particular concern is the fact that a friend told you that she overheard the patient threatening to hurt a former girlfriend. You had been told that what goes on at AA meetings is confidential, and you know how strongly confidentiality is valued in your future profession. You wonder how to uphold your responsibilities to all the parties involved (i.e., the patient, the friend, AA, and your profession), many of which seem to be in conflict, and how to determine which responsibilities take priority over others.

TABLE 13.2
Internship Work Issues Across Professions

Practice Issues

(All Interns)

- Competence in doing the work
- Frequency and focus of supervision
- Consultation
- Education
- Diversity awareness
- Grievance issues
- Limitations in the scope of practice
- Credentialing/license standards and requirements
- Advertising for services
- Dressing for the role
- Relationships with supervisors and staff
- Managing the risks of physical danger and legal liabilities

Integrity Issues

(Helping/Service Professions)

- Dual/multiple status relationships
- Obtaining information
- Disclosure of information
- Recordkeeping
- Informed consent
- Privileged information
- Right to privacy
- Confidentiality
- Upholding the values of benevolence, autonomy, nonmaleficence, justice, fidelity, veracity

- Exceptions to confidentiality, including abuse/neglect cases
- Dangerous-client cases (self and others)
- Third-party payer requests
- Responses to court orders
- Release of information to clients
- Duty to warn and protect
- The integrity of clients
- Attraction/intimate relationships (emotional, physical, sexual)

Intervention Issues

(Direct Service/Helping Professions)

- Clinical issues (transference and countertransference)
- Limitations on scope of responsibilities
- Client's right to self-determination
- Management of referrals
- Size and nature of caseload
- Termination
- Working with special populations
- Abandonment by therapist
- Obtaining information
- Release of information
- Sharing information with colleagues
- Emergency response during nonworking hours
- Differences in legal and ethical practices
- Individual vs. group vs. marriage and family interventions

© Cengage Learning

Integrity Issues

These issues concern the way you approach your work on a daily basis and include such issues as confidentiality, disclosure, and recordkeeping standards. An example of a dilemma in this category is your being out to dinner and overhearing the conversation of a technician who works at your placement site who does not know you but whom you recognize. The conversation is about a customer, and the technician identifies enough data so that you recognize the customer from a meeting earlier that week. You know from your orientation period that the organization has a policy that company personnel are not supposed to talk about customers outside of the office, or at the very least, they must not disclose any identifying information about the customer's profile. The person with whom the technician is speaking is your child's teacher. Neither of them saw you, and you are not sure what to do. You realize that something inappropriate has occurred, but you are hoping to be hired at the end of your internship. You are hesitant to pursue the issue for fear of making the wrong move.

Professional Issues: A World of Responsibilities and Relationships

Once your concerns focus on developing competence, it is quite common to pay attention to aspects of your professional deportment as well as and in relation to that of your supervisors and co-workers. You are discovering another layer of what it means to be a professional.

Focus on THEORY

Just What Is a Professional?

Professional is a loosely defined term that originally referred to "the honorific occupations of medicine, the bar and the clergy" (Sullivan, 2005, p. 35). Since then, of course, the term has expanded to refer to matters that pertain to many other occupations, all of which have three common features: specialized training and codified knowledge acquired through formal learning and apprenticeship; public recognition that the practitioners have certain autonomy to regulate their profession's standards of practice; and, very importantly, a commitment by the individual to provide service to the public that surpasses their personal economic gain (Sullivan, 2005, p. 36). As noted previously, an *issue* refers to a point that is in question or in dispute. So, when we talk about professional *issues*, we are referring to some aspect of how one goes about doing one's professional work that has become a matter of debate among others.

Questioning the Professional Conduct of Others

You may not have even been aware of when you began noticing how others manage themselves in their professional roles, but at some point, it started to take on meaning for you. For example, you may find yourself paying a lot of attention to the ways staff members go about their work, deal with clients, or conduct themselves with colleagues. Perhaps you are beginning to look at others not just in terms of their roles but as professionals with moral, ethical, and legal, responsibilities to their work, to clients, to the organization, to the community, and to you. You may even be tuning into the subtleties of their overall conduct or their specific behaviors and becoming aware of possible improprieties in how they go about their work.

The improprieties that you observe did not begin when you first noticed them; in all likelihood, it is you who have changed and you can now see what has been there all along. There are a number of possible reasons for this change. First, staff members tend not to disclose questionable or surreptitious aspects of themselves so readily with interns or new employees but rather tend to act as they are expected to act in their roles (Kanter, 1977). Consequently, they tend to shield questionable

---

THINK About It

Not in a Million Years!

There are a number of situations that lend themselves, conditionally or not, to questionable behaviors. Some situations are so obvious that mentioning them seems absurd. However, they do need to be mentioned because aspiring professionals, like seasoned professionals, are people too, and they have personal frailties that compromise their ethical standards at times. We both have known and worked with individuals who committed improprieties that neither we, nor they, ever would have expected—not in a million years. As you peruse the following list (Royse, Dhooper, & Rompf, 2012, in part), to which improprieties, if any, could you or someone you know be vulnerable?

- Being sexually intimate with supervisors and clients
- Being dishonest or fraudulent in your actions
- Libelous or slanderous actions against clients
- Threatening or assaultive behaviors against clients or co-workers
- Misrepresenting one's status or qualifications
- Abandoning a client in need of services
- Failing to warn or protect appropriate parties of a violent client
- Failing to use reasonable precautions with self-injurious clients

behaviors and attitudes from interns until they get to know them better. Another explanation is that you have been so busy with your increasing responsibilities that you have not had time to notice these behaviors or even think that there was something to notice. A third possibility is that you held stereotypes that needed to change before you were willing or able to see situations for what they were.

A Word to the Wise... Regardless of the reason, you are bumping up against issues now that did not concern you in the past. And, they can become complicated enough to be potential pitfalls for any aspiring or seasoned professional. When the issues are of an ethical or legal nature (we discuss the legal aspects in an upcoming section), the situation needs to be managed with reason and sensitivity. What is at stake in the least is someone's feelings and at most someone's reputation, career, and integrity. For now, it is important to realize that when the stakes are that high, they are high for you, too, because your opportunities for employment and the future of your career could be compromised if your concerns are not justified and handled professionally.

Questioning Your Own Professional Conduct

I recall not long ago having an ethical issue. I was attracted to another worker in the program. I knew it was not an ethical thing to do and that it was against work policy to date someone within staff. I must admit it was hard because we were attracted to each other. We both made the decision to date even though we knew we were not supposed to. Personally, I would not put myself in that predicament again.
STUDENT REFLECTION

When it comes to your own behavior, you have a number of dimensions to consider. Two especially important ones are illustrated in the example that follows: direct service and civic professionalism. Suppose you are a frontline staff member working with female victims of domestic violence. The agency is fully funded by corporate foundation money and the agency is housed in a well furnished and comfortable space donated by the company. The person who was instrumental in securing this support is a good friend of your family. As you spend some time exploring information about the corporation, you learn that it has investments that you consider to be socially and environmentally harmful: perhaps investments in Sudan or in manufacturing plants that are known polluters or child-labor abusers. This is not an integrity issue as you are used to thinking about them. However, you certainly are pausing to think about the situation you find yourself in, where the corporation supports work you believe in but also supports endeavors you believe are harmful. What will you do?

In terms of your own behavior, it comes down to whether you are able to recognize issues as ethical ones when dealing with them. In this instance, the situation just became more complicated. A client of yours at the center bought a newspaper subscription for you for daily delivery to your office (what a convenience!); she also has been bringing pastries from her bakery for her weekly meetings with you (how tempting!). Do you see a problem with these situations? Do you struggle for the "right answer" after realizing this is an ethical issue?

We find that the areas of professional *relationships* and professional *confidences* are particularly challenging for some interns. Perhaps they are for you also. We'll be exploring those areas in the next section.

Respecting Relationships

When an intern is engaged in more than one relationship (role) at the same time (or sequentially) with someone or someplace connected to her internship, the situation is referred to as *dual or multiple relationships*

Focus on THEORY

Personal Competence & Professional Values: How Compatible?

If you believe that there is a problem with the examples of the gift of a daily delivered newspaper or weekly pastry deliveries described previously, what would you say if you learned that even though you think it is "wrong," there is more than a fleeting possibility that you might accept the daily deliveries and the weekly pastries? It seems that if guidelines about such situations are not clearly spelled out, practitioners rely on their personal value systems and interpretations of the profession's ethics documents, and therein lies a recognized problem: Among practitioners, there is a discrepancy between knowing "what's right" to do and actually doing it (Jennings, Sovereign, Bottorff, Mussell, & Vye, 2005). So, if you know what's right to do and do not do it in this instance, you probably would not be alone in your failure to act. It appears that if professionals believe there is an infraction of a clearly articulated professional code, they tend to do what they thought they should do, especially if legal precedent exists. But if professionals don't believe that to be the case, then they tend not to act consistently according to what they thought was the right thing to do (Bersoff & Koeppl, cited in Jennings et al., 2005; Bernard et al., and Smith et al., cited in Jennings et al., 2005). So, why, if people rely on their personal value systems in such situations, might they not do "the right thing"? Well, some professionals lack honesty and integrity (Smith et al., cited in Jennings et al., 2005), and might not have the courage to act on what is "right" (Rest, cited in Jennings et al., 2005). The central issues here are personal values and personal character and whether there is compatibility between personal competence and the values of the profession.

(Corey, Corey, & Callanan, 2011). The nature of these roles can vary from being an acquaintance, friend, or business client, to being an intimate—emotionally, physically, or sexually (Dorland, 1974, cited in Malley & Reilly, 2001; Gordon & McBride, 2011; Royse, Dhooper, & Rompf, 2012). Such relationships can be difficult to manage as can dual relationships that involve clients. Another group of dual relationships that interns may have to deal with are the *collegial* ones, which refers to your relationships with the co-workers, peers, supervisors, and yes, professors who are involved in your internship (Malley & Reilly, 2001).

All of these relationships need to be managed so that your professional obligations are upheld. You may want to take a moment to think about what you *could* do, what you *would do*, and what you *should* do if your commitment to the profession was being compromised by how you were handling one of these relationships (Rothman, 2002). All such relationships are fraught with complexity and ethical issues. If you are in a potentially compromising relationship, it is best to seek consultation immediately.

Of Special Relevance to The Helping and Service Professions

Safeguarding Confidences

Another area that tends to challenge interns' own professional conduct is that of *confidences*. This issue is rather pervasive and involves clients' rights (to privacy, privileged communication, and confidentiality) and *information disclosure*. These principles are ones that you are obliged to honor in your role as intern, so it is important that you understand them. Although confidentiality and disclosure are professional issues, they can also be ethical and even legal issues, depending on the laws that govern the work of the profession (Eisenstat, personal communication, Sept. 2012).

Privacy, Privileged Communication, and Confidentiality

The clients' rights that you deal with on a daily basis probably include privacy, privileged communication, and confidentiality. *Privacy* refers to the constitutional and statutory rights of your clients to decide when, where, how, and what information about them is disclosed to others by you (Corey, Corey, & Callanan, 2011; Eisenstat, personal communication, Sept., 2012). *Privileged communication* is a legal concept, and the right to such communication belongs exclusively to the client. This concept protects your client from the forced disclosure of information in legal proceedings (Corey, Corey, & Callanan, 2011). You need to check with your supervisor to determine whether clients who see you in a helping or service capacity lose their right to privileged communications because of your status as an intern. If so, the client needs to know that, and your supervisor needs to guide you in informing the client. *Confidentiality* is a legal, professional, and ethical matter that protects the client in a therapeutic relationship from having information disclosed by you without explicit authorization (Corey, Corey, & Callanan, 2011). Your status as an intern also has direct implications for whether the clients you work with are protected by confidentiality statutes. Again, you need to discuss this matter with your supervisor, and your clients will need to be informed.

Disclosure of Information

The issues of what information can be disclosed and to whom can become complicated when you consider disclosing information to your colleagues (while on-site or socially) or to your peers (in the seminar class). In both instances, you have a responsibility to the privacy rights of the client and, at the very least, to be familiar with and follow the policies of the agency, disguise all identifying data, and give much thought to the question *"Why do I need to tell this information to someone?"* The issue

---

Focus on PRACTICE

HIPAA and the Privacy Rule

If you are interning in the health care industry, including mental health service organizations or in companies that provide support or contract services to health care providers, then you might well be affected professionally by the U.S. Department of Health and Human Services Privacy Rule (August 2002). This rule, which is administered by the Office for Civil Rights in the Department of Health and Human Services, is a set of standards that address the use and disclosure of an individual's health information and also standards ensuring that individuals can understand and control how their health information is used. You will be required to become familiar with and trained in these standards and those select circumstances under which such information can be used and disclosed without an individual's permission. Despite all the limitations of the privacy rule, there are a substantial number of exceptions that you will need to know, such as those concerning the public interest and national priority purposes, including but not limited to public health matters; victims of abuse, neglect, or domestic violence; health oversight activities; judicial/ administrative proceedings; law enforcement purposes; decedents; research; cadaveric organ, eye, or tissue donation; serious threats to health or safety; essential government functions; and workers' compensation. Be sure to ask your supervisor if your company or agency is affected by HIPAA's Privacy Rule and if so, in what way that will affect your work. (www.cms.hhs.gov/HIPAAGenInfo/Downloads /HIPAALaw.pdf.)

of disclosure in the world of the helping and service professionals also applies to cyberspace and its information highway. Be it e-mail, chat rooms, bulletin boards, blogs, text and instant messages, tweets, or postings on social media sites, interns need to have an informed command of this aspect of their workplace. Also, remember that when you use these electronic means of communication to discuss clients or other sensitive issues, the sites are not always secure.

The electronic obtaining and releasing of information, along with maintaining that information for the agency, can be difficult for the intern to negotiate appropriately. Whether you are working with databases or releasing information, it is important that you make informed decisions and do so carefully. The agency has its policies to safeguard the maintenance of information as well as forms to ensure that all criteria are met before information is released by you. If you have not yet reviewed them, this is a good time to do so.

Ethical Issues: A World of Principles and Decisions

I try to confront ethical decisions by first asking myself what I think is right. If this gets me nowhere, I ask my coworkers, friends, relatives, professors, or anyone who I feel might have knowledge in that area or who has been in a similar situation.... Then I choose what I think is the best solution.
STUDENT REFLECTION

You might be wondering how you have managed to survive for so long without knowing about ethical issues! This is exactly how our students feel after studying them in a semester-long course.

Chances are that your basic values have served you well. However, the list of potential issues can be overwhelming, even after several readings. Most likely, there are many that you have never heard about and many that seem remotely familiar. Reading through them is a good start. Now that you are becoming familiar with the language of ethics, even if you do not yet know much of what it entails, it is a good time to go ahead and identify the situations that already may have surfaced in your internship and how satisfied you are with your handling of them.

A Word to the Wise...   It would be helpful to your understanding of the discussion that follows if you have a copy of the ethics document that guides the professionals in your agency.

Talking the Talk

In order to have a useful discussion about these issues, there must be a shared language for communicating and a common understanding of the problem. In the section that follows, we identify some of the terms that are frequently used in discussions of ethical matters and their working definitions. It might take you some time to take in the essence of the meanings, so do not be concerned if there seems to be too much to grasp the first time you read through this section. The more you return to it for guidance, the more familiar it becomes and the more you will learn.

Let's start with the term *standards*. When used generically, the term refers to guidelines or codes that govern the behavior of members of a given profession. *Ethical* suggests that someone is acting in accordance with professional standards, codes, guidelines, or policies, and *legal* suggests that someone is acting in accordance with the law. *Values* refer to what is considered intrinsiclly good, useful, and desirable; *moral* refers to what is considered right or wrong conduct in its own right, based on broad mores such as religious principles; and *ethics* refers to the moral principles or rules of conduct of a particular profession (G. Corey, Corey, & Callanan, 2011; Pollock, 2012), such as the ethics of business, service professions, communications media, and so on.

Rules of the Trade

In addition to having a common language, it is important to have access to common resources. For one thing, although your understanding of the issues in your professional field has just begun, you are still responsible for acting in accordance with the values and standards of that profession. These values and standards are embodied in ethical documents variously referred to as *guidelines, standards, regulations, policies, principles,* and *codes.*

It is precisely these types of documents that are reviewed when conflicting or questionable situations arise. Unfortunately, they don't necessarily lend themselves to clear interpretations or resolutions of ethical or professional conflicts. However, they do provide guidelines for behavior and discussions. Technically speaking, there are differences between these documents, which are described in the section that follows. If you are not interested in that level of detail, skip this section and go to "What You Need to Know Now," on p. 374.

Differences in the Documents

If the document is titled *guidelines*, it reflects recommendations from professional groups for acceptable behaviors for the profession. If the title reads *standards*, the statements reflect the rules of behavior for the profession that are drawn up by members of the profession itself and that often carry civil sanctions and set the parameters for ideal behaviors. If you

---

Focus on PRACTICE

Principles and Standards

There are four documents, listed below, that we suggest you take the time to examine. They will help clarify responsible ways to carry out the work of the internship or the profession and will also provide considerable insight into the next section, where we identify many of the issues that could potentially become problematic for you or any practitioner. Your site or campus supervsiors can provide you career specific documents.

- NSEE Principles of Best Practices in Experiential Education (National Society for Experiential Learning, www.nsee.org.)
- NSEE Guiding Principles of Ethical Practice. (www.nsee.org.)
- Ethical Standards of Human Service Professionals (National Organization for Human Services (NOHS) and the Council for Standards in Human Service Education (CSHSE) (www.nationalhumanservices.org; www.cshse.org.)
- The CAS Professional Standards for Internship Programs and The CAS Professional Standards for Service-Learning Programs (The Council for the Advancement of Standards in Higher Education (CAS), 2009). (www.cas.edu.)

---

have heard your supervisor or field instructor talking about accreditation teams or visits, they are referring to an assessment of the site or academic program based on designated standards of behavior for the profession.

If, on the other hand, the document is called *regulations*, it contains dictates typically from governmental authorities and often specifies sanctions for not complying with them. Undoubtedly, you were given a copy of your site's policy manual by the end of the first week in the field. A *policy* refers to the procedures or course of actions set forth by an organization to ensure expediency and prudence in getting the work done. All organizations have policy manuals that they give employees and hopefully interns to read as part of their orientation to the work and workplace. *Principles* are fundamental doctrines of the profession that are rooted in commonsense morality.

Finally, if the document is called a *code*, its statements reflect beliefs about what is right and correct professional conduct. Codes often include standards of practice along with statements that embody the values of a profession. Like the Learning Contract, codes tend to be living documents that continually evolve. They promote professional accountability and facilitate improved practice by protecting the professional from ignorance (e.g., from malpractice suits so long as the professional acts in accordance with acceptable standards), protecting the public from the profession (i.e., protects the consumer from harm), and protecting the profession from the government (i.e., the profession governs and regulates itself, protects itself from internal struggles, and establishes agreed-upon standards of care) (VanHoose & Kottler, 1978, cited in Bradley, Kottler, & Lehrman-Waterman, 2001).

What You Need to Know Now

It is important that you know the resources of your future profession. Your site supervisor can be very helpful in identifying them for you. Another very helpful resource is the book *Codes of Ethics for the Helping Professions*, which includes the full text of codes for fifteen professional specialties (Corey & Corey, 2011). Membership in such organizations can be expensive, but it is well worth the fee (most have a student category of membership with lower fees). The majority of national organizations can provide members with much information and have access to many resources. In addition to a variety of support services, such as casebooks and libraries of instructional videos, many also offer legal counseling and services when necessary.

If you have not done so already, ask your supervisor for a copy of the ethical document that guides your co-workers, for that document should also guide you. We suggest that you take the time to read it. Some of you will be looking at a two-page document; others will be reading 20 pages of professional rules! Reading the shorter documents is a reasonable exercise. However, if you are bound by the American Counselor Association's (ACA, 2005) code or that of the American Psychological Association (APA, 2010), reading such lengthy documents is perhaps not a reasonable undertaking. What is expected of you, though, for the purposes of the internship is to become *familiar* with the categories of codes and those categories relevant to your specific field experience. Bring the document to seminar class. Unless you and all your peers are in the same specialty (e.g., the helping professions or business work), the documents you collectively bring to class will be varied in specialty area, length, and intent.

Ethical Principles and Ethical Values

I believe that it is better to overthink than to be impulsive and regret a choice later on. I do know that the ethical decisions and choices I make in life are a reflection of my values and character. I know that not every ethical decision I make in life is going to be easy or the right choice. But I believe that I have good values, and that will help me with my decisions.
STUDENT REFLECTION

Regardless of how detailed the codes may be or how many times you read them, there may be no answers forthcoming to help you resolve the conflict you experience. When ethical codes fail to provide a direction toward a solution, then the *ethical principles* of the profession can help guide your decisions. *Ethical principles*, as mentioned earlier, are fundamental doctrines of the profession rooted in commonsense morality. If you are not in the helping professions, you may want to locate those principles that guide the work in your field and spend some time thinking about them.

How might you answer the question "What ethical values are relevant to your profession?" Would they be the values that you read about in your academic coursework preparing you for your profession? Presumably so. Would they be the values that you see in practice at the internship site—the ones you are encouraged to demonstrate? Again, presumably so. There is an operating assumption that the ethical values you see in practice are the ones that the professionals believe best inform them when it comes to making ethical decisions; we also know that they will draw upon their personal value systems at times.

THINK About It

How Do Your Ethical Values Measure Up?

In recent years, eight ethical values used by master therapists in the helping professions were identified. These ethical values take on added importance when you think about how practitioners use codes of ethics and their personal value systems when making decisions about ethical issues and dilemmas (Jennings et al., 2005). How do the values cited here figure into the decisions you made earlier about the daily newspaper delivery and the weekly delivery of baked goods?

Building and Maintaining Interpersonal Attachments

- Relational Connection
- Autonomy
- Beneficence
- Nonmaleficence

Building and Maintaining Expertise

- Competence
- Humility
- Professional Growth
- Openness to Complexity and Ambiguity

Of Special Relevance to The Helping and Service Professions

The Six Principles to Guide Your Work

For those of you in the helping professions, we list below the six ethical principles that are commonly accepted as reflecting the highest level of professional functioning (Corey, Corey, & Callanan, 2011). The principles are based on the works of Kitchener (1984) and Meara, Schmidt, and Day (1996) and probably look very familiar to you. These principles have more than one purpose: they can guide your work as well as your decision-making. When using these principles to guide your decisions, be as honest with yourself as possible, as this type of authenticity will be invaluable to the quality of the decisions and the personal insight you develop.

- *Autonomy* refers to the clients' freedom to control the direction of their lives by making decisions that reflect their wishes; this principle affirms the clients' right to self-determination.

- *Beneficence* refers to commitment to do "good," as demonstrated by carrying out work with competence and without prejudice. This principle affirms the clients' dignity and promotes the clients' welfare.

- *Justice* refers to treating others with fairness, regardless of gender, ethnicity, race, age, religious affiliation, sexual orientation, ability, religion, cultural background, or socioeconomic background. This principle affirms the clients' right to equality in services.

Focus on THEORY

Principle Ethics and Virtue Ethics

Kitchener (1984) identified the first five of the six ethical principles as a way for practitioners to make responsible ethical decisions in their work when the ethical codes that are there to guide them fail them because of their broadness or narrowness. Meara, Schmidt, and Day (1996) differentiated Kitchener's principle ethics from virtue ethics, which focus on character traits and ideals (Meara et al., cited in Jennings, 2005, p. 33). Virtue ethics were considered more useful in providing a sense of an individual's moral life than principle ethics; instead, principle ethics were grounded in prescribed rules that were formal and obligatory in nature. Together, the principle and the virtue ethics provide a seminal framework to guide helping practitioners in their ethical decision making, which is rarely a simple matter (Jennings et al., 2005).

- *Nonmaleficence* refers to avoidance of doing harm; this principle affirms the clients' right to respect.
- *Fidelity* refers to having a trustworthy relationship of honest promises and honored commitments; this principle affirms the clients' right to informed consent before committing to interventions.
- *Veracity* refers to being truthful in dealings with clients; this principle affirms the clients' right to full disclosure.

Legal Issues: A World of Laws and Interpretations[1]

An important part of making an ethical decision is taking an engaged approach to knowing the laws that are relevant to and affect your work. Some of you are developing a familiarity with the law, especially if you are interning in legal settings or your work is closely directed by statutes and legal guidelines. Others of you may know little about this aspect of your work. In this section, we will give you a way of thinking about legal matters so you can make better sense of this aspect of your field experience.

All internships are affected to some degree by legal issues. For those of you interning in the criminal justice system, legal mandates govern much if not all of your work. For those of you working with dependent individuals, such as minors, elders, and those with special needs, the intent and extent of your work are largely affected by legal statutes, especially in the area of protection; i.e., abuse, neglect, and exploitation. If you are interning with a legislator, advocating for clients in class-action suits, or interning in hospitals, human resource departments, or in mediation services, you are working with laws. If you are interning in a government agency, your work is affected by the statutes or laws that govern the agency.

A number of legal issues are particularly relevant to interns who work directly with clients, most of which have ethical dimensions as well (see Kiser, 2011). Such issues include, but are not limited to, liability and malpractice; confidentiality, privileged communication, and privacy; disclosure of information; end-of-life decisions; consultations with specialists; crisis intervention; suicide prevention; termination of interventions; intimacy with clients; duty to protect intended victims from violence; and

---

[1]Appreciation is extended to Professor Steven Eisenstat of Suffolk University School of Law, who specializes in civil tort law, for his assistance with the section of this chapter on legal issues.

informed consent (Birkenmaier & Berg-Weger, 2010; Kiser, 2011). The list does not stop here, but we will.

Your responsibility is to know the legal basis, if there is one, for your organization; the laws that affect and govern your work; and the ways in which you are bound by those laws in carrying out your responsibilities (Berg-Weger & Birkenmaier, 2010; Gordon & McBride, 2011). The best way to learn about these matters is to bring your questions and concerns to your supervisors. The following sections will help to prepare you for those discussions.

Talking the Talk

As was the case with ethical matters, there is a terminology specific to legal matters in fieldwork that you need to know. Again, we will take some liberty and use working definitions where possible so that you have a sense of the language and implications. We know that the information in this section is typically what interns most want to know about legal matters. However, the information is very technical, and it is not possible to describe it without a great deal of detail. So, on the one hand, we risk oversimplifying a complex body of information, and on the other hand, we risk boring or causing you undue concern. We will do our best to choose a middle ground. We advise you, though, throughout this discussion and throughout your internship, to bring all matters to your supervisors if you do not have a working understanding of them.

Negligence

A tort is a civil wrong or injury done to another that is not based on an obligation under a contract. There are three types of torts: *intentional torts, negligence torts,* and *strict liability torts.* For the purposes of your internship, it is the negligence tort that is of most concern to you, your site supervisor, and your campus instructor.

For an act to be a negligence tort, there must be a legal duty, owed by one person to another, a breaking (*breach*) of that duty, and harm caused as a direct result of the action. For example, if you are interning at a home health care agency and you voluntarily assume responsibility for an elder in the community, you are then in what is referred to as a *special relationship* with that person. Your duty to your client would be considered breached if you fail to provide the standards of care of the home health care profession; you could do this either by failing to take certain required actions or, if you did act, doing so in a way that does not reflect the standards of care for the home health care profession. It makes sense to raise the issue of *breach of duty* with your supervisor so you can better understand how you could be at risk for such lawsuits (Eisenstat, personal correspondence, Sept. 2012). Negligence torts, then, can result when you fail to exercise a reasonable amount of *care* (standard of care) in a situation that causes harm to another person or to a thing. The basis for the negligence tort can involve doing something carelessly or failing to do something you are supposed to do.

There are two forms of negligence: ordinary negligence (i.e., failing to act as a reasonable person would) and aggravated negligence (i.e., reckless or willful behavior). It is the *ordinary negligence* tort that is the more likely concern in your internship. An example of ordinary negligence occurs when officers (e.g., police, probation, corrections, parole) fail to perform duties owed to clients or inmates or when they perform duties inadequately (Eisenstat, personal correspondence, Sept. 2012). For example, correctional officers have a duty to check regularly on the inmates under their care. If a correctional officer fails to do so and an inmate commits suicide, then the officer could be found negligent in terms of his or her supervisory responsibilities. So, it makes sense to raise the issue of negligence and the potential pitfalls you face with your supervisor so that you can better understand how you could be at risk for such lawsuits.

Malpractice

The term *malpractice* is one you are sure to have heard and know enough about that you do not want it to be a part of your field experience. Malpractice, a form of ordinary negligence, refers to an act that you perform in your professional capacity for which you are being sued. This type of lawsuit charges professional misconduct or unreasonable lack of skill on your part that results in injury or loss to your client. For example, if you are an intern at a residential facility for emotionally challenged adolescents and you fail to take the precautions that are ordinarily provided by other residential facilities/workers in the profession, and your actions or lack thereof result in one of the residents committing suicide, then you most likely would face a malpractice lawsuit (Eisenstat, personal correspondence, Sept. 2012). Again, it is important to bring up the issue of malpractice with your supervisors so you can better understand how you could be at risk for such lawsuits.

It is important to know that there are situations that tend to increase your liability for a malpractice lawsuit. (Being *liable* or having a *liability* means a breach of duty or obligation to another person.) For example, if you fail to use acceptable procedures or you use interventions for which you were not trained, or you fail to choose a reasonable form of intervention, your risk of liability increases significantly. And it does not stop there. If you fail to warn others about or protect others from potential danger, or you fail to secure informed consent appropriately, or you fail to disclose to your client the possible consequences of services and interventions, then your risk of liability could increase. Such disclosures may violate the client's right to confidentiality; deciding whether to disclose or not often places the discloser between a rock and a hard place: disclosing and facing a suit from the client for violating his or her rights to confidentiality or not disclosing and facing a lawsuit if harm comes to the client or third party. Again, it is important to discuss potential malpractice situations with your supervisors.

Rules of the Trade

In addition to having an understanding of the terminology, it is important to have an understanding of the legal framework. Let's start at the beginning with laws. A useful way of thinking about laws is how they are classified. For example, laws can be classified according to how they come into existence. Laws in the United States derive from our constitutions (*constitutional law*, from state and federal constitutions), our legislatures and governmental agencies (*statutes* and *regulations*, respectively), and our *common law* (*case law* from prior decisions by trial courts or appeals courts).

Another way laws are classified is by the nature of their focus: criminal or civil. *Criminal law* refers to a group of laws that seeks to resolve disputes between the government and people. Criminal law seeks punitive measures such as imprisonment and fines to right a wrongdoing (Eisenstat, personal correspondence, Sept. 2012). Many interns work with criminal law on a daily basis; for example, those in criminal justice settings, legal offices, domestic violence agencies, and protective work with dependents.

On the other hand, *civil law*, of which tort law is an example, seeks to resolve disputes between people by enforcing a right or awarding payment or what is referred to as *damages*. Its primary intent is to repair rather than punish behavior, as is the intent of sanctions for criminal matters (Eisenstat, personal correspondence, Sept. 2012). An aspect of civil law that many interns work with is mental health law. This body of law regulates how the government takes care of or responds to people with mental health challenges. If you are interning in a mental health clinic, hospital, residential setting, or a community shelter, your work is affected by this body of law.

In some instances, students deal with both civil and criminal law. For example, interns at offices of the American Civil Liberties Union deal with constitutional rights, which involve both civil law as well as criminal law.

Another way of thinking about law and the helping professions is to separate the laws according to the aspects of the work (Garthwait, 2011). For example, there are laws that regulate the *services* or *actions* that you can give to a client. There are laws that regulate the *work of the service agency*, such as working with youth, working with elders, and working with mentally ill individuals. And there are laws that regulate the *practice* of the profession, such as deportment, licensing, and supervision issues.

Relevant Legal Matters[2]

I just wished I had known exactly what the supervisor is legally supposed to do…. Then I would have been able to point to something in writing. It was very hard knowing the supervisor was wrong but not having anything to point to and say "this is what you are supposed [sic] to be doing for me."

STUDENT REFLECTION

The legal matters of most relevance to interns come under the general categories of *standards of care* and *supervisory malpractice*. Again, these are complex areas of inquiry, and it is not possible for us to address them adequately in one chapter on ethics and laws. However, we hope to give you a way of thinking about them so that you can discuss these matters with supervisors and become better informed about these important areas of practice.

Standards of Care

A matter that is both ethical and legal in nature and affects the work of the intern in the helping professions on a daily basis is that of *reasonable standard of care*. Interestingly enough, this area of potential legal matters is neither universally nor directly addressed in the ethical standards for the helping professions. Kiser (2000, p. 122) has described the components of a reasonable standard of care as including, but not limited to, knowledge of the clients and services being given; delivery of services and interventions based on sound theoretical principles; reliability and availability of services to clients; taking the initiative and acting on behalf of client and public safety; adherence to ethical standards of the profession in relation to client care; and systematic, accurate, thorough, and timely documentation of client care.

Supervisory Malpractice

The second legal matter is that of *supervisory malpractice*. In this instance, it is the behavior of the supervisor that comes under legal as well as ethical scrutiny. As you may be aware, your supervisor is liable for your work because when your supervisor agreed to supervise you, he or she accepted responsibility for all of your work, including your work with clients and for your behavior (deportment) during the internship. If this responsibility sounds pretty serious, it is.

---

[2]We based the discussion that follows on the work of a number of writers and experts in this field, including Birkenmaier & Berg-Weger (2010); Champion (1997); Corey, Corey, & Callanan (2011); Eisenstat (personal correspondence. September, 2012); Falvey (2002); Garthwait (2011); Gordon & McBride (2011); Malley & Reilly (2001); Oran (1985); and Pollock (2012).

Quality of Supervision   Failure to supervise the professional staff appropriately has been the cause of a growing number of malpractice suits (Sherry, cited in Falvey, 2002). This type of lawsuit concerns the quality of supervision given to the intern. The legal scrutiny a supervisor faces in such a lawsuit results from alleged negligence in carrying out supervisory responsibilities and from subsequent injury or damages. You, the intern, along with whomever may have been injured as a result of improper supervision, become the *plaintiffs* (i.e., the ones who bring the complaint), and your supervisor becomes the *defendant* in such a lawsuit for negligence. For example, you have been directed to conduct an in-home assessment to determine the removal of a child based on alleged neglect by the parents. In the process of conducting the interview, the mother becomes despondent, leaves the interview, goes into the bathroom, and slashes her wrists. During the invention that ensued, you are cut by the same instrument the mother used on herself, and you require medical care. If your supervisor did not prepare you adequately to respond to and manage the range of possible reactions to such an interview, your supervisor's risk for liability for failing to train you adequately increases substantially. Such preparation could include, but not be limited to, having you observe and/or conduct such an interview under the direct supervision of an experienced worker or talking with you about the potential for self-destructive reactions to such interviews. In this scenario, the client could sue both the intern and her supervisor (and potentially the agency and the campus) for negligence, and the intern could sue the supervisor and agency and campus for their negligence in properly training her, which led to her injuries. (Eisenstat, personal communication, Sept. 2012). We hope you will never have such memories as part of your internship.

Vicarious Liability   Another type of negligence liability on your supervisor's part is *vicarious liability*. Under vicarious liability, your supervisor could be held responsible for your negligence even if your supervisor did not act negligently. This area of law can become very complicated.

Ordinarily, there must be some form of salary to create an employer-employee relationship, as in cooperative education placements. Practica or service-learning placements, however, typically do not pay for the student's work; internships can be paid, but many are not. Ultimately, the question comes down to whether there is sufficient oversight of the intern—how to do the work, the hours of the work, salary, and so on—that one could argue the level of control that the company exercises over the intern is sufficient to create an employer-employee relationship. This is decided upon by the facts of the specific case.

In the case of paid internships where the employee acts negligently, it is the agency, not the supervisor (since it is the agency paying the salary), who is vicariously liable. Of course, if the supervisor also acts negligently, the agency can be vicariously liable for the supervisor's actions as well. In the case of unpaid internships, vicarious liability can still exist, but once again, it would exist between the "employer" and the intern/employee and not the supervisor and the intern/employee, unless the supervisor is also the employer. If there is a campus supervisor involved in the internship in addition to the organization's employee who oversees the intern's day-to-day work, the agency could argue it lacks sufficient control over the student because a second supervisor is involved and that it is the college or university that should be held vicariously liable for failing to adequately supervise the intern. This scenario of vicarious liability is not likely to arise in a typical internship. The more likely scenario would be the injured party suing the field site and/or the campus for their negligent supervision of the intern. However, both the agency and the campus could be sued through their own negligence and vicariously through the intern's negligence. The two lawsuits are not mutually exclusive; both can be brought (Eisenstat, personal correspondence, Sept. 2012).

Obligations to the Intern   At this point in your understanding of liability, you may be wondering under what circumstances does your supervisor incur potential liability because of her or his failure to meet obligations to you? Four major sources of such liability have been identified (Harrar, et al., 1990, cited in Falvey, 2002). We think it's important to note them so that you are more informed about what you can request of the supervisory relationship.

- If the supervisor is derelict in carrying out supervisory duties for planning your internship, the direction of your internship, or the outcome of your work, then your supervisor's liability can increase.

- If your supervisor gives you inappropriate advice about a treatment intervention that you use, and the intervention is to the detriment of the client, the supervisor's liability can increase.

- If your supervisor fails to listen attentively to your comments about a client and in turn fails to understand the needs of the client, the supervisor's liability can increase.

- If your supervisor assigns you tasks that the supervisor knows you are not trained adequately to perform, then the supervisor's risk of liability can increase.

All these conditions make good common sense, and our experience is that students know intuitively when they are being shortchanged or otherwise not being given quality supervision. However, seeing them in print can be most affirming for an intern. Similarly, it is helpful for practicum students to be aware of these rights as they go about their work in the field.

THINK About It

The Rights of Interns

These rights should look familiar to you. You were first introduced to them in Chapter 3 as your internship was beginning. You've been an intern for a while now. Give some thought to the ways these rights are being respected in your placement. Which rights do you believe are not being respected?

- The right to a field instructor who knows how to supervise, i.e., has been adequately trained and is skilled in the art of supervision
- The right to a supervisor who supervises consistently at regularly designated times
- The right to clear criteria when being evaluated
- The right to growth-oriented, technical, and theoretical learning that is consistent in its expectations

Munson, cited in Royse, Dhooper, & Rompf, 2012.

Grappling with Dilemmas

This is how I try to decide what to do: Think about it, talk about it, and try to look at it from every angle. Then I make a decision.
STUDENT REFLECTION

One of the most wrenching aspects of working with clientele is dealing with a dilemma that involves the welfare of another individual, family, or community. A *dilemma*—be it ethical, legal, professional, or all three—refers to a struggle over alternative courses of action that might resolve a situation. To complicate things, the choices of courses of action tend to be correct in their own right, but they conflict with each other. A dilemma, then, is a situation in which you can find yourself facing more than one justified course of action; that is, two "right" ways of responding.

Our interns tell us that once they develop an understanding of the issues and become comfortable with the language, they begin to see the issues in their daily work. It just so happens that you tend to develop this awareness at the same time that you are moving into a collegial-like relationship with your supervisors. Consequently, you are much more apt to talk about incidents, behaviors, and concerns at this point in your field experience than you were when you began.

Recognizing Dilemmas

The next hurdle is to recognize a dilemma when you see one, which is no easy feat. Our experience tells us that recognizing dilemmas as such is quite challenging for both undergraduate and graduate students and for experienced professionals as well. Often, students cannot readily name ethical issues when they see them, and they do not necessarily see them in a given situation. Nor do they have a language to describe situations that might be ethical in nature or to discuss them. A course in applied ethics taught in a professional studies program at the very least will help the student become aware of the ethical issues of the profession, learn the "language" of applied ethics so issues can be named as such and discussed, and begin to develop the reasoning and problem-solving skills necessary to succeed in the profession. For example, professional studies programs in human services education sometimes *require* a course in professional issues but more often will offer it as an elective or cover the content in other coursework (M. DiGiovanni, Sept. 2012; CSHSE, 2012). In the field of criminal justice education, a study of the effects of a one-semester, elective course in criminal justice ethics suggests that students' values do appear to shift in one measurable way: from having concerns about their personal gain to having concerns about issues of social justice and the welfare of others (Lord & Bjerregaard, 2003). Given that much of criminal justice work is a service profession, this particular shift in values is noteworthy. If you have not studied or discussed ethical issues academically, you may feel particularly underprepared for these challenges. We hope this discussion will help you to frame your understanding of the issues and learn what questions to ask.

Focus on PRACTICE

Got to Know Them!

Typical Types of Dilemmas

There are three basic types of ethical dilemmas that you may encounter in your work as an intern as well as in future careers.

- Dilemmas that result from *your own decisions, behaviors, or attitudes* (e.g., *considering whether to engage in a dual relationship with a client, such as knowingly working with a client whose friend you have dated*)

- Dilemmas that result from another person's *decisions, behaviors, or attitudes* and directly affect you (e.g., *working with a client who, unbeknownst to you, is your cousin's intimate partner*)

- Dilemmas that *you pay attention* to from a distance but that do not directly affect you (e.g., *observing or hearing about a co-worker dating a relative of his or her client*)

Focus on SKILLS

EXPECT the Issues!

One way of dealing with an ethical issue is to anticipate it. A useful and effective way of doing this is to rehearse in your mind the best possible response (how you would like to respond to it), the worst possible response (your worst nightmare in responding), and a more realistic response (found somewhere between the two extremes). Then discuss with peers or your supervisors the implications of the more realistic response and spend additional time rehearsing it.

Walking the Walk

You certainly will be exposed to ethical issues during your fieldwork. You may even experience an ethical dilemma. If that happens to you, remind yourself that you are no stranger to facing difficulties and that you have what it takes to work your way through yet another challenge. And, like the challenges you faced in the past, an ethical issue can become a problem if you do not manage it effectively.

Regardless of the situation, you will need a way of thinking—a framework—for making ethical decisions. Thinking critically about ethical situations is important for making responsible decisions about them. Although most of you have heard this term constantly throughout your education, it is quite possible that you do not know the skills necessary to think critically about an issue, a situation, a decision, or an action. To think critically means to use standards of reasoning such as clarity, precision, accuracy, relevance, depth, and breadth. It involves evaluation techniques, weighing alternative perspectives, and genuine efforts to evaluate all views objectively. Examples include such skills as articulating ideas, asking significant questions, problem solving, and openness to contradictory ideas (Alverno College Productions, 1985). *Why so much thinking* you ask? Thinking deeply and in demanding ways will help you make wise choices—choices that most likely will help your clientele reach their goals (Gibbs & Gambrell, 1996, p. 6).

The Ten Reasoned Steps to Resolving Dilemmas

I find that when I think about the whole situation (ethical decision), I realize that unethical decisions have a lot of repercussions.... People lose their jobs, and I don't want to get into trouble.
STUDENT REFLECTION

There are many decision-making models to guide you in developing critical thinking skills. Some are specifically intended to deal with ethical or legal matters in the helping professions (see, for example, Corey &

Corey, 2011; Corey, Corey, & Callanan, 2011; Gibbs & Gambrill, 1996; Kenyon, 1999; Neukrug, 2013; Rothman, 2002; Mandell & Schram, 2012; Tarvydas, 1997, cited in Tarvydas, Cottone, & Claus, 2003; Woodside & McClam, 2011). We offer you an engaged model that we use when teaching about ethical and legal issues. It is based in part on the *Eight Steps to Creating Change* model in Chapter 9 and incorporates adaptations of other models as well (Close & Meier, 2003; G. Corey et al., 2011). The model, summarized in table 13.3 and detailed below, offers opportunities to think critically in practical ways and to develop a reasoned response to and action plan for the presenting problem.

Step 1: Name the Problem

Collect as much information as you can about the situation. Clarify the conflict. Is it moral? Professional? Ethical? Legal? Given that there are no right or wrong answers to the situation, anticipate ambiguity and challenge yourself to consider the problem from multiple perspectives.

Step 2: Narrow the Focus

Once you have gathered as much information as is reasonable, list the issues you are confronting. Some are more important than others. Describe the critical issues and players; discard the unimportant ones.

Step 3: Consult the Codes

Review the ethical documents of your profession, the policies and regulations of your agency, and related laws to determine whether possible

---

TABLE 13.3
The Ten Reasoned Steps to Resolving Dilemmas

1. Name the Problem
2. Narrow the Focus
3. Consult the Codes
4. Consider the Laws
5. Consult with Colleagues
6. Determine the Goals
7. Brainstorm the Strategies
8. Consider the Consequences
9. Consult the Checklist
   - Principles: Autonomy, Beneficence, Nonmaleficence, Justice, Fidelity, & Veracity
   - Duties to Care and Civic Responsibility
   - Responsibilities Laws, Ethical Documents, Policies, Procedures, & Regulations
10. Step Ten: Decide Diligently

solutions are suggested. Identify aspects of the documents that apply. How compatible are your personal values with those of the profession?

Step 4: Consider the Laws

Chances are you are just becoming familiar with the laws—both civil and criminal—that are relevant to your work. Consult those laws. Once familiar with them, you can contact the legal counsel for your field site (your supervisor should be made aware of this first!) or a law librarian for guidance.

Step 5: Consult with Colleagues

Consult with informed colleagues to discover other ways of considering the problem. Given the responsibility to make a reasoned decision, consulting with colleagues is one way to "act in good faith" and test your justifications. Choose your colleagues wisely.

Step 6: Determine the Goals

Think through what change you hope to bring about in the attitudes, behaviors, or circumstances in question. Question your motives carefully and repeatedly. Is your client's voice heard in the goals you want? Talk with a colleague about whether there may be motives on your part of which you are not aware. Choose your colleague wisely.

Step 7: Brainstorm the Strategies

Identify all possible courses of action, including the absurd. Some may prove useful, although unorthodox. Consider the client's perspective. Is the client's voice heard in your list of options? Discuss options with others. Choose your colleagues wisely.

Step 8: Consider the Consequences

Think about the consequences of each strategy for all involved in the situation. Thoughtfully assess plans. Identify consequences from various perspectives, and question each of the consequences. Remember to include the client's perspective among those you consider.

Step 9: Consult the Checklist

Use the following checklist to evaluate potential areas of ethical and legal misconduct. The questions are based in part on a model of ethical decision making that identifies six fundamental principles of moral behavior: autonomy (*self-determination*), beneficence (*in the best interest of the client*), nonmaleficence (*to do no harm*), justice (*fairness to all*), fidelity (*honest promises and honored commitments*), and veracity (*being truthful*). This model includes such qualities of ethical acts as universality, morality, and reasoned and principled behaviors (G. Corey et al., 2011; Kitchener, 1984; Meara, Schmidt, & Day, 1996; Pollock, 2012).

- Is the action in the best interest of the client? *Consider the six fundamental principles of moral behavior.*

- Does the action violate the rights of another person? *Consider constitutional rights as well as your duty to justice.*

- Does the action involve treating another person only as a means to achieve a self-serving end? *Consider the end-in-itself motive and the utilitarian perspective.*

- Is the action under consideration legal? Is it ethical? *Consider the laws and your legal duties; consider your civic and ethical duties and the components of an ethical act.*

- Does the action create more harm than good for those involved? *Consider the principles of nonmaleficence and beneficence.*

- Does the action violate existing policies, regulations, procedures, or professional standards? *Consider the duty to your professional role.*

- Does the action promote values in culturally affirming ways? *Consider the principles of nonmaleficence and beneficence and the duty to care.*

Step 10: Decide Diligently

Consider carefully the information you have. The more obvious the dilemma, the clearer the course of action; the more nebulous the dilemma, the more difficult the choice. Although hindsight may teach you differently, the best decision under these circumstances is a well-reasoned decision— one with which you can live.

Managing a Professional Crisis

One of the potential pitfalls to a chapter such as this is the tendency, regardless of experience, to worry unnecessarily about your vulnerability to the issues raised in the reading. The fact is that you would not be in the field if your academic program did not prepare you adequately for the experience and you would not be doing the work you are doing if your site supervisor and campus instructor did not believe you were academically and professionally ready for the experience. Even so, issues and situations will present themselves, and they can arise quite quickly.

Sometimes, a situation develops through no fault of your own, and you find yourself at the unenviable end of allegations, complaints, or legal charges. When that happens, it is a very stressful time for all involved, especially you. You are well aware of what you have invested in your field experience and the importance of your grade and learning to your academic work and/or career plans. However, you may not be aware that such situations do arise and that *how* you handle the situation is very important to your supervisors' assessments of you in a crisis.

Knowing how to respond to such a crisis can make all the difference in you having a future in the profession. Managing this type of crisis is no different from managing other crises you have lived through or half-expect to occur at some time. There are many approaches to managing crises. You should know what works and what does not work for you. To add to your growing toolbox for problem management, we offer you this four-pronged engaged approach called HELP (see Table 13.4).

Have Resources in Place

Regardless of the approach you use, it is very important that you know what resources are readily available to you and those that you need to develop. The resources should include, but not necessarily be limited to, legal, emotional, physical, academic, and professional supports. Knowing beforehand what, who, and where the resources are and how to mobilize them is essential to a healthy and effective response. Otherwise, you can find yourself responding in ways that are not helpful to yourself or the situation.

For example, becoming so upset in times of professional crisis that you do not know which way to turn may be an understandable, but not a very useful way of responding. Drinking alcohol, taking drugs, or engaging in other self-destructive behaviors is neither professional nor helpful to the situation. What is useful, though, is identifying your resources beforehand so you know who to call, where to go, and what you can expect from them. Your supervisor can be a wellspring of information and advice as to who should be on your list and what information you should have beforehand. Also, having membership in the profession's national organization (which usually is at a substantially reduced rate for students) allows access to information and often legal advice. It may be helpful at this point to review the sections in Chapter 3 on support systems and identify supports that you need to develop for a professional crisis, which may be quite different from supports for a personal crisis.

TABLE 13.4
HELP: Self-Care in a Professional Crisis
• Have Resources in Place.
• Expect to Learn from the Crisis.
• Lay Out a Crisis Response Plan.
• Practice Self-Awareness.

© Cengage Learning

Expect to Learn from the Crisis

Make a resolution to do the best you can under the circumstances and to learn from a crisis in your internship. It is very important to think through the value of such a resolution now, when you are not in a storm

---

A Case in Point

Tested on the FIRST Day!

On the first day of a practicum in a courthouse, one of our students was asked to assist in gathering information in interviews resulting from a drug sting. Those arrested were of the same ethnicity and spoke the same first language as did our student. The student was approached by those arrested and asked to help them evade court processing. On that same day, a convicted offender asked another student in the same class for a date within the first hour of her practicum at a day reporting center. Both students were shocked at how quickly the situations happened. However, neither of them experienced dilemmas, although they were surprised to learn how easy it was to find themselves in potentially compromising situations.

© Cengage Learning

of emotions that makes it very difficult to appreciate the benefits. Not only does a resolution to learn provide you with an understanding of how you function in a crisis within a professional context, but it gives you insight into an aspect of the profession that you otherwise would only read about. Taking care of yourself legally, emotionally, physically, academically, and professionally is your responsibility in a crisis. It is also an essential factor for riding out the storm in ways that leave you the stronger for it.

Lay Out a Crisis Response Plan

In addition to knowing your resources and knowing yourself in a crisis, it is critical that you have a crisis plan of action; i.e., a plan that allows you to be most helpful to yourself and the situation even in times of high anxiety. An important piece of such a plan is a crisis team for a professional crisis—the people you can call on in an instant to give you the help you need, whether it is legal council or chicken soup, literally and figuratively. It is a team of first responders that you create for yourself. In putting together your critical support team, be sure to identify how to reach them when you need them (i.e., electronic, cell, land line, and postal). Next, you need to think through what you must do to take care of yourself emotionally, physically, and professionally throughout the storm so that you stay afloat when the waters get rough. Maybe a physical outlet is best for you, like running, biking, or workouts at the local gym. Perhaps it is mindfulness meditation or yoga that makes a difference in your ability to cope under pressure. Some find prayer or other meditative activities most helpful. Many find counseling to be comforting and affirming. Whatever it may be, you need to be aware of it, keep it foremost in your mind, and make it part of your agenda in a crisis.

Practice Self-Awareness

It is very important that you understand your reactions, strengths, and weaknesses in a crisis. How you go about solving crises, what works most effectively for you, and what is ineffective in such situations are all informative. Thinking these through before, as opposed to during a crisis, is very important because your objectivity is not compromised by the pain of the situation. This may also be a good time to revisit Chapter 4, especially the parts that help you understand how you function under stress.

Conclusion

You have done a lot of reading about a lot of issues in this chapter, most of which are probably new to you. At best, you have become familiar with some terminology and have become aware of areas of concern that you could face in both the field and the profession. This chapter is intended to give you a way of organizing your thinking about the wide variety of issues—professional, ethical, and legal—that are part of the profession and about ways to respond should a crisis develop in these aspects of the work. We encourage you to use this chapter as a resource throughout your field experience and to consider taking related academic coursework if you have not already done so.

As you continue to develop your sense of competency and professional identity and deal with your concerns about professional behavior, you are fast approaching the last portion of your journey. We discuss that experience in the next chapter and the feelings interns have when their good-byes are bittersweet.

For Review & Reflection

Checking In

Personally Speaking. This is a good time—after making your way through this chapter—to think about your personal system of ethics. Keep in mind that yours might be similar to another person's in some ways, and that you will always find differences. Your system of ethics is unique, regardless of the shared ethics that your profession mandates. What values best describe your system of ethics and to what do you attribute their basis? What is it about your system of ethics that you like, and what would you like to change?

Personal Ponderings (Select the More Meaningful)

- When you consider how you approach ethical decisions, what best describes your style: Thoughtful? Impulsive? Expedient? Other ways of responding? Combinations of styles? What is it about your style that you like—are proud of and why is that? What would you like to change about your style

- Think about how you might react to an allegation, complaint, or legal charge brought against you. Would it matter if you were *not* responsible for the resulting damages? Would it matter if you *were* responsible but denied being so?

Experience Matters (For the EXperienced Intern)

You've had your share of observing the kinds of issues described in this chapter during your history of employment and careers. What similarities and differences have you noticed in this field experience compared to previous work or other internships? What are the implications of those similarities and differences given your career interests? What are your next steps if the implications are not favorable to your future endeavors?

Civically Speaking

As you think about your site's civic connections to the community, give thought to those areas of particular strength—ethically or professionally—in those connections. Once you have a sense of those strengths, there are two areas of inquiry.

First, *So What?* Spend some time with this question because its answer can be more than insightful.

The second question has to do with the legal ways in which your site connects to the community. Do any of those challenge you in an ethical sense, and if so, in what ways? Does that surprise you? Regardless of how you answered this question, once you have a good sense of the answers to both questions, the next question gets tougher: *Now What?* What are your next steps knowing what you know? And, finally, *Then What?* What are the implications for your future?

Seminar Springboards

Practice Makes Easier. Think about an ethical issue, not necessarily a dilemma, that you would not want to face in your internship but one that you know could occur—it's just a matter of time. Share the issues with peers in class, and select one of particular interest to role-play. Rehearsing potential ethical issues will not only be informative but affirming and satisfying once you get through them. It's the getting through that gets tough. Practicing now makes reality a lot easier later.

For Further Exploration

Corey, G., Corey, M., & Callanan, P. (2011). *Issues and ethics in the helping professions* (8th ed.). Belmont, CA: Brooks/Cole.

Seminal text with a focus on the legal, ethical, or professional dimensions of the helping professions.

Corey, G., & Corey, M. (2011). *Codes of ethics for the helping professions* (4th ed.). Belmont, CA: Brooks/Cole.

A compilation of the full text of seventeen ethical documents for the helping professions.

Cottone, R. R., & Tarvydas, V. M. (2007). *Ethical and professional issues in counseling* (3rd ed.). Upper Saddle River, NJ: Pearson.

Comprehensive and informative text focusing on professional, ethical, and legal issues in counseling.

Dolgoff, R., Loewenberg, F., & Harrington, D. (2009). *Ethical decisions for social work practice* (8th ed.). Itasca, IL: F. E. Peacock.

Comprehensive treatment of the topic and resources for social work students, useful glossary of terms, and listing of Internet sources of information about ethics and values.

Goldstein, M. B. (1990). Legal issues in combining service and learning. In J. C. Kendall & Associates (Eds.), *Combining service and learning* (Vol. 2, pp. 39–60). Raleigh, NC: National Society for Experiential Education.

Offers an excellent guide to the legal issues relevant to service-learning.

Goldstein, M. B. (Undated). *Legal issues in experiential education.* Panel Resource Paper #3. Raleigh, NC: National Society for Experiential Education.

Identifies key legal issues in experiential education for academic administrators

Kenyon, P. (1999). *What would you do? An ethical case workbook for human service professionals.* Belmont, CA: Brooks/Cole.

Comprehensive and useful workbook that uses actual field situations to inform student awareness of ethical issues and develop decision-making skills to respond in ethically responsible ways.

King, M. A. (In press). Ensuring quality in experiential education. In G. Hesser, (Ed.). *Strengthening experiential education within your institution* (Revised). Mt. Royal, NJ: National Society for Experiential Education.

A chapter in a classic sourcebook that includes current documents to guide experiential education practices to ensure quality in learning and teaching.

Levine, J. (2012). *Working with people: The helping process* (9th ed.). New York: Longman.

Thoughtful consideration of a variety of professional, moral, and ethical issues in the helping professions.

Rothman, J. C. (2002). *Stepping out into the field: A field work manual for social work students.* Boston: Allyn & Bacon.

A useful and comprehensive guide to varied professional issues for social work students.

Steinman, S. O., Richardson, N. F., & McEnroe, T. (1998). *The ethical decision-making manual for helping professionals.* Belmont, CA: Brooks/Cole.

Offers students a pragmatic and focused approach to awareness of ethical issues and responses and developing decision-making skills.

References

Alverno College Productions. (1985). *Critical thinking: The alverno model.* Milwaukee, WI: Author.

American Counseling Association. (2005). *ACA code of ethics.* Alexandria, VA: Author. Retrieved from www.counseling.org

American Psychological Association. (2010). *Ethical principles of psychologists and code of conduct.* Washington, DC: Author. Retrieved from www.apa.org

Baird, B. N. (2011). *The internship, practicum, and field placement handbook: A guide for the helping professions* (6th ed.). Upper Saddle River, NJ: Prentice Hall.

Birkenmaier & Berg-Weger. (2010). *The practicum companion for social work: Integrating class and field work* (3rd ed.). Boston: Allyn & Bacon.

Bradley, L. J., Kottler, J. A., & Lehrman-Waterman, D. (2001). Ethical issues in supervision. In L. J. Bradley & N. Ladany (Eds.), *Counselor supervision: Principles, process, and practice* (3rd ed.). Philadelphia: Brunner-Routledge, pp. 342–360.

Champion, D. J. (1997). *The Roxbury dictionary of criminal justice.* Los Angeles: Roxbury Publishing Company.

Chiaferi, R., & Griffin, M. (1997). *Developing fieldwork skills.* Belmont, CA: Brooks/Cole.

Close, D., & Meier, N. (2003). *Morality in criminal justice.* Belmont, CA: Wadsworth.

Collins, D., Thomlison, B., & Grinnell, R. M. (1992). *The social work practicum: A student guide.* Itasca, IL: F. E. Peacock.

Corey, G., Corey, M., & Callanan, P. (2011). *Issues and ethics in the helping professions* (8th ed.). Belmont, CA: Brooks/Cole.

Corey, M. S., & Corey, G. (2011). *Becoming a helper* (6th ed.). Belmont, CA: Wadsworth.

Corey, G., & Corey, M. (2011). *Codes of ethics for the helping professions* (4th ed.). Belmont, CA: Brooks/Cole.

Council for the Advancement of Standards in Higher Education (2009). *CAS professional standards and guidelines for internship programs.* Washington, DC: Author. Retrieved from www.cas.edu

CSHSE (2012). Council for standards in human service education program standards. Retrieved from http://www.cshse.org/standards.html

Falvey, J. with Bray, T. (2002). *Managing clinical supervision: Ethical practice and legal risk management.* Belmont, CA: Brooks/Cole.

Garthwait, C. (2011). *The social work practicum: A guide and workbook for students* (5th ed.). Upper Saddle River, NJ: Pearson.

Gibbs, L., & Gambrill, E. (1996). *Critical thinking for social workers: A workbook.* Thousand Oaks, CA: Pine Forge Press.

Goldstein, M. B. (Ed.) (1990). Legal issues in combining service and learning. In J. C. Kendall & Associates (Eds.), *Combining service and learning* (Vol. 2, pp. 39–60). Raleigh, NC: National Society for Experiential Education. Retreived from www.nsee.org

Gordon, G., & McBride, R. B. (2011). *Criminal justice internships: Theory into practice* (7th ed.). Cincinnati, OH: Anderson Publishing.

HIPAA/Standards for Privacy of Individually Identifiable Health Information (August 2002). Washington, DC: U.S. Department of Health and Human Services.

Jennings, L., Sovereign, A., Bottorff, N., Mussell, M. P., & Vye, C. (2005). Nine ethical values of master therapists. *Journal of mental health counseling, 27*(1), 32–47.

Kanter, R. M. (1977). *Men and women of the corporation.* New York: Basic Books.

Kenyon, P. (1999). *What would you do? An ethical case workbook for human service professionals.* Belmont, CA: Brooks/Cole.

Kiser, P. M. (2000). *Getting the most from your human service internship.* Belmont, CA: Brooks/Cole.

Kiser, P. M. (2011). *Getting the most from your human service internship* (3rd. ed.). Belmont, CA: Brooks/Cole.

Kitchener, K. S. (1984). Intuition, critical evaluation and ethical principles: The foundation for ethical decisions in counseling psychology. *The counseling psychologist, 12*(3), 43–45.

Lord, V. B., & Bjerregaard, B. E. (2003). Ethics courses: Their impact on the values and ethical decisions of criminal justice students. *Journal of criminal justice education, 14*(2), 191–211.

Malley, P., & Reilly, E. (2001). Ethical issues. In Boylan, J., Malley, P., & Reilly, E. (Eds.), *Practicum & internship: Textbook and resource guide for counseling and psychotherapy* (3rd ed.). Philadelphia: Brunner-Routledge, pp. 93–128.

Mandell, B. R., & Schram, B. (2012). *An introduction to human services: Policy and practice* (8th ed.). Upper Saddle River, NJ: Pearson.

Martin, M. L. (Ed.). (1991). *Employment setting as practicum site: A field instruction dilemma.* Dubuque, IA: Kendall/Hunt.

Meara, N. M., Schmidt, L. D., & Day, J. D. (1996). Principles and virtues: A foundation for ethical decisions, policies and character. *Counseling psychologist, 2*(1), 4–77.

National Organization for Human Service Education (2000). Ethical standards of human service professionals. *Human service education, 20*(1), 61–68. Retrieved from www.nationalhumanservices.org

National Society for Experiential Education. (1998). Standards of practice: Eight principles of good practice for all experiential learning activities. Presented at the Annual Meeting, Norfolk, VA. Retitled. (2009). NSEE principles of best practices in experiential education. Mt.Royal, NJ: Author. Retrieved from www.nsee.org

National Society for Experiential Education. (2010). NSEE guiding principles of ethical practice. Mt. Royal, NJ: Author. Retrieved from www.nsee.org

Neukrug, E. (2013). *Theory, practice and trends in human services: An overview of an emerging profession* (5th ed.). Belmont, CA: Brooks/Cole.

Oran, D. (1985). *Law dictionary for nonlawyers* (2nd ed.). St. Paul, MN: West.

Pollock, J. M. (2012). *Ethics in crime and justice: Dilemmas and decisions* (7th ed.). Belmont, CA: Wadsworth.

Rothman, J. C. (2002). *Stepping out into the field: A field work manual for social work students.* Boston: Allyn & Bacon.

Royse, D., Dhooper, S. S., & Rompf, E. L. (2012). *Field instruction: A guide for social work students* (6th ed.). New York: Longman.

Schultz, M. (1992). Internships in sociology: Liability issues and risk management measures. *Teaching sociology, 20,* 183–191.

Sullivan, W. M. (2005). *Work and integrity: The crisis and promise of professionalism in America* (2nd ed.). The Carnegie Foundation for the Advancement of Teaching. San Francisco: Jossey-Bass.

Tarvydas, V., Cottone, R., & Claus, R. (2003). Ethical decision-making processes. In Cottone, R., & Tarvydas, V. (Eds.). *Ethical and professional issues in counseling* (2nd ed.). Upper Saddle River, NJ: Prentice Hall.

Taylor, D. (1999). *Jumpstarting your career: An internship guide for criminal justice.* Upper Saddle River, NJ: Prentice Hall.

Wilson, S. J. (1981). *Field instruction: Techniques for supervisors.* New York: Free Press.

Woodside, M., & McClam, T. (2011). *An introduction to human services* (7th ed.). Belmont, CA: Brooks/Cole.

CHAPTER 14

And in the End:
The Culmination Stage

*I've been taking in as much information as I can before
"it's all over" and have been concerned with career goals.
The past couple of weeks have just been filled with an
overwhelming amount of anxieties and mixed emotions.*
STUDENT REFLECTION

As incredible as it must seem to you at times, the end is in sight. For most of you it is the end of something special, although for some of you it may not have been the experience you hoped for. In either case, the beginning of your internship may seem like yesterday; it may seem like years ago; or, you may go back and forth between those two extremes. Regardless, this is a time to look forward and be proud of what you have accomplished; it is also a time to reflect on the experience that you have had in the field.

Making Sense of Endings

We refer to this last stage as the *Culmination* stage because everything in this stage is reaching a final crescendo. Another stage, another set of risks and opportunities, and another potential crisis, await you. Endings are a necessary part of anyone's development (Kegan, 1982). Like other critical junctures discussed in this book, this one is normal, as are the concerns and issues associated with it. Normal does not mean easy, though. Remember the distinction between difficulties and problems discussed in Chapter 8? You could think of endings as difficulties that can be solved, resolved, or aggravated into problems (Watzlawick, Weaklund, & Fisch, 1974), depending on the approach you take.

A Word to the Wise... As your internship ends, you may feel more than ever the pull to do more and more. There may be a lot that can or needs to be finished, but becoming too involved in a frenzy of activity can also be a way to avoid facing your feelings about the ending of your field experience. Once again, we remind you—and it is especially important now—that you need to take time for reflection, even if that means jotting down a time on your calendar for it to happen. You will have to protect that time more jealously than ever, as the external pressures mount and the temptation to avoid your feelings grows stronger.

A Myriad of Feelings

In our experience, interns approaching the end of their experience report many different feelings. There are often feelings of pride and sometimes of mastery. Others report relief and anticipation of freedom. A different kind of relief is expressed by interns who have struggled or had a disappointing experience. However, we also hear about sadness, anger, loss, and confusion, regardless of how satisfying the internship has been. In a survey of psychotherapy interns, Robert Gould (1978) found that many of them reported increased anxiety and depression as the end approached as well as decreased effectiveness. The interns in his study also reported feeling moodier. That has been our experience as well; you would not be unusual if you found yourself experiencing several—or even all—of these feelings, simultaneously or sequentially, with your emotional landscape shifting by the hour. What is going on here?

Changes and More Changes

For one thing, you have a lot on your plate right now. Your internship is ending, and you may or may not be ready. The calendar says the internship is almost over, but you may feel as though you have just gotten started, especially if it took some time to find your rhythm (Suelzle & Borzak, 1981). In addition, you may or may not have completed the work you set out for yourself. Projects may not be finished; clients may not be quite where you would like them to be; customers may not be quite satisfied. On the other hand, some of you may feel as though you have been finished for a while.

The web of relationships that has been the social context of your internship is changing yet again. Many relationships are ending; others will be redefined as you leave the role of intern. Endings are part of most relationships and a necessary part of helping relationships (Levine, 2012). The more the relationships mean to you, though, the harder it will be to end them: *Now I am really sad. I loved my internship and don't want to lose the relationships I've formed here. I'm trying to make plans with everyone so we can keep in touch.*

The external context of your life is shifting as well. There may be new demands on your time and energy as you feel the pressure of endings and beginnings (Suelzle & Borzak, 1981). Your attention may be pulled toward papers and final projects for your internship seminar as well as to papers and exams in other classes. Your summer or holiday plans may need to be finalized. If you are graduating and have not yet found a job, your job search is occupying more and more time and attention. If you have a family, they may be dealing with their own endings at school or at work, and they may be dealing with the endings in your life as well. Of course, your internship is not the only thing that is ending. Your seminar class, the semester, the school year, even your college career may all be drawing to a close. That's a lot of endings, a lot of beginnings, a lot of good-byes, and a lot of hellos. Good-byes are never easy, and for some they can be very stressful. The prospect of beginning in a new school, a new town, a new job, or all of the above is daunting as well as exciting.

In Their Own Words	Voices of Culmination

The most important part of this internship was how it affected me personally… my thoughts, feelings, and beliefs. I went into this internship thinking I knew myself, but learned very quickly that was not true. I am going to miss learning that kind of learning.

I am amazed just how much my life can change over the course of just a few months. I am no way near the same person I was back at the beginning of my internship. I am proud of who I have become and where I am headed. I just feel sad it's all ending.

Now that my internship has come to an end, I feel excited to be moving on. I feel inspired and more prepared. I feel I have been challenged and I have proven to myself that I am able to handle those challenges. It has been a beautiful ride and one of unexpected fulfillment.

I went from "that college student" to a professional. It took some time to adjust and it wasn't glamorous at times, but I made it through. They took me under their wings; they didn't take me by the hand. They let me have responsibility as soon as I walked through the door. They treated me as a professional and now I'm ready for what's ahead.

I look forward to the future. My internship experience will continue to help even after it's over. I will be able to draw from the confidence and skills I learned. I can honestly say that I am not nervous about the future. Rather, I have a sense of confidence about it and a feeling that whatever I end up doing, I will be able to do it well if I work hard enough at it.

Thinking About Endings

As you head into this phase of your internship, take some time to think about yourself. All these endings and beginnings are likely to tap a variety of emotional issues for you. Some of the issues you thought about in Chapter 4 may resurface or surface for the first time. You may also discover some new issues; don't let them catch you by surprise.

Although this may be your first internship, don't be so sure the experience of culmination is all new for you. You have had endings, separations, and losses before, and these experiences can leave you with some unfinished emotional business. Your experience of ending the internship will be colored by those experiences, as well as by your response patterns, your cultural norms, and even your family patterns.

Think back on the experiences of separation and loss in your life. Going away to camp or to college, having close siblings leave home, moving, divorce, ending an intimate relationship, and being fired from a job are examples of separation experiences you may have had. Those experiences are never easy and in some cases they can be traumatic. Perhaps some of the hurts from those experiences have not healed. Perhaps you wish you had behaved differently or that others had.

Now think about how you say good-bye and how people have said good-bye to you. Some people just leave and don't say a word; others write long letters or schedule good-bye lunches or dinners. The way you say good-bye probably depends somewhat on the nature of the relationship that is ending, but perhaps you can detect some patterns in how you approach this task. Remember the discussion of dysfunctional patterns in Chapter 4? This is an area in which many people have those patterns, and they can surface in an internship. Here is one we have seen many times:

> *Saying good-bye to someone I don't care that much about is pretty easy. When it's someone I'm really invested in, though, I don't handle it well. Usually what I do is get really busy. I keep promising myself to go to lunch, or dinner, or something with the person, but I never seem to make the time. Then all of a sudden there is no time, and I end up saying a hurried good-bye. I know I have hurt people's feelings that way.*

Scott Haas (1990), in a book about his internship, notices himself falling into a similar pattern as the end of his internship approaches:

> *I create obstacles to avoid thinking about the end. I distract myself: I make lists. I ruminate about minor inconveniences (like wondering for days whether the gas company will correct their bill). I develop new projects and interests. I go shopping, and then in the store can't remember why I ever wanted the thing I'm about to buy. When all else fails, I pretend there is no end. I'm just imagining it: it really isn't happening. (p. 171)*

Finishing the Work

As you enter the last weeks of your internship, taking an engaged stance is very important to a satisfying ending. You need to think about what work remains and what you want to and can do about it. The nature of the work makes a difference here. If you have been given a series of small, concrete projects, such as a report to write or an event to plan, and they are complete or near completion, then ending the internship will be somewhat easier (Suelzle & Borzak, 1981). If you are part of a large project that will not end until after you have gone, then feeling some closure about your involvement can be harder. Perhaps there is a component of the project you can complete or a summary of your work that you can write. Finally, you may be involved in a complex project that must be completed before you leave. Perhaps, for example, you have done some research for your supervisor and have agreed to summarize your findings. There may be the temptation to read one more article or interview one more person; in some cases, there is always more research you can do. At a certain point, however, you have to stop generating and start summarizing.

The work needs to be completed in such a way that your supervisor has all the necessary information to carry the work forward, whether it is project-based work, work with community groups, client or customer work, or research-based work. Additionally, it is important that you offer recommendations for how the work should be continued, and be sure all loose ends are tied up before leaving the work for someone else.

The Tasks at Hand

As you can see from the following chart, there are three general areas of focus at this point in the internship. They can apply to the *work* you are doing, the *people* you work with, the internship *site* in general, and of course, to *you*. As is often the case, we describe them separately but as you experience the end of the internship they will often blend together. The first is *reaching closure*—identifying and addressing any tasks or issues that you have not yet addressed or completed, and addressing them wherever possible (Levine, 2012; Shulman, 1983). The second is *redefining relationships*. Your internship has unfolded in the context of a web of relationships—with supervisors, co-workers, and in some case clientele—and these relationships are going to change now. Some of them will end; some may continue, but in a different form. Even when the relationship ends, though, in many cases you will still think (and feel) about it; dealing with those feelings is part of the redefinition. The final area of focus is *planning for the future* (Levine, 2012). You cannot control the future, of course, and plans need to be flexible, but planning will help ease your anxieties and feel more empowered.

TABLE 14.1
Culmination

Associated Concerns	Critical Tasks	Response to Tasks	
		Engaged Response	Disengaged Response
Saying good-bye **Transfer of responsibilities**	Endings and Closure	Proactively seeks closure in key relationships	Looks to others to initiate closure; waiting for the end
Completion of tasks **Next steps**	Redefining Relationships	Identifies and deals with feelings in key relationships	Avoids any unpleasant feelings
Multiple endings **Closing rituals**	The Future	Makes intentional and realistic plans for self and clients	Does not plan, merely stops

Sweitzer and King, 2012

Remaining Engaged

As we have emphasized throughout this book, and through every stage, these issues can be addressed in an engaged or disengaged way. At the risk of being repetitive, we take a moment now to remind you of some key features of engagement:

- *Stay Active* We once again encourage you to be as active as possible in facing and working through these issues. You can look to others to set direction for you, and you can accept whatever direction they set, but you will learn more and learn more deeply if you take an active role in ending the internship well.

- *Be Responsible* Remember that you are responsible for your learning and you have learned a great deal about how to be an engaged and empowered learner. Some supervisors are very active in focusing on these issues and helping you to do so; some campus instructors are active in these ways as well. If so, you are lucky; in any case you are still responsible.

- *Acknowledge Feelings* You may find strong feelings about supervisors, peers, and others coming to the surface. It may or may not be appropriate to express these feelings directly to the person who has engendered them, but it is important to acknowledge them, if only to yourself. At the very least, you will want to find a place where you can express them to someone else and say whatever you want to say before worrying about just what to say to the people involved. Often, interns find ways to avoid facing and expressing these feelings, particularly the negative ones; this is one kind of disengagement in the culmination stage. Avoidance behaviors may include joking, lateness, or absence.

THINK About It

The Importance of Rituals

Some of the tasks discussed in this section will be taken care of almost by themselves. They will happen naturally, or someone else will initiate them. Others will not, though, and in those cases you will need to be more proactive. In writing about ending an internship, some authors have discussed the need for rituals (Baird, 2011; Rogers, Collins, Barlow, & Grinnell, 2000). That may sound like a strange term to you, even calling up religious or pagan images. However, any formal way of marking an event or passage (such as a going-away luncheon) can be considered a ritual. In this case, rituals can help provide a sense of completion or closure. Rituals can add a sense of "specialness" to the ending; they can help you recall what was significant and important. They can also help ease the transition by connecting the past to the future. Finally, rituals can create a formal, structured opportunity to experience and express emotions that might otherwise be repressed (Abramson & Fortune, 1990; Baird, 2011).

- *Use Resources*   If you are like most interns, you will find that some of these issues are resolved easily while others are more stubborn. Some tasks are harder to finish than you may have thought, and some of the people you work with, especially clients, may actively resist your efforts to end well; it is important not to give up and we remind you that you have an array of tools at your disposal for dealing with difficulties, as well as a support system.

Handling the Slips and Slides...at Journey's End

As we pointed out in Chapter 2, your internship is going to come to an end regardless of how you approach the tasks of ending. Although internships rarely fall apart in the final stage, it is still possible for you to experience a fundamental shift in the way you feel about the experience. You may find yourself thinking that this experience hasn't been so great after all, in spite of all that you know you have achieved. Some interns need to avoid the painful feelings associated with ending to such an extent that they, unintentionally and without awareness, "sabotage" the experience. They start showing up late, or not at all, doing sub-par work, or engaging in other behaviors that lead to warnings or worse as well as escalating negative feelings. For these interns, it is almost as if it is easier to walk away angry than to walk away sad. Others report feeling numb or empty, not experiencing any strong emotions but also losing their sense of accomplishment, competence, and success that they have worked so hard to achieve. This is the face of disillusionment in this stage, and it is a *crisis of closure.*

As we have stressed throughout the book, interns who arrive at this sense of disillusionment can do so through multiple paths. Sometimes it is a case of taking a passive stance too often or giving up too easily. Sometimes the experience of ending touches personal issues in an unexpected way, especially if endings have been hard for you in the past. And it is possible for external events—at the internship or in your life—to pull you down this path, especially when a crisis of personal significance occurs. For example, the loss of a loved one, a divorce (yours, someone in your family, a close friend), or the loss of your employment can trigger a serious slide. So, too, can a big project that fails or is halted through no fault of yours, or the abrupt departure of your supervisor or the arrival of a new one who is far less skilled and sensitive. There are many more examples of the "dangers" of this particular crisis. Most do avoid disillusionment or climb out of it by maintaining an engaged approach and embracing the opportunity such situations present, such as employing the range of problem solving tools at your command, including in those in the Essentials chapters (3 and 9). We will discuss specific opportunities as we move through the discussion of this stage's tasks.

Closure with Supervisors

Many interns have one or more campus supervisors or instructors as well as more than one supervisor at the site. Such an arrangement calls for closure with each of these supervisors. How you pursue closure with the campus supervisor or instructor will depend on the nature of the relationship. Touching base by e-mail, social media, or text message with a campus supervisor once or twice during the internship is a different relationship than meeting regularly with the supervisor or communicating frequently by electronic means. In some instances, you may want to ask the site and campus supervisors to be references for you as you begin your employment search. How you decide to bring closure on those relationships certainly will affect the supervisors' decisions to do so.

For most interns, the site supervisor is the person who has most powerfully affected the experience. Your site supervisor is in a position of power, at least with regard to you, and often is in a position of some leadership in the organization. Usually the site supervisor is the person with whom you have worked most closely, shared the most, and learned the most. Now it is time to bring closure on that relationship or perhaps move it to a new phase of relationship. There are two dimensions to this closure process: the formal and the informal phases (Birkenmaier & Berg-Weger, 2010). In the formal phase of the process, there is a final evaluation with the site supervisor and perhaps an evaluation of that supervisor by the intern as well. In the informal—and equally important—phase, the supervisory relationship is processed and discussed, as are the feelings about the ending of that relationship.

The Final Evaluation

You have probably had evaluations at your internship before this point; at least we hope so. Perhaps you are less nervous about them than you once were; perhaps not. In any case, you have one more to go through, and this can be the most important one of all. Many internship programs and placement sites use a written final evaluation of some sort. Most of them have their own format, so we are not going to provide one here. However, if you would like to see some examples, we refer you to work by Wilson (1981), Stanton and Ali (1994), Baird (2011), Garthwait (2011), and Birkenmaier and Berg-Weger (2010). As was the case with the initial or mid-semester evaluation, we recommend that you become familiar with the form used at your field placement or reach an agreement about what it will be before the evaluation is actually completed. That way you will have a better idea of what to expect and there will be far fewer disappointing surprises. And if the form does not seem like a good fit for you or your situation, now is the time to bring that up with your supervisor and see whether there are alternatives.

Regardless of what form you use, or whether you use one, it is important that you have a final evaluation conference with your supervisor. This conference can either precede or follow the written evaluation. Some of the time can be used to prepare for that evaluation or to go over it. In any case, be sure to schedule an adequate block of time, at least an hour (Baird, 2011; Faiver et al., 1994; Shulman, 1983). There may be lots of tempting reasons for both you and your supervisor to avoid this session. You may be nervous about the feedback; your supervisor may not be good at endings either; both of you may be pretty busy trying to wrap up projects or cases. This is one of those times when a formal, scheduled time—a ritual—will ensure that the task actually gets done.

The Final Conference

You may not realize it, but this could be one of *the* moments in your education that you have been preparing for over time, especially if your internship is a degree requirement. This moment takes on even more weight and additional significance if it determines whether you successfully complete your academic program and graduate. Ideally, if your supervisor and you have had ongoing assessments and evaluations, there should be no surprises as you approach the final meeting with your supervisor(s) at the site. But, that is not always the case, so this conference will undoubtedly be more productive if you spend some time preparing for it.

Getting Ready

First, before you even consider what is on the form, take some time to review the internship experience with all its joys and frustrations. You have been reflecting on a regular basis, but it is easy to miss some important moments or broader themes; so, it is time to think reflectively once again, and to think introspectively as well.

Reflecting on the Work and the Learning   Reread your journal and make some notes for yourself. If you had a mid-semester evaluation, look it over as well. Stanton and Ali (1994) have suggested dividing your reflection into two sections: work performance and learning. As you will see, these two areas have some overlap, but they are worth considering separately. In discussing work performance, you and your supervisor should cover the areas where you seem to have been especially effective, such as the clients with whom you have worked most successfully, the aspects of the project or research where your strengths were most obvious, and the service areas where you have demonstrated the most skill (Faiver et al., 1994). Both your and the supervisors' perceptions of these issues are important. In discussing what you have learned, use your written goals and objectives as references and consider how well you met each goal, the reasons why, and how what you have learned will be useful in your professional and personal life.

Managing the Emotional Stakes   It is also important to prepare for the affective aspect of this experience (Stanton & Ali, 1994). Being evaluated is always an emotional experience to some degree, and for some interns the emotional stakes are very high. Brian Baird (2011) has suggested completing a self-evaluation beforehand, perhaps using

---

THINK About It

Growth, Knowledge, and Interests

If you are stuck in reflecting on your work performance and on what you learned, here are some topics from Birkenmaier & Berg-Weger's work (2010) that you may find helpful:

- Areas of personal and professional growth
- Knowledge and skills you gained
- New or confirmed areas of interest
- Goals met, unmet, and partially met
- Your level of confidence
- Your ability to function as part of a team
- Your ability to make good use of supervision

the agency form, in which you downplay your strengths and call more attention to your areas for improvement. As you do so, monitor how you feel and how satisfied you are with your reaction. We also suggest you spend some time thinking about your emotional reactions to praise and criticism, as you did earlier. Some interns have a difficult time listening to praise; others chafe at criticism. Still others find it hard to tell other people how they really feel about them. Anything you can do to help yourself participate honestly and openly in this process will pay dividends.

Engaging the Evaluation

How you handle yourself in the evaluation process may make a difference in the evaluation you receive, regardless of how you performed throughout the internship. That is because you are expected to take an active role in the evaluation discussion; in other words, you are expected to continue to be engaged in this aspect of the internship as well. Nearly everyone remembers the final evaluation conference, and we want to ensure that you leave that meeting with pride, regardless of whether you earned high ratings across all factors on which you are evaluated. Here's how.

During the evaluation, it is important to balance the positive and the critical (Baird, 2011; Shulman, 1983; Wilson, 1981). It is also important that the feedback be productive and constructive. It is normal to want to focus on the positive and bask in the glow of your accomplishments and your supervisor's praise, but constructive feedback or criticism is just as important. There are bound to be things you need to work on and perhaps areas where you didn't perform as well as you or your supervisor would like. Helping professionals are working with clients on just this issue. Just as their clients need to know where they still have work to do, so do they. Ignoring these areas is a little like saying, "I'm perfect. I learned everything I need to know to be a professional in this field in just a few short weeks." That sounds more like a quote from a late-night TV advertisement than the stance of a reflective, thoughtful professional. In some cases, you may need to be proactive by asking your supervisor to identify areas for growth.

It is also important, during the evaluation interview, to listen as carefully and critically as you can, clarify things as you need to, and ask questions (Birkenmaier & Berg-Weger, 2010). If you don't understand what your supervisor is saying or on what it is based, it will be hard to learn from the positive or the negative, and the process will not be productive or constructive.

You may not like some of what you hear or you may think it is unfair. It is important, though, to try and keep those two issues separate (Rogers et al., 2000). We suggest you listen quietly and respectfully, ask for clarifications and examples, and then go away and think about it for a while before responding. Take some time to vent your feelings, and then consider whether some or all of the criticism is valid. Try to reframe the criticisms into opportunities to learn and grow and remember that this is a professional issue. It is neither a personal attack nor a comment on your overall worth and value (Munson, 1993). If after careful reflection you believe that comments have been made about you that are not true or not grounded in facts, you should discuss this with your campus instructor or supervisor.

Feedback for the Supervisors

The end of an internship may also be a time to offer some feedback to your site supervisors (Baird, 2011; Faiver, Eisengart, & Colonna, 2004; Rogers, Collins, Barlow, & Grinnell, 2000). Some supervisors will request this from you, and they may or may not give you advance notice. There are even forms available to offer written feedback (Rogers et al., 2000). You probably have mixed feelings about this prospect, and we don't blame you. On the one hand, this is an opportunity to tell your supervisor what went well for you in the relationship and where there may be areas for improvement; and it is an opportunity for both of you to learn more about yourselves. You also have an obligation to future interns (Rogers et al., 2000). On the other hand, it is important to think carefully before offering criticism, regardless of how constructively put, unless it is requested (Baird, 2011); this is a case in which being proactive may not be a good idea. After all, your supervisors have and may continue to have power over you in the form of grades, letters of recommendation, and word of mouth in the community. Your campus instructor or supervisor may be able to help you decide what and whether to share with your site supervisor. If you decide to give some feedback, we encourage you to practice or role-play beforehand with your peers or your campus supervisor (seminar class tends to be a good way to do this) and to keep in mind the principles of effective feedback (see p. 63) to guide what you do.

Ending the Supervisory Relationships

The informal part of the closure process with your site supervisor(s) is at the same time easier and more difficult. For better or worse, there is no form to fill out, no big meeting, and no high-stakes evaluation; just the world of feelings. For most interns, the relationship with the site supervisor is among the most significant of the internship. For many, the relationship has been close and positive, both intellectually and emotionally. For others, the relationship has not been close or satisfying, but still one in which a lot was learned. Now that the relationship is ending, the ending can engender a variety of feelings for both of you. If your relationship with your supervisor has been mostly an intellectual, dispassionate one, it may make saying good-bye easier (Haas, 1990). Some supervisors will initiate a conversation with you about these feelings; some will avoid it; and some will engage but need to be prompted by you. If possible, it is important to try to give voice to your feelings.

Some interns find these conversations disappointing (which does not mean they are not important). Many interns develop powerful feelings of closeness, respect, and affection for their supervisors. In some cases, the feelings are mutual. But remember that while this is a new and unique experience for you, it may not be for your supervisor. It is more typical that the supervisor has enjoyed working with you and watching you grow through the experience; and the supervisor might well be sorry to see you leave. However, the supervisor may not feel personally close to you and has no desire to continue the relationship after the internship is over. If this is the case, try not to let it change the value of your experience, or even your feelings about your supervisor.

THINK About It

Getting Letters of Recommendation

If you have had a good experience in the field, you will probably want a letter from your site supervisor and maybe from a co-worker, either for your next internship or for future employment. It's a good idea to ask for it now, while you and the internship are still fresh in the supervisor's and co-worker's minds, rather than call or write months later. Baird (2011) offers some important guidelines to follow when asking for a recommendation:

- *Don't Assume* Before requesting the letter, ask your supervisor if he or she is comfortable writing a supportive letter for you. This may seem silly, especially if you have a good relationship. Many supervisors, if they do not feel they can write a positive letter, will simply suggest that you ask someone else. However, you don't want to take the chance that the letter will be less positive than you expected, especially if it is sent directly to a school or employer.
- *Know What You Need* Make clear what your future goals are. A letter for graduate school may look somewhat different than a letter for employment. There may also be some jobs for which your supervisor feels you are better qualified than others.
- *Plan Ahead* Give plenty of notice and let the reference know of the time frame; provide whatever forms and envelopes are needed.

Quick Tip: If the letters are not sent electronically, then use preaddressed, stamped envelopes as they show courtesy and consideration.

Of Special Relevance to The Helping and Service Professions

Saying Good-Bye to Clients

Just when I'm really starting to feel comfortable with the kids it's almost time to end.

STUDENT REFLECTION

Termination is the word used to describe the ending of a therapeutic relationship. To some of you, this is a familiar term. You may have studied it in class, and if your agency does one-to-one counseling, you may have heard the term used there as well. For others, the term may be new, somewhat strange, and harsh sounding. Not all of you are involved in counseling or psychotherapy in your internship, but if you are working directly with clients, you need to think carefully about ending your work with them and saying good-bye, regardless of the label you apply to the process.

Furthermore, as an intern you are saying good-bye under unusual circumstances. Usually when a relationship between a patient or client and a helping professional ends, it is because the work is done or the client decides to end it. However, you are dealing with what is called a *forced termination* (Baird, 2011; Gould, 1978). It is the calendar that dictates the ending. The therapeutic work is not necessarily over, and the clients are not necessarily ready to stop or start over with another helping professional. Ending relationships with clients is bound to evoke emotional reactions for them and for you. Dealing with those feelings and ending well is an important part of the internship. For some of you it will be relatively easy. For others, it may be one of the biggest challenges you face.

Part of the reason interns in the helping and service professions have such varied experiences in coming to closure with clients is that there are such great variations in the internship sites. Some sites offer a lot of one-to-one counseling. Others, such as adolescent shelters, offer some, but in the context of daily living and recreation, in the *milieu.* Other sites offer none at all. There is also great variation in the characteristics of the clients. Some client populations are much more emotionally vulnerable and may take your departure that much harder. Some clients are more autonomous than others. Clients who drop into a senior services center, for example, or a town recreation department are probably functioning pretty well on their own. They appreciate you, but they really don't *need* you. Finally, some clients are simply more aware of your leaving than others. Interns working with clients who have Alzheimer's disease often report having to tell each client over and over that they are leaving. Many of the clients did not remember the interns from day to day, although they seemed glad to see them each day. Leaving was difficult for the interns, but probably not for the clients. In any case, there are four important issues for you to think about when bringing closure with clients:

- Deciding when and how to tell them
- Addressing the unfinished business
- Dealing with feelings, both yours and theirs
- Planning for future needs

Timing and Style

There are many different theories and opinions about when and how to tell clients you are leaving. Because internship sites and client populations vary so greatly, you must look to your site, your site supervisor, and your campus supervisor or instructor to guide you in this area; there is no prescribed framework that applies to all or most interns. Some agencies recommend that you tell clients when you first meet them that you will be leaving at the end of the term or the year; in fact, some agencies tell the clients for you. Some clients are used to seeing interns come and go; as soon as you say you are an intern, they know you will not be there for very long. Other agencies recommend that you begin discussion of termination one week, two weeks, or more in advance. Some recommend that you not mention it until right before you go.

There is also a variety of methods for discussing termination, from individual conferences, to group meetings, to letters, or any combination of these approaches. If you do not know how your agency handles termination, you need to take the initiative to ask. Depending on your needs and attitudes about ending, you may feel that the agency's policy is too casual or forces you to focus on something that does not seem like a big deal. Remember that the clients are the primary focus; the decision needs to be made based on what will work best for them, not for you. Still, this is a good opportunity for discussion with your supervisor if you disagree with the agency's policy, or if it does not match a theory you have studied. Both of you might learn something, and at the very least you will be more likely to accept and follow through with the agency's procedures.

Unfinished Business

As you approach the end of your relationship with clients it is a good time to reflect, certainly in private and perhaps with clients, on their progress. If they had specific goals, this is the time to review them. Even if they did not, your memory and your journal are good sources for remembering what your clients were like when you first arrived, as well as what your relationship with them was like. Both clients and workers can become so caught up in the problems and challenges of the present that they don't think much about the change that has already occurred. Add to that the difficult feelings that termination can bring and you can see why it is important to take time to reflect on the positive.

It is equally important, though, to be clear with yourself and with clients about the work that remains. All of this is especially important if the client is going to continue at the agency, which is usually the case. You may want to use some of your supervision time to review each of your clients with your supervisor, or you may want to discuss them as a group. You will want to talk about how you feel about each of these terminations, as well as how you think each client is going to react. Baird (2011) discusses a termination scale developed by Fair and Bressler (1992) that has 55 items covering emotional responses and planning. You may want to make use of this resource (Fair and Bressler, 1992, cited in Baird, 2011, pp. 172–173).

Overarching Feelings

In many cases, the termination process can be an emotional one for everyone involved. For both you and the clients, your leaving may recall echoes of past experiences of separation and loss. In addition, many interns report feeling nervous and apprehensive about discussing termination with clients because they are afraid of what the clients' reactions might be. Given the variety of client personalities and the separation experiences they have had, you can expect an equal variety of reactions to the news that you are leaving. It makes sense if both you and your clients approach the termination process with some trepidation. Of course, termination can be a learning opportunity as well. Clients can learn that good-byes, while often painful, do not have to be traumatic, angry, or hysterical. And, the responsibility is yours to model for them positive and empowering ways of dealing with their feelings.

The Client's Experience

Clients may be especially vulnerable to feelings of separation and loss. For many clients—especially those who have been subjected to abuse, neglect, or parents with substance abuse problems—separations have been arbitrary and unpredictable, such as when a parent abandons a child or goes on a drinking binge and returns days later. For some people, separation is associated with anger and hysteria, as in the loud and even violent arguments that can lead to the disruption or ending of a marriage or other intimate relationship. You may also find that clients have had particularly traumatic separations, as in the death of a parent, sibling, or close friend. Finally, many of your clients will have worked with several agencies, workers, or several residential settings and have had many termination experiences. If some or all of these experiences have been painful, this termination may bring up those feelings again (Baird, 2011), and you are right to be concerned. This student poignantly observed just that issue.

> *'Why do you have to go?' is probably the most difficult [question].*
> *Many of these children have had people who come and go in their lives*
> *and I don't want them to look at me as one of those people.*

Clients also may feel especially vulnerable to feelings of abandonment. After all, you are leaving because of your school calendar, not because they no longer need to work with you; in fact, they may wonder how much you really cared about them in the first place (Baird, 2011; Stanziani, 1993). Some clients may feel angry and believe their trust has been betrayed, even if you have been clear from the beginning that you will be leaving at the end of the semester. It is important to remember that whatever reaction you receive may be only part of the picture (Penn, 1990). Termination, like the end of any significant relationship, usually creates feelings of anger, sadness, and appreciation. Your client may be expressing only one of those feelings, and indeed that may be the only feeling he or she can access. However, it is important for you to remain open to the other feelings as well, and perhaps to help your client be open to them too. It is also important to separate what clients choose to show you from what their deeper feelings may be.

Sometimes these feelings are expressed indirectly, not unlike your own feelings about ending the internship (Penn, 1990). Clinically, this is referred to as *resistance.* Some clients may be genuinely indifferent to your leaving. But others will feign indifference or be indifferent because their real feelings are too hard to accept. Another indirect way of dealing with termination issues is to demand that termination come immediately. In these cases, once you tell clients you will be leaving, they may stop coming to the agency or, in residential settings, ask for another worker to be assigned right away. Other clients may begin to exhibit older, more problematic behaviors, as if to say, "You can't leave me; I still need you." Finally, clients may begin to devalue the work you have done together, since it is easier to say good-bye to something or someone who, after all, wasn't that important anyway.

The Intern's Experience

For most interns, saying good-bye to clients is an emotional experience as well, sometimes in unexpected ways. You may find yourself feeling guilty that you haven't been able to do more for some or all of your clients, and again, this feeling is often unrelated to how much you have actually accomplished (Baird, 2011). In other cases, you may worry that you are the only one who can work successfully with a particular client or group of clients. It may be that you are the first person to really "get through" to a client. That is a wonderful feeling, but it places extra weight on the termination process. As one intern said, *they do not receive the same help from others in certain areas…. It just makes me sick inside.* Remember that although the client may regress some when you leave, chances are that at least some of the progress that was made will remain. It can, however, be hard to remember or believe that in the moment. If a client starts to regress, you may begin to question the value of your work or your effectiveness in general (Gould, 1978).

Focus on PRACTICE

Knowing the Pitfalls

All helping professionals are vulnerable to client reactions, but as an intern, especially if this is your first intensive experience in the helping field, you may be particularly vulnerable. The closeness that can happen between worker and client is a heady, heartwarming experience, and it is especially affecting the first time (Baird, 2011). Both of us recall with special vividness our first few clients and the special difficulties that we faced in saying good-bye. There are pitfalls to watch out for as you take in and process clients' reactions to your leaving, and we've identified several common ones for you.

- *Personalizing Reactions*   It may be easy for some of you to take quite personally the client's reactions, as if they were statements about *you*. Of course, they have something to do with you, but remember that you may also be hearing expressions of their current struggles with themselves and echoes of old wounds. Work with your site supervisor to prevent such personalizing.

- *Confusing Needs and Issues*   It is important to separate the clients' needs and issues from those of your own. If, for example, you struggle with a need to be liked, then you may be especially vulnerable to a client who turns on you in anger or devalues the experience. Work with your site supervisor to prevent such confusion.

- *Mixing Feelings and Decisions*   It is important to deal with the feelings and issues separately from the decisions you make about how to respond to your clients. You may need to be liked, but liking you in that moment may not be the best thing for your client. If a client has never been able to express disappointment with anyone, then the ability to do that with you is far more important than your need to be reassured. Work with your site supervisor to ensure that you keep both separate.

- *Seeking Affirmation from Clients*   Some interns unconsciously try to get clients to make the termination easier for them. They seek reassurance that they have done a good job or that whatever problems remain are not their fault (Baird, 2011). In these times, it is especially challenging to keep the client's needs in the foreground. You may very well need some support and some focused attention to help you process your feelings and needs. That is what your campus instructor, site supervisor, and peers are there for; don't turn to your clients for that support. Instead, discuss your support needs in supervision.

- *Promising to Return*   Another pitfall is promising to come back and visit. Clients may ask you to come back or to maintain contact in some other way. They may be avoiding the difficulties of separation, they

(continued)

may not understand the boundaries of a professional relationship, or they may just be doing what they would do with anyone they have come to care about. Even if they don't ask, you may find yourself wanting to reassure them that you will visit or keep in touch. Use extreme caution before making any such promises. Consult with your supervisor and your campus instructor. Above all, if you do make commitments, make them with care; the worst thing you can is promise to keep in touch and then fail to do so.

The Future

The last issue to think about in terminating with clients is the future, both for them and for your relationship with them. Just as this is a time to review goals and accomplishments, it is also a time to set or reestablish goals for the future. You also need to think about who will be dealing with your clients after you leave.

Transitioning Clients In many agencies, intern supervisors plan the transition for clients that will occur when you leave, and you should be part of that planning. If your supervisor is not raising this issue, we suggest that you do. This transition, often referred to as *transferring clients*, requires time and attention if it is to be done well (Baird, 2011; Faiver et al., 2004; Rogers et al., 2000). The nature of the transfer will depend a great deal on your specific setting. You may actually have some clients that have been assigned to you. For example, there may be individuals who come to a clinic for counseling, a group you have been assigned to lead or co-lead, children and families whom you visit at their homes, or community groups that you meet with on a regular basis. In residential settings, it is not unusual for each resident to have a primary staff member and sometimes interns move into this capacity after they have been there for a few weeks. In other settings, there will not be clients who have been officially assigned to you, but you may have developed especially close relationships with certain individuals, and it may be clear to you that you are the one to whom they turn most often. In any case, the issue of who is going to take over is one that should be discussed with your supervisor, with your co-workers, and of course with your clients. You should also be sure to reserve some time to work with the person who will be taking over, and you should begin this process well before your last day.

Relationships after the Internship Ends   What sort of relationship will you have with your students, patients, residents, or clients after you leave? What sort should you have? In many cases, the answer to both questions is "none." At most, you might hear from them by mail or e-mail, which is not initiated by you. In some instances, either because the client asks for more contact or because you want more contact, the issue becomes less clear. As we discussed in Chapters 7 and 13 in the instance of the helping and teaching professions, sexual/intimate relationships with clients while you are an intern are unacceptable and are a clear violation of the ethical standards of virtually every professional organization. Social relationships with clients are also discouraged, although in some settings they are unavoidable and may even be beneficial. Now, however, you are leaving the internship site, and there may be clients with whom you want to pursue a friendship or a romance. Some of you, particularly those working with adolescents, may find this absurd and out of the question. But for those of you who are working with clients your own age, such a situation is not out of the question; the issue does come up.

If you are in the helping professions, it is important to know that some of the ethical codes of those professions are quite specific about the amount of time that must pass before such relationships can develop. This issue is complicated once again by the wide variety of placements and clients that interns deal with. Relationships after the internship ends have been the subject of a good deal of discussion in the counseling field over the years. So if you were thinking that maybe you could just pick right up with a different kind of relationship with one of your clients, you may want to think again.

The internship site may also have policies in place about relationships with clients. For example, if a friendship develops and you subsequently decide to apply for a position with the internship site a few years later, you might be surprised to learn that the site considers the relationship a conflict of interest and may not hire you. If you find yourself in this position, it's best to be fully informed before you leave the agency so you have a clear understanding of how your decisions affect your future options at the site and in the professional field where you interned. If the supervisor doesn't raise it in the final days of your internship, you need to do so if you think you'll be continuing a relationship(s) after the internship ends.

Another consideration is social media and what's okay and not okay in terms of using social networks to stay connected with former clients. Clients may ask you to join one of their social networks, such as Facebook or LinkedIn. Remember that if you accept such an invitation, you are giving the client access to a good deal of information about you that you might not normally share. Again, it would be prudent to find out what the agency policy is about connecting with former clients in these ways. You may be surprised to learn that policies are already in place. And it's likely that such policies are being discussed at the national level of the professions when their ethical codes are ready for updates, especially when it comes to the helping professions.

The major concern about such relationships is the unfair power differential between you and your former patient, student, resident, or client (Salisbury & Kiner, 1996). Through working with clients, helping professionals often have knowledge of their most sensitive issues and areas of vulnerability. Those being helped, on the other hand, usually have no such knowledge of the professionals' lives, and this puts them in a vulnerable position in a friendship or romance.

A Word to the Wise... Not all of you do crisis counseling, nor are you necessarily working with clients who have come to you or the agency because of mental health issues. If you are working in a recreation center, doing community organizing, or providing health education for college students, the power differential referred to may be quite different. So, unfortunately, there are no hard-and-fast rules to follow, just some important issues to think about. If you are considering pursuing any sort of relationship with someone you've worked with in a helping capacity, you ought to talk that over very carefully with your campus and site supervisors. Whatever you decide, remember that even though you are no longer working with them, the primary concern in your decision making still must be their welfare, not yours.

Bidding Farewells

I think to myself this is the last time I am going to be doing this and seeing these people. It's been a busy and scary week.
STUDENT REFLECTION

The Placement Site

The last week of an internship is often hard in many ways. You have had your final evaluation meeting with your supervisor; you have had meetings about the status of the projects you have been working on; if you have clients, you have prepared them for your departure; and, when applicable, you have notified human resources accordingly. It becomes more difficult as you listen to the future of the work you have been doing, work that you will not necessarily see materialize. All of this is leading up to the final day, the last day of your field experience.

Rituals and Remembrances

Think about the pace of life at your site. Most agencies in the helping and service professions are understaffed and extremely busy, just as are some areas of the private sector. In residential settings, large and small, crises

How You Don't Want Your Last Day to Go

Your last day, a day you have been looking forward to with an incredible mixture of feelings for weeks, is approaching fast. You begin to imagine what it will be like when everyone says good-bye to you and what you will say as you leave. As you move through the last day, though, no one says a thing to you about leaving. Your supervisor doesn't seem to be around, and for your co-workers, it seems to be business as usual. Conversations focus on the work, and during quiet times, it's the usual office small talk.

You go about your business, feeling a little confused but certain that someone will say something soon. Maybe they're being sneaky, and there is a little party planned for you later. No such luck. As you are leaving for the day, you say good-bye to people, pretty much as you always do. One of your co-workers looks up from what she's doing and says, *"See you next week."* "Well, no," you respond. *"Today is my last day."* *"Oh my gosh,"* says your colleague. *"I totally forgot! Well, hey, it's been great. Good luck to you."* You start to reply, but the phone rings, a client calls, or a crisis erupts, and your colleague swirls away, back into the normal pace of life at the site.

When driving home, you feel differently than you thought you would. Yes, there is joy and some relief, but also some hurt and a vague sense of emptiness. You try not to focus on that feeling, but it gnaws at you. What's wrong with those people anyway? Maybe you didn't mean as much to them as you thought. Maybe they're just rude and not quite the people or professionals you thought they were. You feel as if you deserve better from them after all you have done for the organization.

One way of looking at what has happened here is that you took a relatively passive, disengaged approach to achieving closure, and now your feelings are sliding and need to be stopped.

erupt all the time. And it's not the last day for anyone else; it is *your* last day! For that matter, it may not even be the end of their day or week. So it is understandable that they might forget about your departure.

The point here is that you may need to be proactive in assuring you get the sense of closure you want. This is another time when a ritual may be in order. Some agencies have fairly elaborate rituals when someone leaves. There might be time set aside at a staff meeting, or there might be a party or a farewell lunch. Other sites don't have any of these rituals in place. You won't know what your site typically does unless you take the initiative and raise the issue. For example, talk with your supervisor about the best way to mark the end of your internship. Ask what the norms are and be clear about what you need. It may be that there is just no time for a group activity, but at least you will know that, and you can schedule individual 15-minute appointments with some of your co-workers. In our experience, most supervisors are glad to help you make something happen, but they might not think of it on their own or may want to know what you prefer.

Even if there is going to be a formal good-bye celebration, there may be individual co-workers to whom you have grown especially close. They may take the initiative to have a final conversation with you, or they may not for any number of reasons. Here again, you need to take the initiative and schedule some time with them. This is yet another opportunity to practice your feedback skills; it is a time to let them know specifically what you have learned from them, what you appreciate about working with them, what memories you are left with from the experience, and how you feel about leaving or about them as supervisors and mentors. They may, in turn, do the same for you. Remember, though, the main point is for you to say what you want and need to say, and if you do, the conversation is a success. You don't want to go away with that nagging feeling that you wish you had said such and such to so-and-so. Whatever you get back from them is a bonus.

Planning for the future also means considering the nature of your involvement, if any, with the site and its work after the internship is over. Some interns are offered part-time work, relief work, or even full-time jobs. We often tell our students to be flattered, but be wary. Make sure that accepting is really your best option and not just a way to avoid bringing things to an end. In other cases, you may want to come back for a visit or to see and help with an event you have been working on. Seeing a project, event, or contract you have worked on come to fruition can be wonderful as well. Just take care not to promise more than you can deliver. If the site is counting on you and you get caught up in your life and don't follow through, you leave a bad feeling about you and possibly about your campus or program.

Planning for your future includes a reassessment of the role of the internship in your career plans, the role you see yourself taking as a civic professional, and concrete steps to move your life forward, such as preparing your résumé, developing your portfolio, compiling your professional networking list, and adjusting your out-going message on your voice mail to reflect the professional you have become!

The Seminar Class

It provided me a comfortable environment where we could meet new people, learn new things, and share thoughts. It allowed me to grow as a student, an individual, and a professional. I really looked forward to it...we all went for the ride together. My "family"...the seminar group— after you share so many experiences, you kind of move past friendship and end up in the realm of "family."

STUDENT REFLECTION

THINK About It

Ready to Move On? Tools for the Task

Regardless of whether you are staying on at the site in some capacity or transitioning out when the internship ends, there will be a change in your status once the internship is over. The following may help you be ready.

- What is most pressing for you as the end of your internship draws near? Do you have a plan in place to manage those issues and move forward?
- Think about these five words: *completion, over, termination, culmination,* and *ending.* Do you feel differently about any of them? What can you do to staying with the positive feelings as you move through this period?
- For those transitioning from role of intern to that of employee, what it is about the transition that leaves you feeling good? Not so good? What can be done to bring about the needed changes?
- For those of you transitioning out of the site when the internship is over, what do you expect to be most challenging and most manageable about that transition?

If you were lucky enough to have a seminar class as part of your internship, you will have to find a way to say good-bye to your peers and campus supervisor as well.

Seminar class is usually a source of significant support during the internship. It is the one place where you know that others understand what you are describing, appreciate what you are experiencing, and share your concerns, questions and expectations as well. It is also the one place where you can raise issues and express feelings about the internship without being judged. Often, it is not until the final class approaches or during the final class itself that interns appreciate the uniqueness of the seminar and realize the impending absence of this source of support and guidance. And then, those who have difficulties with endings often find themselves responding in old, familiar ways— ways we addressed earlier in this chapter. It might make sense to review that material (p. 401) at this point if this issue resonates with you.

Whether your seminar is held on campus or on-line, you will want to think about the final class and what shape you would like it to take, and discuss your needs as well as the needs of your peers and instructor. For on-campus seminars, students sometimes want to have the class held off campus at a nearby restaurant, sharing memories of the class and what they are taking away from the experience of being a community of interns. Other times, for both on-campus and online seminars, interns want to spend the final class involved in experiential activities focused on "endings." These activities, often a combination of the interns' favorites and some of the campus supervisor's favorites, are designed for reflecting on the highlights of the internship, the ways in which the interns have grown and changed through the experience, and how the class made a difference in the journey of the internship. For both models of seminar, there are times when interns prefer that the class be held "as usual" and that the topic of endings be addressed both in experiential ways through activities as well as through conversations about their own best ways of bringing closure and saying good-byes. Of course, your campus supervisor also has expectations for the final class, and those remain to be seen.

This is a good time to reflect on what you *need* from the final class, what you *want* to happen during the final class, and what your *plan* is to ensure that your needs are met by the time the final class ends. Campus supervisors welcome the interns taking this engaged approach to it and working with the interns to develop a final seminar that meets everyone's needs.

Looking Around and Moving On

Your internship is just about over now, and your future is beginning already. Probably, there were times when you thought this moment and this day would never come or times when you were amazed and unnerved at how fast it was approaching. New challenges await you, of course. Some of you may be headed for another internship, some for a job in the field, and still others for the next level of studies and more field-based learning.

Planning for your future includes a reassessment of the role of the internship in your career plans, the role you see yourself taking as a civic professional, and concrete steps to move your life forward, such as preparing your résumé, developing your portfolio, compiling your professional networking list, and adjusting the outgoing message on your voice mail to reflect the professional you are becoming!

Reflecting on Growth

The moments in life when we feel uncomfortable are when a person grows the most. I truly feel I am a different person in April than I was in September. I feel wiser. I feel better rounded, competent, humble.
STUDENT REFLECTION

Now is the time to consider the next steps in your personal, professional, and civic development and the obvious place to begin is the document that guided you through your journey and brought you to this point in your experience. Indeed, it is time for one last look at your Learning Contract and to think about what lies beyond it for you. Take pride in having met your learning goals and take time to reflect on your growth and accomplishments, including those that were not planned. There is always unplanned, unexpected learning that takes place in an internship, and sometimes it is the most memorable of the experience.

You have learned a great deal about yourself in the journey you are completing. Importantly, you have learned that your formation as a person is never complete. We have tried to emphasize this dimension of learning, growth, and development in this book; interns often tell us that this is the most powerful learning of all for them. You know more about how you respond to challenges and why. And, you know that self-understanding is a process, not an accomplishment. You have more tools to pursue your self-understanding, and the practice you have had in the field, if you continue it, will make those tools more and more second nature to you.

Your formation as a professional will never be complete. Now, however, you are more likely to be aware and in charge of that growth process. You can expect that additional knowledge, skills, and even values, may be required or encouraged in your next situation, whether that is another internship, a more advanced degree program, or the next step in your career. We hope you have come to appreciate the importance of development through the pursuit of new knowledge and skills and further understanding of your values; and that you have *internalized* a commitment to your own growth as a person and as a professional.

Remember, though, that the professional arena is only part of the picture. We have tried to help you look at yourself not just as a professional, but as a civic professional, with a clear sense of the human and community context of the work you do and the *public relevance* of your profession. We have also encouraged you to use the internship as a vehicle for your civic awareness and development. We hope that you continue to identify and pursue the values, skills, and knowledge that make you an engaged and effective citizen.

Preparing a Professional Portfolio

Many interns have found value in creating a career portfolio that includes the internship experience. This type of portfolio focuses primarily on *presenting* your experience, skills, and competencies. This is a reflective tool that is a powerful asset as you seek employment or further education. There are different categories of content you can consider including, such as reflections on your work, sample papers, case studies and summaries of projects (being mindful of confidentiality), performance reviews, media coverage of your work, and so on. Whatever time you spend on this endeavor will be time well spent. We suggest that you follow a process for portfolio development, which can be found online on the numerous websites that inform and guide the portfolio process. We offer one example below.

- *Consult, Consult!* Exchange ideas about portfolios and the resources and websites being used to inform its development with peers, supervisors, colleagues, and friends.

- *Create the Vision* Draft a list of potential contents and an introductory statement. Think about how you want to use your portfolio and what you expect it to accomplish for you. It is a marketing tool for you to get in the "door" of where you want to go. Keep that in mind when you "see" the final product in your mind at this point in the process. Be sure it is what you need and want it to accomplish.

- *Construct the Framework* Brainstorm all the possible organizing lenses or metaphors you might use to create and organize your portfolio. Examples include being the humanitarian student, the emerging professional, the ethical decision maker, the community member, the effective organizer, the civic professional, the caring advocate, and so on. Then, begin collecting the documents you need to substantiate your self-identified strengths and pay attention to the pool of resources you are using to further develop your framework. This can take time, so be sure to build that factor into your plan so you aren't disappointed when you need your portfolio and you are waiting for essential materials to arrive, such as letters of reference and transcripts.

- *Convey Your Profile* Here is where you choose the actual format and design of the portfolio that includes variations on both web-based and paper models. For example, some students choose an online portfolio with a personal website, a video of personal statement of intent, and links to essential documentation. Other students choose a traditional paper CV model, with a résumé and a file of documentation to support interview-specific paperwork. Many students are choosing a combination of these two models. The important thing is that your portfolio conveys the profile you want it to convey.

A Final Word to the Wise... Keep in mind that the developmental stages of an internship are not phases you go through just once. In your new position or in another field placement, you are going to go through them again. The concerns about expectations and acceptance and the challenges of keeping yourself moving and growing, of confronting problems, and of ending well will all visit you again and again in your work and career. When we tell our students this piece of news, some of them roll their eyes and hold their heads. When they remind each other of how they aptly handled the challenges they faced and how much they learned, they realize that their learning will travel with them into the future.

You have learned valuable skills. If you continue to use them, they will grow sharper and more integrated over time. Even if your life and career

---

THINK About It

Am I REALLY Ready to Move On?

You know well the concrete steps that must be taken to move into the future. If that future is the beginning of your career, then here's a practical to-do list...in case you need a reminder!

- Update your résumé to include the internship.
- Polish your interviewing skills.
- Fine-tune your networking list.
- Develop your portfolio.
- Contact your references.
- Change your outgoing voice mail messages.
- Visit the career services office on campus.
- Join your professional organization and its list serves.
- Start building your library of professional resources.
- Begin the search for your first career position.
- Create your personal website.
- Think about what's next to learn.
- Think about your next steps.

---

change directions now, you can still take a lot of what you have learned with you. You have learned what to expect from an internship. The challenges will have a different shape and pace because they will be happening in a different place. You have grown and changed so much that it may feel like these challenges are happening to a different person. And, of course, you will handle the issues in a new way. However, the concerns themselves should seem familiar to you if you can get enough distance and perspective from your everyday activities. Perhaps you will continue to keep a journal...and pause frequently for reflection on what you have written.

A Fond Farewell...

It may seem odd for us to say this, since we have never met you, but we wish you well in your life's work and in your development as a civic professional. We are always glad to hear from former students and welcome hearing from you about how this book may or may not have been useful to your journey in experiential learning. And if there are topics we didn't discuss and you think we should, then let us know, and we'll give attention to them. We leave you with these thoughts from one of our students:

> *It's like moving out of your house. You're leaving something that has had such an impact on your life and starting something unknown and exciting. You are leaving a foundation of growth and a place filled with memories. You're going into unknown territory, and that is exciting.*
> STUDENT REFLECTION

For Further Exploration

Shumer, R. (In press). Evaluating and assessing experiential learning. In G. Hesser, (Ed.). *Strengthening experiential education within your institution* (Revised). Mt. Royal, NJ: National Society for Experiential Education.

A chapter in a classic sourcebook that considers trends in the approaches to and the focus of evaluation of experiential learning, as well as their impact on individuals, institutions, and communities. Tools and frameworks are described that include evaluation and assessment as an integral part of the process of experiential learning.

References

Abramson, J. S., & Fortune, A. E. (1990). Improving field instruction: An evaluation of a seminar for new field instructors. *Journal of Social Work Education*, 26(3), 273–286.

Baird, B. N. (2011). *The internship, practicum and field placement handbook: A guide for the helping professions.* Upper Saddle River, NJ: Pearson Prentice Hall.

Birkenmaier, J., & Berg-Weger, M. (2010). *The practical companion for social work: Integrating class and field work* (2nd ed.). Needham Heights, MA: Allyn & Bacon.

Fair, S. M., & Bressler, J. M. (1992). Therapist-initiated termination of psychotherapy. *The Clinical Supervisor*, 10(1), 171–189.

Faiver, C., Eisengart, S., & Colonna, R. (1994). *The counselor intern's handbook.* Pacific Grove, CA: Brooks/Cole.

Garthwait, C. L. (2011). *The social work practicum: A guide and workbook for students.* (5th ed.). Needham Heights, MA: Allyn & Bacon.

Gould, R. P. (1978). Students' experience with the termination phase of individual treatment. *Smith College Studies in Social Work*, 48(3), 235–269.

Haas, S. (1990). *Hearing voices: Reflections of a psychology intern.* New York: Penguin.

Kegan, R. (1982). *The evolving self: Problem and process in human development.* Cambridge, MA: Harvard University Press.

Levine, J. (2012). *Working with people: The helping process.* New York, Longman.

Munson, C. E. (1993). *Clinical social work supervision* (2nd ed.). New York: Haworth Press.

Penn, L. S. (1990). When the therapist must leave: Forced termination of psychodynamic psychotherapy. *Professional Psychology: Research and Practice,* 21, 379–384.

Rogers, G., Collins, D., Barlow, C. A., & Grinnell, R. M. (2000). *Guide to the social work practicum.* Itasca, IL: F. E. Peacock.

Salisbury, W. A., & Kiner, R. T. (1996). Post termination friendship between counselors and clients. *Journal of Counseling and Development,* 74(5), 495–500.

Shulman, L. (1983). *Teaching the helping skills: A field instructor's guide.* Itasca, IL: F. E. Peacock.

Stanton, T., & Ali, K. (1994). *The experienced hand: A student manual for making the most of an internship* (2nd ed.). New York: Caroll Press.

Stanziani, P. (1993). *Practicum handbook: A guide to finding, obtaining, and getting the most out of an internship in the mental health field.* Cambridge, MA: Inky Publications.

Sweitzer, H. F., & King, M. A. (2012). *Stages of an internship revisited.* Paper presented at the National Organization for Human Services, Milwaukee, WI.

Suelzle, M., & Borzak, L. (1981). Stages of fieldwork. In L. Borzak (Ed.), *Field study: A sourcebook for experiential learning* (pp. 136–150). Beverly Hills, CA: Sage Publications.

Watzlawick, P., Weaklund, J. H., & Fisch, F. (1974). *Change: Principles of problem formation and problem resolution.* New York: Norton.

Wilson, S. J. (1981). *Field instruction: Techniques for supervisors.* New York: Free Press.

 Index